What Is Nanotechnology and Why Does It Matter?

The Authors

Fritz Allhoff is an assistant professor in the Department of Philosophy at Western Michigan University, where he also holds an affiliation with the Mallinson Institute for Science Education. He has held fellowships at the Center for Philosophy of Science at the University of Pittsburgh and the Centre for Applied Philosophy and Public Ethics at the Australian National University. He has edited more than 20 books, including two on the social and ethical implications of nanotechnology.

Patrick Lin is the director of Ethics + Emerging Sciences Group at California Polytechnic State University, San Luis Obispo, and holds academic appointments at Dartmouth College, US Naval Academy and Western Michigan University. He is also lead author of a recent major report funded by the Department of Navy entitled *Autonomous Military Robotics: Risk, Ethics, and Design* (2008), as well as other publications in technology ethics, including edited anthologies on nanoethics.

Daniel Moore is a research scientist on nanoscale semiconductor solutions for IBM. He served on the Georgia Institute of Technolog's honor committee and has received numerous fellowships, including the Molecular Design Institute Fellowship, the GT Nanoscience and Technology Fellowship, the School of Materials Science and Engineering Advanced Publication Award, the NSF STEP Fellowship, and a Sam Nunn Security Fellowship. His professional experience includes nanoscale research in other leading industry laboratories.

What Is Nanotechnology and Why Does It Matter?

From Science to Ethics

Fritz Allhoff, Patrick Lin, and Daniel Moore

WILEY-BLACKWELL

A John Wiley & Sons, Ltd., Publication

This edition first published 2010
© 2010 Fritz Allhoff, Patrick Lin, and Daniel Moore

Blackwell Publishing was acquired by John Wiley & Sons in February 2007.
Blackwell's publishing program has been merged with Wiley's global Scientific,
Technical, and Medical business to form Wiley-Blackwell.

Registered Office
John Wiley & Sons Ltd, The Atrium, Southern Gate, Chichester, West Sussex,
PO19 8SQ, United Kingdom

Editorial Offices
350 Main Street, Malden, MA 02148-5020, USA
9600 Garsington Road, Oxford, OX4 2DQ, UK
The Atrium, Southern Gate, Chichester, West Sussex, PO19 8SQ, UK

For details of our global editorial offices, for customer services, and for information
about how to apply for permission to reuse the copyright material in this book
please see our website at www.wiley.com/wiley-blackwell.

The right of Fritz Allhoff, Patrick Lin, and Daniel Moore to be identified as the
authors of this work has been asserted in accordance with the Copyright, Designs
and Patents Act 1988.

Library of Congress Cataloging-in-Publication Data
Allhoff, Fritz.
    What is nanotechnology and why does it matter : from science to ethics / Fritz Allhoff,
Patrick Lin, and Daniel Moore.
        p. cm.
    Includes bibliographical references and index.
    ISBN 978-1-4051-7545-6 (hardcover : alk. paper) – ISBN 978-1-4051-7544-9
(pbk. : alk. paper) 1. Nanotechnology. 2. Nanotechnology–Moral and ethical aspects.
I. Lin, Patrick. II. Moore, Daniel, 1979– III. Title.

    T174.7.A496 2010
    620'.5–dc22

                                                                                    2009032168

A catalogue record for this book is available from the British Library.

Set in 10/12pt Sabon by Graphicraft Limited, Hong Kong

01   2010

Contents

Preface

New and emerging technologies both excite and worry us – academics, policy-makers, businesses, investors, ethicists, journalists, futurists, humanitarians, the global public – and nanotechnology today is certainly a flashpoint for irrational exuberance and fears. By definition, there is a knowledge gap during the nascent stages of any new technology, and nanotechnology is no exception: researchers and engineers are still learning about nanoscience and its applications. But, in the meantime, hope and hype naturally and irresistibly fill this vacuum of information.

In nanotechnology, we find an entire range of speculative possibilities, from limitless life to the end of days. Depending on whom you ask, nano-technology is predicted to help solve the world's energy and hunger issues. Or it can help us easily ride into orbit on a space elevator or evolve into cybernetic beings. Or by playing with this fire – by manipulating the basic building blocks of nature – nanotechnology may scorch the earth and fulfill a prophecy of Armageddon. We may finally cause our own undoing by unleashing a powerful technology that we do not yet fully understand and thus may not be able to control.[1]

So which is it: will nanotechnology usher in a bright era for humanity, or its reckless demise? Or perhaps neither: nanotechnology could be much ado about nothing, given the admittedly humdrum products it enables today from longer-lasting tennis balls to stain-resistant pants. Of course, none of us knows the answer, despite a continuous flood of predictions. But what

---

[1]   In addition to forthcoming discussion in this book, see Fritz Allhoff, Patrick Lin, James Moor, and John Weckert (eds), *Nanoethics: The Social and Ethical Implications of Nano-technology* (Hoboken, NJ: John Wiley & Sons, 2007). See also Fritz Allhoff and Patrick Lin (eds), *Nanotechnology and Society: Current and Emerging Ethical Issues* (Dordrecht: Springer, 2008).

we do know is that nanotechnology is rapidly entering the marketplace today, and ongoing research reveals risks (e.g., environmental and health harms) as well as fantastical innovations (e.g., invisibility cloaks and 'smart dust'). It is also becoming clear that nanotechnology has the potential to profoundly change the world, even if today's products are uninspired and only incrementally better than previous ones. Therefore, understanding what nanotechnology is and why it matters is the first step in a roadmap toward our future: it is the next generation for industries, financial markets, research labs, headlines, and our everyday lives.

In this book, we hope to tame the riot of speculation with an informed and balanced look at nanotechnology and its issues. To that end, this book is organized in three units. First, we discuss the science behind nanotechnology, including the nanoscale, tools of the trade, nanomaterials, and applied nanotechnology. In the second unit, we provide a general framework to evaluate the particular ethical and social issues that nanotechnology raises (which are then covered in the third unit); for instance, we will discuss the different ways to understand risk, regulation, and fairness in the use and dissemination of nanotechnology. In the third unit, we dedicate chapters to some of the most urgent and contentious applications of nanotechnology: environment, military, privacy, medicine, and enhancement. In choosing this list, we have focused on near- and mid-term issues rather than more speculative ones – such as life extension, space exploration, and molecular manufacturing – though these latter are mentioned when appropriate. We should also note that given the unusually broad range of issues that arise in connection with the social and ethical implications of nanotechnology, we have made specific choices regarding our coverage; to this end, we have focused on core scientific and philosophical issues and ones on which we are qualified to write. Other important topics – such as existing (and potential) laws and regulations, economic impacts, and so on – can be pursued elsewhere by the interested reader.

While all of the chapters were the result of collaboration, one or other of us bore the primary responsibility for each. Having one scientist and two philosophers on the project undoubtedly makes it stronger and, while our disciplinary backgrounds undoubtedly come through, we hope the presentation and styles are well integrated; editing a project like this takes as long as writing it! Some of the chapters have extended scientific or philosophical primers, which we include for those with less background and/or for pedagogical reasons as this book stands for various course adoptions. Those with the relevant professional training can certainly skip these discussions.

New technologies are not easy to understand, nor are the public policy questions they engender. Therefore, we commend you, the reader, not only for your patience in sorting through these issues, but also for your foresight in recognizing how essential nanotechnology will be to society and our collective futures.

We further thank the institutions and organizations with which the authors are – or have been during time of writing – affiliated for their support: The Australian National University's Centre for Applied Philosophy and Public Ethics, California Polytechnic State University (San Luis Obispo), Dartmouth College, Georgia Institute of Technology, IBM, The Nanoethics Group, University of Oxford's Future of Humanity Institute, and Western Michigan University.

Parts of this book are also based upon work supported by the US National Science Foundation under Grant Nos 0620694 and 0621021. Any opinions, findings, conclusions, or recommendations expressed in this book are those of the authors and do not necessarily reflect the view of the National Science Foundation or the aforementioned organizations.

Finally, we thank our editor at Wiley-Blackwell, Jeff Dean, for his vision and boundless patience; our project managers, Barbara Duke and Sarah Dancy; our editorial assistants, Marcus Adams, Tim Linnemann, and, especially, Lindsay Mouchet; our copyeditor, Brian Goodale; three anonymous external reviewers; and our supportive families.

Fritz Allhoff, PhD
Patrick Lin, PhD
Daniel Moore, PhD

Unit I

What Is Nanotechnology?

1

The Basics of Nanotechnology

1.1 Definitions and Scales

Before delving into the depths of nanotechnology and nanoscale science, we should be clear what we mean when we use terms such as 'nanotechnology,' 'nanoscience,' and 'nanoscale.' It is a basic nomenclature, used to describe certain attributes of certain systems, and such nomenclatures are typically designed to eliminate confusion and encourage accurate communication among those discussing the system. To be sure, there will be several other variants of words with the prefix 'nano-' used in this book and it is important to make sure that we have our meanings straight if any sort of meaningful discussion is possible.

The prefix 'nano-' is derived from the Greek word *nannos*, meaning "very short man." Most of the measurement prefixes used today originate from Greek and Latin words used in measurements. For example, 'kilo-' is from the Greek word *khilioi* meaning "one thousand" and 'milli-' is from the Latin word *mille* meaning "one thousand." Greek and Latin words for numbers cover the everyday level of measurements from one-thousandth to one thousand. Beyond that, it gets interesting. To describe one billion (1,000,000,000) of something we use the prefix 'giga-' which is from the Greek word *gigas* meaning "giant." We also get the word "gigantic" from this root. On the other end of the spectrum, to describe one millionth (0.000 001) we use the prefix 'micro-' from the Greek word *mikros* meaning "small." To describe one trillionth (0.000 000 000 001) we use the prefix 'pico-' from the Spanish word *pico*, which can mean both a "beak" and a "small quantity." These prefixes are extremely useful when discussing very large or very small values. For example, we could refer to the radius of the Earth at its equator as being 6,378,100 meters, but it is more useful (and requires less effort) to refer to it as 6,378.1 kilometers. The most common scientific prefixes and their derivations are shown in Table 1.1.

Table 1.1 Etymology of scientific prefixes

Prefix	Language of origin	Word	Meaning of word	Value
Zetta	Latin	Septem	Seven	$(10^3)^7 = 10^{21}$
Exa	Greek	Hexa	Six	$(10^3)^6 = 10^{18}$
Peta	Greek	Penta	Five	$(10^3)^5 = 10^{15}$
Tera	Greek	Teras	Monster	10^{12}
Giga	Greek	Gigas	Giant	10^9
Mega	Greek	Megas	Great	10^6
Kilo	Greek	Khilioi	One thousand	10^3
Hecto	Greek	Hekaton	One hundred	10^2
Deca	Greek	Deka	Ten	10^1
Deci	Latin	Decem	Ten	10^{-1}
Centi	Latin	Centum	One hundred	10^{-2}
Milli	Latin	Mille	One thousand	10^{-3}
Micro	Greek	Mikros	Small	10^{-6}
Nano	Greek	Nannos	Dwarf	10^{-9}
Pico	Spanish	Pico	Beak, small quantity	10^{-12}
Femto	Danish	Femten	Fifteen	10^{-15}
Atto	Danish	Atten	Eighteen	10^{-18}
Zepto	Latin	Septem	Seven	$(10^{-3})^7 = 10^{-21}$

At its root, the prefix 'nano-' refers to a scale of size in the metric system. 'Nano' is used in scientific units to denote one-billionth (0.000 000 001) of the base unit. For example, it takes one billion nanoseconds to make a second. In everyday practical use, the term 'nanosecond' is not very useful in describing time accurately. Imagine discussing time in these terms: we would say things like "dinner will be ready in 300,000,000,000 nanoseconds." Instead, the term 'nanosecond' is mainly used to refer to a very short period of time. (A nanosecond is to a second as one second is to approximately 30 years.)

When we are talking about *nano*technology, we are talking about a scale – an order of magnitude – of size, or length. We are making a reference to objects that are sized on a scale that is relevant when we discuss *nanometers* (nm). Using this terminology makes it easier to discuss the size of objects that are the main attraction in nanotechnology, namely atoms. If we were to describe the size of atoms and molecules in feet or meters, we would have to say that a hydrogen atom (the smallest atom) is 7.874×10^{-10} feet or 2.4×10^{-10} meters. Instead, we can use nanometers and say that the hydrogen atom is 0.24 nm.

The nanoscale, then, is the size scale at which nanotechnology operates. Though we have a lower limit on this scale size (the size of one atom), pinning down an upper limit on this scale is more difficult. A useful and well-accepted convention is that for something to exist on the nanoscale, at least one of its dimensions (height, width, or depth) must be less than about 100 nanometers. In fact, it is these limits to the nanoscale that the National Nanotechnology Initiative (NNI) uses for its definition of nanotechnology: "Nanotechnology is the understanding and control of matter at dimensions of roughly 1 to 100 nm, where unique phenomena enable novel applications."[1] To this, it is useful to add two other statements to form a complete definition. First, nanotechnology includes the forming and use of materials, structures, devices, and systems that have unique properties because of their small size. Also, nanotechnology includes the technologies that enable the control of materials at the nanoscale.

Though we have established that the 'nano-' in 'nanotechnology' signifies a particular scale, it is important to get a good idea of what that scale is – that is, what size the nanoscale is in relation to our everyday experience. There are several analogies that we can use to explain the size of a nanometer in relation to sizes that are more commonly known. For example, it takes 50,000 nm to make up the width of a single strand of human hair. Another example is as follows: a nanometer compared to the size of a meter is roughly of the same proportion as a golf ball compared to the size of the Earth. Perhaps the best way to illustrate the nanometer scale is by describing the range of length scales from the centimeter down to the nanoscale. An ant is on the order of 5 mm (10^{-3} meters) in size. The head of a pin is 1–2 mm. A dust mite is 200 μm (10^{-6} meters) in size. A human hair is about half the size of a dust mite, 100 μm wide. The red blood cells that flow in our veins are about 8 μm in diameter. Even smaller than that, the ATP synthase of our cells is 10 nm (10^{-9}) in diameter. The size of the double helix of DNA on the nanoscale is about 2 nm wide. Finally, atoms themselves are typically less than a nanometer in size and are sometimes spoken about in terms of angstroms (10^{-10} meters).

1.2   The Origins of Nanotechnology

Nanotechnology, like any other successful technology, has many founders. In one sense, the very field of chemistry has been working on nanotechnology since its inception, as have materials science, condensed physics, and solid state physics. The nanoscale is not really all that new. But investigating and designing with a specific eye on the nanoscale is new – and revolutionary.

The term 'nanotechnology' can be traced back to 1974. It was first used by Norio Taniguchi in a paper entitled "On the Basic Concept of 'Nano-Technology'."[2] In this paper, Taniguchi described nanotechnology as

[1]   "What is Nanotechnology?" National Nanotechnology Initiative. Available at http://www.nano.gov/html/facts/whatIsNano.html (accessed October 11, 2008).
[2]   Norio Taniguchi, "On the Basic Concept of Nanotechnology," *Proceedings of the International Conference of Production Engineering, London, Part II*. British Society of Precision Engineering, 1974.

the technology that engineers materials at the nanometer level. However, nanotechnology's history predates this. Traditionally, the origins of nano-technology are traced back to a speech given by Richard Feynman at the California Institute of Technology in December 1959 called "There's Plenty of Room at the Bottom."[3] In this talk, Feynman spoke about the principles of miniaturization and atomic-level precision and how these concepts do not violate any known law of physics. He proposed that it was possible to build a surgical nanoscale robot by developing quarter-scale manipulator hands that would build quarter-scale machine tools analogous to those found in machine shops, continuing until the nanoscale is reached, eight iterations later. As we will see, this is not exactly the path that nanotechnology research has actually followed.

Feynman also discussed systems in nature that achieve atomic-level pre-cision unaided by human design. Furthermore, he laid out some precise steps that might need to be taken in order to begin work in this uncharted field.[4] These included the development of more powerful electron microscopes, key tools in viewing the very small. He also discussed the need for more fundamental discovery in biology and biochemistry. Feynman concluded this talk with a prize challenge. The first challenge was to take "the infor-mation on the page of a book and put it on an area 1/25,000 smaller in linear scale, in such a manner that it can be read by an electron micro-scope."[5] The second challenge was to make "an operating electric motor – a rotary electric motor which can be controlled from the outside and, not counting the lead-in wires, is only 1/64 inch cube."[6] He ended the talk by saying "I do not expect that such prizes will have to wait very long for claimants."[7] He was right about one of the prizes: the motor was built fairly quickly and by a craftsman using tools available at the time. However, it was not until 1985 that a graduate student at Stanford named Tom Newman reduced the first paragraph of Charles Dickens' *A Tale of Two Cities* to 1/25,000 its size.

In his paper on the basic concept of nanotechnology, Taniguchi devel-oped Feynman's ideas in more detail. Taniguchi stated, " 'Nano-technology' is the production technology to get the extra high accuracy and ultra fine dimensions, i.e., the preciseness and fineness of the order of 1 nm

[3] Richard P. Feynman, "There's Plenty of Room at the Bottom," *Journal of Micro-electromechanical Systems* 1 (1992): 60–6.
[4] Physicists, especially, had not explored this field much. Most physicists at the time were focused on high-energy physics, probing into the atom to look at quarks and subnuclear reac-tions, astrophysics, or nuclear physics. Much of what fell in between was left to chemists and engineers, as many of the fundamental equations in this area of the natural world were believed to be explained already.
[5] Feynman, "There's Plenty of Room," p. 66.
[6] Ibid.
[7] Ibid.

(nanometer), 10^{-9} m in length. With materials the smallest bit size of stock removal, flow, or design is one atom (generally about 1/5th of a nanometer), so the limit of fineness of materials is on the order of one nanometer."[8] In the paper, Taniguchi discussed his concept of 'nanotechnology' in materials processing, basing this on the microscopic behavior of materials. Taniguchi imagined that ion sputtering would be the most promising process for the technology. As we will discuss later in this section, many tools and techniques are used for the development of this type of nanotechnology.

Then, in 1987, K. Eric Drexler published his book, *Engines of Creation: The Coming Era of Nanotechnology.*[9] Aimed at a non-technical audience while also appealing to scientists, Drexler's book was a highly original work describing a new form of technology based on molecular "assemblers," which would be able to "place atoms in almost any reasonable arrangement" and thereby allow the formation of "almost anything the laws of nature allow." This may sound like a fanciful and fantastical idea but, as Drexler points out, this is something that nature already does, unaided by human design, with the biologically based machines inside our own bodies (and those of any biological species). There has been significant debate about the possibilities, promise, and troubles with what is now called "molecular manufacturing." Even the possibility of these machines is widely debated. Suffice it to say, however, that *Engines of Creation* marks a distinct jumping-off point for nanotechnology and the associative scientific research. Even though much of this research had nothing at all to do with molecular manufacturing, the focus on the scale of the research objects became the most important factor. Tools developed to handle individual atoms, such as the scanning probe microscope at IBM, enabled researchers to study and manipulate individual atoms and molecules with a degree never before possible. In a very famous image, researchers at IBM moved xenon atoms around on a nickel substrate with a scanning tunneling microscope. This image used the atoms to spell out the company's logo, "IBM." Electron microscopes developed to the point that they could be in more and more research environments (including, sometimes, in biological applications as Feynman desired). Able to see individual atoms and their arrangements within materials, researchers began to study developing atomically precise materials and devices.

The discovery of novel materials on the nanoscale notably began with the Buckminsterfullerene (also called the buckyball). The buckyball was so named because of the resemblance to the geodesic domes that the architect Richard Buckminster Fuller popularized. Discovered in 1985 at Rice University, it consists of an arrangement of 60 carbon atoms. In 1991,

[8] Taniguchi, "On the Basic Concept."
[9] Eric Drexler, *Engines of Creation: The Coming Era of Nanotechnology* (New York: Broadway Books, 1987).

nanoscale materials became the focus of intense research with the discovery of the carbon nanotubes by Sumio Iijima of NEC. At a somewhat feverish pace, novel nanoscale material after novel nanoscale material has been reported ever since. Nanotechnology is now recognized as the future of technological development. In 2000, the US government developed the National Nanotechnology Initiative (NNI) in order to administer funding and develop nanotechnology as the major research thrust of the twenty-first century.

1.3   The Current State of Nanotechnology

Having looked at the basic history of nanotechnology, we can now investigate where we currently stand. In particular, how is nanotechnology researched in laboratories across the world today? What is the current direction of nanotechnology research? The answer helps us understand the development, characterization, and functionalization of nanoscale materials and the science that governs them. This involves three main thrusts of research: nanoscale science (or "nanoscience" – the science of interaction and behavior at the nanoscale), nanomaterials development (the actual experimental development of nanoscale materials, including their use in device applications), and modeling (computer modeling of interactions and properties of nanoscale materials).

Understanding the underlying science of nanoscale interactions is extremely important to the development of nanotechnology; these interactions constitute one of the main areas of research in the field of nanotechnology. The laws of physics that operate on objects at the nanoscale combine classical (or Newtonian) mechanics, which governs operations of everyday objects, and quantum mechanics, which governs the interactions of very small things. Though many of the fundamental laws of nature that operate on this level have been discovered, science at this scale is still very difficult. Quantum mechanics works at this scale, but the interplay between the high number – that is, greater than two – of atoms in nanoscale materials can make it difficult to predict the actual outcome of these interactions. Furthermore, classical mechanics works on this scale, but the small size of the materials and the close scale of the interactions can make forces that are well understood at large scales (e.g., friction) and/or powerful at those scales (e.g., gravity) mysterious and/or less powerful at the nanoscale. Understanding the forces and theories at play within nanotechnology is just one aspect of nanoscience.

Another very important aspect of nanoscience is understanding the formation of nanoscale materials and devices. In looking at the nanoscale, traditional (non-nano) materials, structures, and devices are often referred to as "bulk technology." To be sure, this "bulk" style of technology has led to many great accomplishments: we easily make wonderful computing devices, ultra-strong steel, and very pure ceramics. Using bulk technology, we can create exquisitely small devices and materials. However, this creation is still done by cutting, chipping, pounding, extruding, melting, and performing other such bulk procedures to materials to create the new device, structure, or material. The main difference with nanotechnology is this creation process. With nanotechnology, we start on the atomic scale and, controlling atomic/molecular placement and arrangement, we build up the technology into unique devices, materials, and structures. This new type of formation requires new types of synthesis, requiring a new understanding of the formation of materials on the nanoscale. Furthermore, many materials have extremely unique properties when they are developed at a nanoscale. Many materials configure themselves in different atomic arrangements not seen in the bulk form of the same materials. Understanding the changes that these materials undergo as they are formed on a smaller scale is vital to developing the use of these materials in devices.

Nanotechnology today focuses, as we have already mentioned, on the development, understanding, and use of materials at the nanoscale, or nano-materials. Materials are made up by an arrangement of particular atoms – typically in a specific way – which helps define the property of the material. For example, steel is one of the strongest engineering materials, and its strength will increase as carbon is added to it. Steel is made of mostly iron with other elements added: stainless steel, for example, contains 10 percent chromium to protect the material against corrosion.

Actually, materials throughout history have basically defined the technology of the age. We refer to the Stone Age and the Iron Age because of the types of materials that were used or developed during those eras to make the technology that was used in everyday life. For example, the Bronze Age was a period in civilization's development when the most advanced widespread metalworking consisted of techniques for smelting copper and tin from naturally occurring ore and then alloying those metals in order to cast bronze. In more modern times we can see that the development of methods for using silicon and other semiconductors is essential to developing modern computing devices and many other devices that we use on a daily basis.

On the nanoscale, this association between material, device, and structure only amplifies since these are virtually indistinguishable from each other at that scale. Nanomaterials can also be used in conjunction with other materials, thus augmenting the properties of those other materials. In this sense, nanoscale materials have been important to the materials field for some time. For example, nano-sized carbon black particles have been used to reinforce tires for nearly 100 years. Another, more common, example is precipitation hardening of materials. Precipitation hardening is a heat treatment technique that is used to strengthen materials, particularly some metals. It relies on producing fine, impure nanoscale particles, which then impede the moving of defects within the material. Since these defects are the dominant cause of plasticity in materials, the treatment hardens the material. This accidental discovery in 1906 allowed for significant improvements in the strength of aluminum. At the time, researchers could not image these precipitates. It was later discovered that the precipitates were nano-sized particles.

Materials are the essence of technology at the nanoscale. Because of the scale of the technology, the atomic species and structure define not only the properties of the material but also the function of the device. Furthermore, different materials interact differently with their environment when they are sized on the nanoscale. Bulk materials interact with their environment in a certain way because the vast majority of their atoms are inside the volume of the material rather than on the surface; this makes the surface-to-volume ratio very small. Atoms respond to their environment differently when they are surrounded by other atoms than when they are on a surface and do not have atoms surrounding them. And the relative amount of atoms on the surface can greatly influence the properties of the material as a whole. With nanoscale materials, many of the atoms reside on the surface of the material and therefore the surface-to-volume ratio is much larger. For example, a spherical particle that has a radius of 100 micrometers will have a surface-to-volume ratio of about 30,000. This may seem large, but the percentage of atoms on the surface of the material is very small, only 0.006 percent (that is, only 6 out of every 100,000 atoms are on the surface). Compare this with the surface-to-volume ratio of a particle that has a radius of 10 nm. This ratio would be 300,000,000. Here, the percentage of atoms on the surface of the material is a much larger 6 percent (or 6 out of every 100). This can radically change the properties of the material, how it interacts with its environment, and how it can be used in devices.

Another important aspect of nanotechnology is the modeling of nanoscale devices, materials, and interactions. Modeling has an interesting history, through which initially scientists would build actual physical models of molecules and structures in their offices. With these models they would perform the calculations for each atom, move them around on their model, and then start the process all over again – iteration after iteration. John Desmond Bernal, a scientist who pioneered using X-rays to examine the structure of materials, was one such modeler. As he wrote: "I took a number of rubber balls and stuck them together with rods of a selection of different lengths ranging from 2.75 to 4 inch. I tried to do this in the first place as casually as possible, working in my own office, being interrupted every five minutes or so and not remembering what I had done before the interruption."[10] Clearly, another way to perform these calculations was necessary.

---

[10] John Desmond Bernal, "The Bakerian Lecture, 1962: The Structure of Liquids," *Proceedings of the Royal Society* 208 (1964): 299–322.

Though the simplest calculations can be done by hand on a sheet of paper, computers are required for understanding and modeling the behaviors of large systems (including biological molecules and chemical systems). For these, a massive amount of computing power is necessary. Modern modeling systems that are used in predicting the behavior of nanoscale systems all rely on the atom as their fundamental unit.[11] Modeling provides an approximate solution. The reason that an exact solution cannot be determined for many arrangements is known as the "many-body problem." This problem is best illustrated by looking at quantum mechanics. When we are trying to determine the energy of an atom, the electrons play a central role. Let us look at a few equations just to see how much more complicated they become as electrons are added; without any technical knowledge about how to solve the equations, this point should still be readily apparent. For example, if we wanted to determine the energy of a hydrogen atom (with one electron and one proton), the equation looks like this:

$$E_\alpha = \frac{<\psi_\alpha|\hat{H}|\psi_\alpha>}{(<\psi_\alpha|\psi_\alpha>)}$$

where the symbols are standard quantum mechanical symbols.[12] This problem can be solved by hand. Now, if we add an electron to the system as in a helium atom, the equation becomes more complicated:

$$\left[-\frac{1}{2}\nabla_1^2 - \frac{1}{2}\nabla_2^2 - \frac{Z}{r_1} - \frac{Z}{r_2} + \frac{1}{|\overline{r_1} - \overline{r_2}|}\right]\psi(\overline{r_1}, \overline{r_2}) = E_{el}\psi(\overline{r_1}, \overline{r_2})$$

This is still solvable by hand, though it would take a little more work. Beyond this, however, the problem becomes unsolvable exactly; it has to be approximated. The equation for the many electrons looks like the following:

$$\left[-\frac{1}{2}\sum_i \nabla_i^2 - \sum_i \frac{Z}{r_j} + \sum_i \sum_{j>i} \frac{1}{|\overline{r_i} - \overline{r_j}|}\right]\psi(\overline{r_1}, \ldots, \overline{r_n}) = E_{el}\psi(\overline{r_1}, \ldots, \overline{r_n})$$

This equation is unsolvable when the number of electrons is greater than two. The tabulation of a function of one variable requires about a page, but a full calculation of the wave function (the description) of the element

---

[11] This may not seem all that special, but physical and chemical law relies on quantum calculations in which the atom is not the fundamental unit. In quantum chemistry, the electron is the fundamental unit; it helps determine the bonding of the atoms and the electrical/magnetic characteristics of the material.

[12] It is not within the scope of this work, nor is it necessary to get into the details of the calculation. Rest assured that this calculation is relatively simple and can be performed by hand on the back of an envelope to receive an exact answer.

iron has 78 different variables in it. Even if we simplified the number of each variable to a number like 10 (a very crude approximation), the full determination would require 10^{78} entries. That's a one with 78 zeros after it for just one iron atom. Imagine if we want to determine the properties and interactions of more complex systems! Clearly, approximation is necessary.

Computer simulation and modeling are not without their limitations. A simulation is at best as good as any underlying assumptions. Oftentimes the assumptions and simplifications are ill-conceived, which can cause the results to be misleading. Furthermore, long simulations can be ill-conditioned (i.e., not well suited for computation) and will accumulate errors. A better choice of algorithm can help relieve this problem, but it cannot eliminate it completely. Also, many of the functions used for simplification are not very good for large systems (because of the problems mentioned earlier), and the molecular dynamics simulations that are based on them will come out flawed. Overall, the larger the system being modeled, the more problematic the simulation will be and the more its results will deviate from physical reality.

1.4   The Future of Nanotechnology

The future of nanotechnology has been the subject of myriad books and articles, which span across many genres. Non-fiction books have included Drexler's *Engines of Creation*, John Storrs Hall's *Nanofuture: What's Next for Nanotechnology*,[13] *The Spike* by Damien Broderick,[14] Ray Kurzweil's *The Singularity Is Near*,[15] and investment books such as *Investing in Nanotechnology* by Jack Uldrich.[16] Fiction works include *Prey* by Michael Crichton[17] and Greg Bear's *Slant*.[18] These works cover nanotechnology in a wide variety of ways, from alarmism over technology run amok to Pollyanna-like predictions of a future utopia.

The discussion of the science and development of nanotechnology in this book emphasizes near-term issues and applications; our priority is more to characterize these properly than to offer speculative commentary which would very likely end up wrong. However, it is important to keep in mind the

[13]   J. Storrs Hall, *Nanofuture: What's Next for Nanotechnology* (New York: Prometheus Books, 2005).

[14]   Damien Broderick, *The Spike: How Our Lives Are Being Transformed by Rapidly Advancing Technologies* (New York: Forge Books, 2002).

[15]   Ray Kurzweil, *The Singularity Is Near: When Humans Transcend Biology* (New York: Viking, 2005).

[16]   Jack Uldrich, *Investing in Nanotechnology* (New York: Adams Media Corporation, 2006).

[17]   Michael Crichton, *Prey* (New York: HarperCollins, 2002).

[18]   Greg Bear, *Slant* (New York: Tor Books, 1998).

longer-term implications of nanotechnology, especially in terms of thinking about how the technology will be developed and the more ambitious applications that are driving that development. Researchers think about this when they apply for grants and when they deliberate what type of research to pursue. Grant managers consider the future of technology when they decide which projects to fund. Companies consider the future of technology when deciding how to develop their own financing and technology development.

Here, we will briefly examine some of the future of nanotechnology, though we will also return to the prospect of its particular applications in subsequent chapters. We want to keep in mind some of the broader developments toward which nanotechnology is working. The discussion will include nanotechnology's impacts on computing and robotics, medicine, and molecular machining. We will find that there is significant overlap between these categories and developments.

Nanotechnology computing development is foreseen in two forms. The first is in the development of much smaller devices with much better properties of the computing circuits. This evolution of computing allows for higher density of circuitry and newer architectures, giving greater and greater computing power. It follows a similar path to that of current computing technology. Most readers are familiar with Moore's law, which predicts the doubling of computing power every 24 months.[19]

Kurzweil has written extensively about what this law means (and accelerating exponential growth) and what it means in relation to the development of nanotechnology:

> Moore's Law of Integrated Circuits was not the first, but the fifth paradigm to provide accelerating price-performance. Computing devices have been consistently multiplying in power (per unit of time) from the mechanical calculating devices used in the 1890 US Census, to Turing's relay-based "Robinson" machine that cracked the German Enigma code, to the CBS vacuum tube computer that predicted the election of Eisenhower, to the transistor-based machines used in the first space launches, to the integrated-circuit-based personal [computers].[20]

The second way in which nanotechnology could bear on computing is far more revolutionary. Smaller and more powerful computing allows for the development of nanoscale machines (sometimes called "nanobots").

---

[19] Moore's law is variously formulated with differences in the doubling time quoted (from one to two years) and the actual measure that is being doubled (from the number of transistors on a single chip to computing power). The law owes to Intel co-founder Gordon Moore, who used the two-year figure for the doubling of transistors on a chip. See Gordon E. Moore, "Cramming More Components onto Integrated Circuits," *Electronics* 38.8 (1965): 56–9. See also Intel's website at http://www.intel.com/technology/mooreslaw/ (accessed August 17, 2007).
[20] Ray Kurzweil, "The Law of Accelerating Returns," March 7, 2001. Available at http://www.kurzweilai.net/articles/art0134.html (accessed April 25, 2007).

These machines could be autonomous – as envisioned in Crichton's *Prey* – or they could require outside control. The small nature of these machines would provide for significant advantages over larger robots. For example, nanobots could venture to places previously unthinkable for machines, such as into the bloodstream or into cells. They could be used on rescue missions, searching in places that are too dangerous or too small for larger robotics or humans to venture. They could serve as a laboratory on a chip and be sent to study environments that are harmful for humans, such as the inside of volcanoes, tornadoes, and hurricanes or even outside the Earth's atmosphere. Consider the success of the two rovers that have been studying Mars for the past four years (at the time of writing). They have helped us investigate and learn more about the planet Mars than have years of study by telescope. Smaller-scale (nanoscale) robotics could be made with less material, which would ultimately make them cheaper and require less energy. Though a single nanobot may not be able to provide the level of analysis that a larger robot could, networks of these machines working together could use effective parallel processing to study much more.[21]

Nanoscale robots could also assist with medical treatments; we will discuss this in more detail in Chapter 11 (esp. §11.3), but offer some preliminary discussion here. A nanoscale machine could be programmed to flow through the bloodstream and seek out diseased cells. These cells could then be destroyed specifically. Alternatively, new medicines that target only very specific cells or parts of the body (highly selective medicines) could be developed to eliminate many of the adverse side effects of treatments. Chemotherapy is one treatment that could greatly benefit from this. Traditionally chemotherapy essentially involves a race to kill the diseased cancerous cells before the healthy cells are killed as well.[22] Being able to target only tumor cells with treatments would be a great boon to fighting cancer. In fact, much of the research in nanomedicine focuses on cancer treatment. Consider the use of dendrimers. A dendrimer is a synthetic macromolecule comprising branched repeating units, much like a tree that grows in three dimensions from a core root; the properties of this molecule are controlled by the surface functional groups.[23] Using dendrimers, James Baker at the University of Michigan has shown how to transport a powerful drug inside tumor cells. Folic acid is attached to the surface of the material, and in this way the dendrimer is selectively absorbed by the cancer cells. The

---

[21] These nanobots may not even require powering their own locomotion once they are on the surface. As light as and smaller than dust, the "nanobots" could simply be blown about by the winds on the planet. Of course we would lose some control over *where* the nanobots are, but a much greater portion of the surface of the planet could be covered in much less time.
[22] Of course, this is an extreme oversimplification of cancer treatments.
[23] See §7.5 for further uses of dendrimers in medical applications, specifically HIV/AIDS prevention.

surface also has anti-cancer drugs attached to it, thereby achieving selective delivery of the drug.[24] Also, sensor test chips containing nanowires made to selectively detect different diseases and irregularities in blood can be developed to detect, *in vivo* and in parallel, many other different diseases.

Of course, the idea of "nanomedicine" is not limited to curing diseases. Using nanoscale and nano-enabled robots to assist in surgery could allow for more precise and safer surgery; already robots with steadier hands than humans are used to perform certain surgeries. Nanoscale robots could enter the body and perform controlled surgery on individual cells, tissues, or organs without requiring an external incision. Human enhancement is another area where nanomedicine is being developed: mankind has a long history of using tools and technology to enhance and augment the natural ability of humans, and nanotechnology can easily be imagined to enhance many different aspects of human performance, from memory to physical ability. We discuss enhancement in greater detail in Chapter 12 (though also see §9.2 for military applications).

Molecular manufacturing was described in detail by Drexler. In particular, he used the biological assembler as supposed proof of the feasibility of assembling atoms and molecules in extremely precise arrangements. According to Drexler:

> Nature shows that molecules can serve as machines because living things work by means of such machinery. Enzymes are molecular machines that make, break, and rearrange the bonds holding other molecules together. Muscles are driven by molecular machines that haul fibers past one another. DNA serves as a data-storage system, transmitting digital instructions to molecular machines, the ribosomes, that manufacture protein molecules. And these protein molecules, in turn, make up most of the molecular machinery.[25]

This vision, while exciting, nevertheless faces challenges. The primary contention is over the feasibility of molecular manufacturing; i.e., whether it is scientifically possible. A well-publicized "debate" between Drexler and Richard Smalley – a Nobel Prize winner and the discoverer of the buckyball – appeared in the pages of *Chemical and Engineering News*.[26] The two have very different views concerning what nanotechnology is and how it works. The scientific objections to the idea of molecular assemblers are summed up as the "fat fingers problem" and the "sticky fingers problem."

---

[24] Jolanta F. Kukowska-Latallo et al., "Nanoparticle Targeting of Anticancer Drug Improves Therapeutic Response in Animal Model of Human Epithelial Cancer," *Cancer Research* 65.12 (2005): 5317–24.
[25] K. Eric Drexler, "Appendix A: Machines of Inner Space," in B.C. Crandall and James Lewis (eds), *Nanotechnology: Research and Perspectives* (New York: MIT Press, 1993), pp. 325–6.
[26] K. Eric Drexler and Richard Smalley, "Point–Counterpoint: Nanotechnology," *Chemical and Engineering News* 81.48 (2003): 37–42.

The "fat fingers problem" refers to the lack of space available to build a molecular assembly. To assemble a structure one molecule at a time, multiple manipulators are necessary. There simply is not enough room in the nano-sized reaction region to account for all of the atomic-sized "fingers" of all the manipulators necessary to have complete control of the chemistry involved. The "sticky fingers problem" refers to how atoms stick to each other via bonding: this adhesion makes the atoms of the manipulator hands adhere to the atom that is being moved around. Even despite these problems, future nanotechnology may still take a wide range of paths leading to radical new types of technology.

We now turn to familiar forms of nanoscale technology used in the world around us.

1.5   Nanotechnology in Nature and Applications

Natural nanotechnology

Nanotechnology has emerged as a discipline only recently. Like many other technological disciplines, it draws a significant amount of inspiration from nature. With many more years of development than human technology and a very large "laboratory" in which to test new ideas, nature has inspired innovations in technology for a very long time. Going back to the designs of Leonardo da Vinci, inspiration by nature for technologies to help humans are very evident. For example, da Vinci studied in extensive detail the flight of birds to help him design his plans for a helicopter and a hang glider. The wings of many of his hang gliders were based upon those of bats. Modern technology also draws on many concepts observed in nature.

Understanding how nature uses nanoscale forces and materials to achieve some very remarkable things can be used to design engineered devices serving other purposes. Biomimetics (the research field that deals with recreating and mimicking nature's mechanisms in technology) attempts to take advantage of nature's billions of years of evolutionary experience in order to make more effective materials and technology. By looking at how evolutionary pressures have created efficient and adapted traits and structures, scientists can try to use some of those traits and structures in their own designs. Or, as a corollary, failed designs would have been eliminated in nature, so, by copying nature, scientists will avoid at least some of those failures.

One of the more famous examples of biomimetic technology is Velcro. Velcro was developed in 1941 by Georges de Mestral after he noticed burdock seeds sticking to his clothes and his dog's fur. On examining the ball-like seeds, he noticed the hook-ended nature of their fibers that allowed them to stick to other materials. This evolutionary trick allows the seeds to travel and spread: the hooks of the burrs "grab" onto loops of thread or fur. Evolution designed this to help spread the seed, but de Mestral used it in order to aid in adhesion and create the common application, Velcro, that we know and use today.

Scientists have recognized the special functionality of nanoscale materials for some time. In the natural world, the impact of design on the nanoscale is well known and nature has evolved some very interesting uses for nano-materials. For example, some bacteria have magnetic nanoparticles inside them, which are used as a compass and help provide a sense of direction to the bacterium. Even larger creatures have taken advantage of nanoscale design. Gecko foot hair, nanoscale in size, has been shown to be central in the gecko's exceptional ability to climb rapidly up smooth vertical surfaces. Even the most basic building blocks of biological things are an example of nano-scale design. Most forms of movement in the cellular world are powered by molecular motors that use sophisticated intramolecular amplification mech-anisms to take nanometer steps along protein tracks in the cytoplasm.

Gecko feet are an interesting example of nanoscale design providing func-tionality on a larger scale. Geckos can hang out on a wall and effortlessly cling to just about anything. As far back as Aristotle, it was observed how they can "run up and down a tree in any way."[27] Until recently it was not known exactly how the geckos accomplished such feats, but a 2002 research article in the *Proceedings of the National Academy of Sciences*[28] explained it as a dry adhesion mechanism. The mechanism relies on a force known as the van der Waals force. This force will be discussed in more detail in §2.2, but it implies that the gecko's adhesive properties are not the result of surface chemistry or an epoxy. Instead, they are the result of the size and shape of seta tips. Greater adhesive strength is achieved simply by having a greater surface area (that is, by subdividing the setae).

Another example of nanotechnology in nature is magnetotactic bacteria. "Magnetotactic" is the name of a class of bacteria that orient themselves along the magnetic field lines of the Earth's magnetic field, in much the same way that a compass does. These bacteria were first reported in 1963. This ability to orient themselves arises from the presence of chains of magnetic materials inside their cells. This magnetic material is typically either magnetite (Fe_3O_4) or greigite (Fe_3S_4). That nature uses such a chain of materials that look much like nanowires implies that such magnetic nanoscale wires could be used in technological applications as well.

In a greater sense, the mechanisms that control any biological cell oper-ate at the nanoscale and represent a source of inspiration for a whole array of nanotechnology applications. If nanomedicine is to have any impact, we must understand and control movement through cells, bloodstreams, and

---

[27] Richard McKeon and C.D. Reeve, *The Basic Works of Aristotle* (New York: Modern Library, 2001), p. 140.
[28] Kellar Autumn et al., "Evidence for van der Waals Adhesion in Gecko Setae," *Proceedings of the National Academy of Sciences U.S.A.* 99.19 (2002): 12252–6.

biological environments. What better place to draw inspiration than from nature itself, which has been grappling with these problems for eons?

It is important, however, to understand that biology and evolutionarily designed solutions can only take us so far in technological development. Because nature has evolved mechanisms suited for certain purposes (e.g., survival and reproduction), human intuition and guidance are needed in order to use technology for other particular purposes. Furthermore, though nature may have highly optimized solutions for problems, they may not be the *most* optimal solutions. From one generation to the next, nature has a limited toolkit upon which to draw. Humans will not, all of a sudden, develop solid state memory storage simply because it is "better" in some sense than our biological memory storage. We will not develop, through evolution, the ability to walk on walls as geckos have. Where evolution has brought us from has a large impact on where we will go. Our toolkit of biological solutions determines, to a large extent, the toolkit that we will develop through evolution.

Historical nanotechnology

Nanotechnology has not only been present in nature, but has also been used unwittingly in human-made technology for centuries. For example – and as already mentioned in §1.3 – nanoscale carbon black particles, basically high-tech soot, have been used as a reinforcing additive in tires for nearly 100 years. Nanomaterials have also been used unwittingly by artisans for centuries. When gold is significantly reduced in size, it no longer retains its familiar yellow-metallic appearance, but can take on an array of colors. The red paint used on Chinese vases as far back as the Ming dynasty is the result of Chinese artisans grinding up gold particles until they are on the order of 25 nm in size. Separately, medieval artisans in Europe discovered that by mixing gold chloride into molten glass they could create a rich ruby color. By varying the amount of gold put in the mixture, different colors are produced. Though the cause was unknown at the time, the tiny gold spheres were being tuned to absorb and reflect the sunlight in slightly different ways, according to the size of these particles. Nanomaterials have been used unknowingly to make stained glass by grinding up gold and silver nanoparticles to small sizes. Both gold and silver nanoparticles change color significantly according to their size and shape. At 25 nm in diameter and spherical, gold is red, at 50 nm and spherical, it is green, and at 100 nm and relatively spherical, it is orange. Silver is blue at 40 nm when spherical, yellow at 100 nm when spherical, and red at 100 nm when prismatic.

Though these are early examples of using the nanoscale size to enhance or change the property of a material, they at best qualify as accidental nanotechnology. In order to categorize them as nanotechnology, they would have to have been intentional. The main difference between the modern push for technology and these previous historical examples of nanoparticle use is that the modern use is intentional and with an understanding of the underlying mechanisms that produce the new properties.

The multidisciplinary world

We have written a lot about materials because they are central to the nanoscale world and, therefore, to the development of nanotechnology. However, nanotechnology draws from a wide range of disciplines. Because it requires extensive knowledge of chemical interactions to manipulate matter at the molecular scale, chemistry is very important to nanotechnology. Because, on the nanoscale, the laws of nature merge the laws of the very small (quantum mechanics) and the laws of the large (classical physics), physics is very important to understanding nanotechnology. Because the material is vital to the function of device and structure, the science and engineering of materials are vital to developing nanotechnology. Specific applications require even more disciplines. Medical applications such as drug delivery require knowledge of biomedical applications. Even more, because nanoscale devices will have to move through a bloodstream, fluid dynamics becomes important.

The point is that this wide range of disciplines brings interesting opportunities and challenges to researchers. There are opportunities for nearly any scientist to be involved in developing nanoscale science and technology. One of the true revolutionary effects of developing nanotechnology is that scientists of all different stripes are working together on a wide variety of projects. Physicists are providing insight into curing cancer, chemists are developing new types of lights, and many more bridges are built between disciplines that once were separated from each other.

The challenges are also numerous. Each field brings its own expertise and terminology. This makes communication between practitioners in these fields difficult. Consequently, a large amount of overlap in research occurs and insights take some time to be shared and understood. In order to solve this challenge, many multidisciplinary centers are being established. In these centers, biologists and physicists work side by side in hopes that proximity will breed cooperation in research, understanding of the different fields, and innovation from fresh perspectives on the problems of the fields. For example, the University of Chicago established an Interdisciplinary Research Building in 2003 in order to create such a cooperative environment. This environment is necessary because, as we have seen and will see, nanotechnology draws from many different disciplines. In subsequent chapters – and especially in the first unit of this book – we will provide a more scientific context for nanotechnology, thus elaborating on the themes developed in this chapter. The second and third units of this book will build on that context and show some uses to which nanotechnology is applied, as well as the associative social and ethical dimensions of those uses.

2

Tools of the Trade

The development of new technologies generally has been catalyzed by the development of new tools and theories that have enabled us to perceive further and deeper than we were able to before. Nanotechnology is no different. New technologies require new ways to analyze and understand the new phenomena that are occurring. With nanotechnology, these tools are still being developed, but many of the tools and theories in use incorporate developments in physics during the last 100 years. Our understanding of the world and its laws has changed radically in that period, as we have moved from the classical world of Newtonian mechanics and Galilean relativity to the world of quantum mechanics and special and general relativity. These are not small changes in our worldviews.

In classical mechanics, billiard balls bounce off each other and, given knowledge of all the forces on an object and using specific measuring tools, it is possible to determine the fate of that object – which seems charmingly quaint today. In quantum mechanics, the billiard balls may travel through each other or suddenly appear in some other place. It is also impossible to determine with accuracy the position of an object while also knowing its momentum. And we can only give the probability that an object may roll down a hill instead of up. It is theoretically possible – though exceedingly improbable – that all the molecules in the Nile River will suddenly begin to flow south rather than north for a moment. In classical Galilean relativity, if someone were travelling at a certain speed and threw a ball forward, he could simply add the two speeds together to discover the speed of the ball. Now, with the discovery of modern relativity, we have to qualify the speed according to the reference frame from which we are speaking, and we can no longer simply add these speeds to find the speed of the ball. Furthermore, no matter how much we try and no matter how much energy we put behind it, the laws of nature will not let us accelerate an object faster than the speed of light.

The changes that have wrought our understanding of the world in the past 100 years extend beyond the field of physics. Biology has seen huge progress as well. The discovery of the DNA molecule and its structure as a double helix has changed the way the field defines and applies itself, and the new field of genetics has emerged. This genetic understanding of life has revolutionized medical technology and biomedical engineering.

So what do these changes mean for the development of nanotechnology? What tools do they help create that have enabled us to begin to develop technology at a nanoscale? What theories both empower and limit nanotechnology? This chapter will describe some of the basic tools that are used in nanotechnologies and discuss the theories that impact them.

2.1   Seeing the Nanoscale

In his 1959 talk (see §1.2), Richard Feynman said that if the physics community wanted to contribute something important to biology, it should develop an electron microscope that is 1,000 times more powerful than the electron microscopes then available. Electron microscopes today represent one of the most important tools for understanding processes and materials at a nanoscale. As one professor has put it, "if you aren't using [electron] microscopes, you aren't doing nano."[1] But what is an electron microscope? How is it used to create images? We are unable to see atoms and molecules with our eye. How do electron microscopes help us see them? Why are light microscopes not good enough for this task? Understanding these questions will elucidate the size and nature of nanotechnology, the difficulty in designing, utilizing, and measuring matter on such a small scale, and the challenges that face scientists working with nanotechnology. Microscopes have always been an important outlet for developing new technologies and improving new scientific techniques. Learning about the tools that are used in nanotechnology will allow us to understand how it is being developed and what its limitations are.

The nature of light

In order to answer those questions it is first necessary to understand the nature of light. At first glance, this may seem like a simple issue, but it is one of the oldest in science and one that we have only recently begun to fully understand. In Hellenistic theories, light was emitted from a person's eye, traveled infinitely fast, and illuminated objects. Ptolemy, who wrote about the refraction of light through materials, postulated this along with

---

[1]   Zhong Lin Wang, "Nanotechnology and Nanomanufacturing," in Hwaiyu Geng (ed.), *Semiconductor Manufacturing Handbook* (New York: McGraw-Hill Professional, 2005), ch. 22.3.

other Greek thinkers. In order to describe how night was dark, it was further postulated that the rays emitted from the eye interacted with rays from the sun. Descartes argued that light was a disturbance of the continuous substance of which the universe is made. By analogy with sound waves, Descartes assumed that light traveled faster in a dense medium (e.g., a solid) than a sparse medium (e.g., a gas). Though much of Descartes's theory has been discarded,[2] it is considered an early version of the wave theory of light.

The modern wave theory of light was first published by Robert Hooke[3] and Christian Huygens in the late 1600s.[4] They proposed that light was emitted in all directions as waves unaffected by gravity. This wave theory predicted that light waves could interfere with each other and that light could be polarized. Different colors of light were caused by different wavelengths (i.e., the distances between successive peaks of those waves). Because the theory required that light travel through a medium, it was presumed that there was an *ether* that permeated the universe.

Light, as we now know, is electromagnetic radiation. This was discovered in the mid 1800s by Michael Faraday who noticed that the angle of polarization of a beam of light could be altered by a magnetic field. Faraday theorized that light is an electromagnetic vibration. This means that it requires no medium in which to travel, no ether. Soon after, James Clerk Maxwell discovered that electromagnetic waves travel at a constant speed. This speed happens to be equivalent to the already known speed of light. Light, therefore, is an electromagnetic wave.

The so-called wave theory was successful in explaining a large array of optical problems and was considered one of the great triumphs of physics. However, by the early twentieth century, several effects appeared that could not be explained by the wave theory of light. The first of these was an effect observed in what is known as the Michelson-Morley experiment. The Michelson-Morley experiment was designed to determine the drift that occurred because of the motion of the Earth in the ether that permeated the universe. However, the experiment instead determined that there was no drift, basically confirming that light travels at a constant speed regardless of the frame of reference in which it is measured. This constant speed of light contradicted the basic mechanical laws of motion. In 1905, Albert Einstein resolved this paradox by revising Newton's laws of motion in order to account for this. This revision resulted in the theory of relativity. However, it did not call into question the fundamental wave-like nature of light.

---

[2] For example, Descartes's theory held that the universe is made from a continuous substance (viz, ether) and maintained certain commitments about the relative speeds of light in substances; both the ether and these commitments have now been rejected.

[3] Robert Hooke, *Micrographia* (1665). Available at http://www.gutenberg.org/etext/15491 (accessed October 22, 2008).

[4] Christian Huygens, *Treatise on Light* (1690). Available at http://www.gutenberg.org/files/14725/14725-h/14725-h.htm (accessed October 14, 2008).

The photoelectric effect did. When light strikes a metal surface, electrons are ejected from that surface, causing an electric current to flow across a voltage. It was observed that the energy of these electrons was dependent upon the frequency and not the intensity of the light, as was predicted by the wave theory of light. Moreover, for a certain metal, there was a certain frequency of light below which no current would flow. Einstein also solved this puzzle in 1905, this time by invoking the particle theory of light. Einstein proposed a description of "light quanta" and showed how these quanta could explain the photoelectric effect.[5]

A third anomaly with the wave theory of light was the problem of black body radiation. Here, the wave theory of light could not explain the measurements of the light spectrum of thermal radiators, called "black bodies." An example of a thermal radiator is the sun or the filament in an incandescent light bulb. In 1900, Max Planck developed a theory based on the idea that black bodies emit light only as discrete quanta of energy. Each quantum of light is called a 'photon,' which has an energy inversely proportional to the wavelength of light emitted:

$$E = \frac{hc}{\lambda}$$

where h is a constant known as Planck's constant, λ is the wavelength of light, and c is the speed of light.[6] The photoelectric effect along with black body radiation led to the establishment of quantum mechanics as a fundamental explanation of light.

The modern theory of light treats light as having a dual nature. In some situations, it is a wave. In other situations, it is a particle. This is known as "wave–particle duality." In fact, all objects have properties of both waves and particles, from large objects like planets (the wavelength is amazingly long) to small objects like electrons.[7] Since science is all about solving problems, whichever explanation fits best for a given problem will determine the property that light exhibits.

For optics and microscopes, the wave nature of light is the property that fits best. When we talk about the spectrum of light, we are talking about the range of wavelengths of light that are used. On one end of the spectrum are

---

[5] It was for this work, published in a paper entitled "On a Heuristic Viewpoint Concerning the Production and Transformation of Light," that Einstein was awarded the Nobel Prize in Physics in 1921. Albert Einstein, "Über einen die Erzeugung und Verwandlung des Lichtes betreffenden heuristischen Gesichtspunkt," *Annalen der Physik* 17 (1905): 132–48. Available at http://www.physik.uni-augsburg.de/annalen/history/papers/1905_17_132-148.pdf (accessed October 24, 2008).

[6] For this work, Planck was awarded the Nobel Prize in Physics in 1918.

[7] For discovering that electrons have a wave–particle duality, Louis de Broglie received the Nobel Prize in 1929.

gamma rays and X-rays. Then, there are ultraviolet rays (with wavelengths on the order of 10 to 100 nm). Longer wavelengths bring the visible spectrum. The visible spectrum of light actually makes up a very small part of the overall electromagnetic spectrum. Blue light is on the lower end of the visible spectrum at around 400 nm. Red light is on the long end of the spectrum at around 750 nm. Then there is infrared light up until 100,000 nm. Next on the spectrum are microwaves, which are on the order of 1 cm long. Radio waves make up the remaining end of the spectrum.

A useful tool that utilizes this wave-like nature of light is called spectroscopy. Spectroscopy is the study of the interaction between light waves and matter. Originally reserved for visible light interaction, spectroscopy is now used with many different types of light wave. Spectroscopy is mainly used to identify substances through the spectrum emitted from or absorbed by a material. Two examples of spectroscopy that are used in analysis are infrared (IR) spectroscopy and Raman spectroscopy.

IR spectroscopy uses the infrared region of the light spectrum, and it works because chemical bonds vibrate when they are illuminated by specific frequencies of infrared light. With IR spectroscopy, a beam of infrared light is passed through a material and the amount of energy absorbed by the material at each wavelength is measured. An absorbance spectrum is plotted up. Materials will then produce a specific signature of absorbed energy at certain wavelengths. This signature is compared against a library of known materials and the makeup of the unknown material can be determined. Raman spectroscopy works in a similar method, but instead of absorbing from a range of wavelengths, it relies on the scattering of a single wavelength of light. This laser beam of light interacts with the material and the energy of the laser light is shifted either up or down. This shift in the energy can be read and gives information about the makeup of the material.

Electron microscopes

What we are most concerned about in microscopes is their resolving power, or resolution. The resolving power of a lens is ultimately limited by diffraction, which comprises the various phenomena associated with wave propagation, like bending, spreading, or anything else that disrupts the wave of light. Light passing through the lens of a microscope interferes with itself. The result can be a blurring of the image. At best, we have an empirical diffraction limit that was determined by Lord Rayleigh in the early 1900s. For an ideal lens (which never really exists) of focal length f, we have a minimum spatial resolution of

$$\Delta l = 1.22 \frac{f\lambda}{D}$$

where D is the diameter of the focused light beam (for example, a laser) used to illuminate the material. Interestingly, this shows that a wide focused light beam can resolve a smaller spot than a narrow one.

To get a rough estimate of the resolution of a particular medium, a good rule of thumb is that the material being imaged should be at least as large as half the wavelength being used to image it. So, if we are to use visible light, the smallest object that can be imaged is on the order of 200 nm. Clearly, this is not small enough to resolve nanoscale materials, atoms, or molecules. Something of much lower wavelength is needed in order to image the nanoscale materials, atoms, and molecules that nanotechnology uses, such as electrons. It was not until the development of electron microscopy that imaging materials at the nanoscale became possible. For a typical low-voltage scanning electron microscope (SEM) with a beam energy of 5 kiloelectron volts (keV), the wavelength of the electron beam is around 0.0173 nm. For a 100 keV transmission electron microscope (TEM), the wavelength is even smaller, 0.0037 nm. For a 400 keV electron beam, the wavelength is 0.00028 nm. Of course, this is the theoretical limit and the actual resolution of the microscope is never as good as this. Aberrations, defects, astigmatisms, and other practical considerations combine to significantly reduce the resolution of the microscopes.[8] However, with certain configurations, these electron microscopes can even distinguish individual atoms. This enables us to discover and investigate nanomaterials and how their atomic makeup impacts their properties.

As mentioned above, there are two types of electron microscope: transmission and scanning. The transmission electron microscope (TEM) was first demonstrated and developed in the 1930s. Transmission electron microscopy involves a high-energy electron beam consisting of electrons emitted from a metal cathode and directed into a beam by magnetic lenses. The electron beam is transmitted through a thin slice of a material; the sample must be very thin so that the electrons can travel through it. The electrons are then magnified by magnetic lenses and focused onto a fluorescent screen or a charge-coupled device (CCD) camera. The TEM acts like a projector, projecting the sample being measured onto the fluorescent screen.

By using electrons to image objects, resolutions of better than a single atom can be achieved. A high-resolution transmission electron microscope image appears on a screen that is activated by electrons interacting with it. In it, the three-dimensional lattice of atoms in a material is projected onto a two-dimensional surface. The spherical imprint of the individual atoms appears as circular patterns, and the lattice arrangement of the atoms can be discerned. Though these images lack color, they can reveal a lot of data about the material.

---

[8] Zhong Lin Wang and Z.C. Kang, *Functional and Smart Materials: Structural Evolution and Structure Analysis*, 1st edn (New York: Plenum Press, 1998).

Transmission electron microscopy does have some limits, though. To begin with, as mentioned above, the sample must be very thin. It must be made thin enough to be transparent to electrons. There are different ways to prepare the material. One way is to cut a very thin slice of the sample. Or it can be fixed in plastic and then sliced. Any of these methods is time consuming and difficult. Furthermore, TEM imaging requires sending very high-energy electrons through a sample. This amount of energy can affect even the staunchest material. Imaging a material can also change the atomic structure of the material. This is evident in zinc sulfide where illumination with a 200 keV electron beam can change the atoms from a hexagonal arrangement to a cubic arrangement.[9] Beam interaction needs to be accounted for with almost any sample. Biological samples, the sort for which Feynman spoke of needing better electron microscopes, have even more difficulty being imaged. Biological specimens contain large quantities of water. Since TEM works in a vacuum, this water must be removed. However, removing the water can disrupt the sample. So the tissue must be preserved with different additives. These molecules cross-link with each other and then they become trapped as stable structures. The specimen is then dehydrated with acetone or a type of alcohol. Typically, it is then embedded in a polymer or plastic and sliced into very thin sections that can be imaged in the TEM.

Despite its limitations, TEM has proven to be one of the most effective tools for researching nanoscale objects. With specialized arrangements, *in situ* experiments can be run to determine all sorts of properties of different materials. In 1999, an ingenious *in situ* TEM experiment was used to determine the elasticity of carbon nanotubes. First, a single carbon nanotube was attached to a needle on a TEM sample holder using an electrical carbon epoxy. And then an alternating current was passed through the sample holder and the needle (and through the carbon nanotube). The nanotube vibrated – or resonated – according to the frequency of the current. From this, the elasticity of the nanotubes was determined in relation to the diameter of the tube.[10] In addition to measuring properties of nanoscale structures, transmission electron microscopy has been used to observe on an atomic level the formation of nanostructures and other atomic processes. Unfortunately, because of the sample preparation limitations, it is very difficult to observe molecular interactions and biological processes.

A very useful type of transmission electron microscope is a scanning transmission electron microscope (STEM). Like a TEM, the electrons still pass through the sample, but the electron beam is focused onto a narrow region, which is then scanned over the sample in a raster. This allows for

[9]   Daniel Moore et al., "Wurtzite ZnS Nanosaws Produced by Polar Surfaces," *Chemical Physics Letters* 385 (2004): 8–11.
[10]   Philippe Poncharal et al., "Electrostatic Deflections and Electromechanical Resonances of Carbon Nanotubes," *Science* 283.5407 (1999): 1513.

chemical analysis techniques such as mapping the sample by energy dispersive spectroscopy (EDS) and partial electron energy loss spectroscopy (PEELS). These data can be obtained at the same time as the image, allowing for real-time correlation.

The second type of electron microscope is the scanning electron microscope (SEM). The SEM uses an electron beam of much lower energy than a transmission electron microscope. Where a TEM uses an electron beam energy on the order of 200 keV, a typical SEM uses an electron beam with an energy on the order of 5–30 keV. The SEM has another major difference from a TEM. Electrons from an "electron gun" are beamed at the specimen. The beam irradiates the specimen, causing it to emit other electrons. The electrons are then detected. As the electron beam continuously sweeps across the specimen at a high speed, an image of the surface of the specimen is formed. The magnification of the beam can be changed by changing the width of the electron beam. Though resolutions on an atomic scale cannot be achieved by SEM, resolutions of up to 1 nm can be achieved. However, SEMs can image a large area of the specimen and can give a very good image of the surface morphology of nanoscale materials and devices.

Scanning electron microscopes typically require samples to be imaged in a vacuum and this limits their uses, especially with biological samples. Biological samples outgas, producing a lot of vapor while in a vacuum, and need to be dried or frozen. Also, samples that do not conduct electricity at all (any insulators and biological samples) need to be coated with a metal, such as gold, in order to be imaged. Environmental SEMs allow samples to be observed in humid and gaseous environments. This enables observing biological samples, phase transitions of materials, and gaseous sensors. Sensing technology is one of the main currently developing areas of nanotechnology, as we will see, and when nanowires are used in an environmental SEM the sensing process can be controlled and studied.

Both scanning and transmission electron microscopes represent very powerful tools for nanotechnology research and tool development. This is due to their versatility; they allow for *in situ* experiments and they enable observation of physical processes at an atomic level.

Scanning probe microscopy

Another tool is known as a scanning probe microscope (SPM) – though calling it a "microscope" is somewhat of a stretch. The general field of scanning probe microscopy involves forming images using a physical probe that scans over the specimen. The interaction between the probe and the surface as a function of position is recorded as the probe scans over the specimen. This interaction can take many forms. In a sense, scanning probe microscopes are not microscopes at all. They are not "illuminated" in any way and no direct image of the specimen is created. Instead, the specimen is "felt" by the probe tip. In contrast to optical and electron microscopes, scanning probe microscopy is a sort of Braille by which different aspects of a specimen can be observed and understood.

The first scanning probe microscope was created by Gerd Binnig and Heinrich Rohrer in 1981 at IBM. It was a scanning tunneling microscope (STM). This meant that the probe (which is usually atomically sharp) is moved over the surface of the specimen while a voltage is applied between the probe and the surface. Even though the probe does not touch the surface, electrons still flow between the surface and the probe tip. This jumping of electrons is called tunneling and it forms an electrical current. The size of this current is dependent on the distance between the probe and the surface. Typically, the probe is moved up and down so as to maintain a constant current as the surface topography changes. By scanning the probe tip over the surface and measuring the height of the probe tip, a display of the surface can be obtained. High-quality STMs reach the resolution of a single atom.

Another popular type of scanning probe microscope is the atomic force microscope (AFM). With it, the probe tip scans across a surface, typically being dragged across the specimen. The top of the cantilever is reflective (often it consists of silicon coated with gold). A laser is aimed at the cantilever and reflected onto a photodiode and the height of the surface is determined. When the tip is brought into contact with the specimen and scanned across the surface, the cantilever is deflected depending on the height of the surface. The AFM can be used to manipulate atoms mechanically. The probe tip can be used to push atoms around on a surface and move them into different arrangements. It also can be used to differentiate atoms of different elements. Because different elements have different chemical bonds that they form with the probe tip, and these interactions can alter the frequency of the tip's vibration, the different atoms can be detected and mapped. The AFM can be fitted with specifically designed tips in order to measure forces other than mechanical displacement. In addition to the AFM, many other types of scanning probe microscopy are used to measure different types of force, including magnetic force microscopy, electrostatic force microscopy, and scanning electrochemical microscopy.

Scanning probe microscopes provide researchers with several new directions in which to study a material or manipulate it. For one, they provide three-dimensional information about the surface of the specimen instead of the two-dimensional information that electron microscopes provide. Samples viewed by some types of SPM do not require any special treatments that irreversibly alter the sample. Furthermore, it is easier to study macromolecules and biological samples with an SPM than it is with electron microscopes. Electron microscopes require a vacuum, while SPMs can typically be used in a standard room environment or even in liquid.

Scanning probe microscopy does have its limits. One major limitation is the scanning rate and the field of view. A typical scan with an AFM requires several minutes for a field of view much smaller than that viewed with an

SEM. This slow rate does more than annoy researchers. It can lead to drifting of both the sample and the probe and confuse the information received. Furthermore, the quality of the image is specifically dependent on the probe tip used, and an incorrect choice (e.g., an old tip that is not very sharp) or atmospheric disturbance (e.g., someone walking down the hallway nearby) can distort the image and lead to artifacts in the findings that are not real. This makes images obtained by SPMs somewhat difficult to interpret. Despite these limitations, scanning probe microscopes are an essential tool in studying and manipulating nanoscale materials.

A note on the engineering problem

In order for products to be made commercially viable (in the end, nanotechnology needs to deliver products that can be sold), they must have certain characteristics that make them profitable for those who make them. On a very basic level, the products must be made with a process adaptable to a production environment. Second, the production process must be able to be monitored to ensure the product will function properly. And, third, the products must have a high level of reproducibility.

Electron microscopes and scanning probe microscopes currently serve as the monitors for nanotechnology. They are also used in the semiconductor industry – which is seeing the size of the devices on its chips shrink and shrink to keep up with Moore's law – to monitor the processes used to make increasingly complex chips at different steps in the path. Using a scanning electron microscope, we can see these features and image them very quickly. Also, we can measure them, such as to determine that a certain line on a certain layer of a chip has a width of 85 nm with an error of 5 nm. As the feature size on the chip shrinks, we will then be able to say that a certain feature has a width of 65 nm with an error of 5 nm. As the feature size shrinks even more, we will say 45 nm with an error of 5 nm. The problem is in the *error*. The processes that are developed to manufacture products have certain tolerances within which they can be off target but still good enough. If too many devices are being made outside these tolerances, then they fail and do not make it to a final product. They have to be reworked and the company loses money and the product becomes commercially non-viable.

We have already noted, with the size of gold particles and their color, that at the nanoscale small changes in the size of a material can have large changes in its behavior. One to two nanometers here or there can make all the difference. Nanotechnology then requires not only newer tools for making things, but also newer measuring techniques with better tolerances. Sometimes all that is needed is a better algorithm to perform image analysis.

It is sometimes very easy to get caught up in what is scientifically possible and ignore the engineering problems that come with it. Certainly, humans knew very early on that it was possible for animals to fly: birds are a direct testament to this. However, airplanes that carry humans cannot possibly be allowed to get tired as frequently as birds do and they certainly cannot fly as sporadically as birds. The engineering solution to a problem very rarely exactly copies nature's solution, as issues of cost, efficiency, and scalability affect technological solutions in a very different manner than they do natural solutions.

2.2   Basic Governing Theories

In addition to actual machines and devices that are used to study the nanoscale, scientists draw upon many different theories in order to predict and explain the phenomena that occur when dealing with nanoscale objects. In a very real sense, theories are the lifeblood of science and technology. They drive research and enable development. They lead to the creation of new tools and new understandings of the world around us. Sometimes, theories far outpace the technological advances that they enable. The electron was first developed theoretically in 1897 as a way of understanding the inner workings of the atom. It was not until the mid twentieth century that electrons were first utilized to create the modern electronic computing device.[11]

Though nanotechnology relies on almost the entire spectrum of scientific theories to drive its full development (from evolution to cellular automata), there are some main theories that are driving and enabling the current level of nanotechnology. Among the most important theories are those of quantum mechanics, chemical bonding, and the structure of materials (crystal structure). Let us consider these in order.

Quantum mechanics

We briefly discussed some of the strange interactions that quantum theories have predicted, making it a significant departure from traditional Newtonian physics. Because nanotechnology operates on a small enough scale that the effects of a single atom (or small groups of atoms) become important, quantum theory is crucial to any explanation of phenomena on the nanoscale. Here, we will delve into some of the more specific aspects of quantum mechanics.

It is necessary to use quantum theories to understand the behavior of nanoscale systems and smaller because Newtonian theories collapse and fail at this scale. If Newton's laws governed the workings of an atom, negatively charged electrons would be attracted to the positively charged nucleus

---

[11]   The invention of the point-contact transistor by William Shockley, John Bardeen, and Walter Brattain at Bell Labs in 1947 marked the first major step in introducing electrons to computing. Before this, computers were based on vacuum tubes.

of the atom and the two would collide. This, of course, does not happen. Quantum mechanics confines electrons to electronic shells of specific radii from the nucleus of the atom. This enables many different technologies, including light emitting diodes (LEDs). Electrons can be made to "jump" from one shell to another, emitting light of a specific wavelength.

Earlier, it was noted that every object is, like light, both a wave and a particle. We can observe different aspects of an object, such as its position or its momentum. The act of observing the object forces the wave to assume a specific state – the one we observe. The best description of this process (and the most well known) is called Schrödinger's cat. Austrian physicist Erwin Schrödinger wrote the following:

> One can even set up quite ridiculous cases. A cat is penned up in a steel chamber, along with the following device (which must be secured against direct interference by the cat): in a Geiger counter there is a tiny bit of radioactive substance, so small, that perhaps in the course of the hour one of the atoms decays, but also, with equal probability, perhaps none; if it happens, the counter tube discharges and through a relay releases a hammer which shatters a small flask of hydrocyanic acid. If one has left this entire system to itself for an hour, one would say that the cat still lives if meanwhile no atom has decayed. The ψ-function [probability function] of the entire system would express this by having in it the living and dead cat (pardon the expression) mixed or smeared out in equal parts.
>
> It is typical of these cases that an indeterminacy originally restricted to the atomic domain becomes transformed into macroscopic indeterminacy, which can then be resolved by direct observation. That prevents us from so naively accepting as valid a "blurred model" for representing reality. In itself it would not embody anything unclear or contradictory. There is a difference between a shaky or out-of-focus photograph and a snapshot of clouds and fog banks.[12]

In other words, after an hour the so-called wave function of the cat has an equal probability of returning a value of "alive" or "dead" when we observe the cat. The cat is both alive and dead at the same time until it is observed. Observing this quantum cat forces the wave function into one specific state – alive or dead – each of which has an equal probability of occurring.

It is this probability that led to Einstein's skepticism about quantum mechanics, often nominally misattributed under the aegis that "God does not play dice": "an inner voice tells me that it is not yet the real thing. The theory says a lot, but does not really bring us any closer to the secret

---

[12] Erwin Schrödinger, "The Present Situation in Quantum Mechanics," trans. John D. Trimmer, *Proceedings of the American Philosophical Society* 124 (1980): 323. Schrödinger's original essay was "Die gegenwärtige Situation in der Quantenmechanik," *Naturwissenschaften* 23 (1935): 807–12, 823–8, 844–9.

of the 'old one.' I, at any rate, am convinced that *He* is not playing at dice."[13]
Einstein's argument against quantum mechanics was basically that even
though a coin flip or a roll of a die *can* be described using probability, this
does not mean that the physical motions are unpredictable; with quantum
mechanics we just do not know the underlying mechanism. However,
quantum mechanics has proven to be very successful in predicting and
explaining phenomena. Einstein believed that, although quantum mechanics successfully explains so many phenomena, a more complete theory will
eventually be developed – and this still remains to be seen. In the meantime,
quantum theory remains a vital tool in explaining and predicting many of
the marvels that occur within nanotechnology, as will be shown.

The basics of quantum mechanics can be summarized in four basic
principles:

1 Heisenberg's uncertainty principle states that it is not possible to know
   the values of the properties of a system at once with any precision. In
   fact, the more certainly we know one property, the less certain we can
   be about another property. This is mathematically described with the
   following equation:

$$\Delta x \Delta p \geq \frac{\hbar}{2}$$

   where $\hbar$ is Planck's constant (6.626×10^{-34} joule seconds) divided by
   2π, and Δx and Δp represent the certainties in the measurement of the
   position and the momentum of an object.[14] This principle claims there
   is no way of knowing where a moving object (particle or larger) is *ever*
   – even given every possible detail. Therefore, it is impossible to predict
   where it will go with absolute certainty.
2 The wave–particle duality is in all matter. Any single experiment can
   show either the wave-like properties or the particle-like properties, but
   not both.
3 Any system (or object) is completely described by a wave function. This
   description is probabilistic and the probability of an event is related to
   the amplitude of the wave function.

[13] Letter to Max Born, written December 4, 1926. In Irene Born (trans.), *The Born–Einstein
Letters* (New York: Walker and Company, 1971), p. 91 (emphasis in original). Another close
call is "Though I am now an old fogey I am still hard at work and still refuse to believe that
God plays dice." Letter to Ilse Rosenthal-Schneider, written May 11, 1945. In Ilse Rosenthal-
Schneider (trans.), *Reality and Scientific Truth: Discussions with Einstein, Van Laue, and Planck*
(Detroit, MI: Wayne State University Press, 1981), p. 33. We thank Tim McGrew and John
Norton for help researching this quote.
[14] Other such pairs have been found including the time coordinate and energy of an object
and the angular position and angular momentum of an object.

4 Quantum mechanical descriptions of large systems closely approximate the classical description.

These four principles do not offer a complete picture of quantum mechanics in the same way that Newton's laws of motion do not completely describe the entirety of classical mechanics. Rather they provide the fundamental basis for understanding the meaning of quantum mechanics. We will see how these principles and quantum mechanics apply to different aspects of nanotechnology.

Chemical bonds

Chemical bonding is responsible for the interactions between atoms and between molecules and is thus very important to the development of nanotechnology. Though the rigorous theoretical explanation of different chemical bonds is the subject matter of quantum theories, theorists usually appeal to more qualitative descriptions that are easier to understand. Chemical bonds rely mainly on the sharing of electrons or electronic interactions in order to bind the atoms together. In general there are four main types of bonds that differ in the strength with which the bond connects the two atoms. The stronger bond types are covalent and ionic bonds. The weaker bonds are hydrogen and van der Waals bonds.

Covalent bonding is characterized by the *sharing* of electrons between atoms. In general, atoms "prefer" to have eight electrons in their outermost shell, the valence shell. Elements that have filled up their outermost shell naturally are known as noble gases because they tend to be inert: they do not interact with other elements. Covalent bonds between atoms attempt to share electrons in such a way that a noble gas configuration is created for each atom. This can happen with atoms of the same element or with atoms of different elements. An example of like elements is hydrogen, which gives rise to hydrogen bonding. Hydrogen atoms have one electron and one proton. The first electron shell on atoms can have a maximum of two electrons. So, a hydrogen atom bonds with another hydrogen atom through covalent bonding and, therefore, each atom can effectively have a full outer shell. This forms the diatomic hydrogen molecule. An example of different atoms forming a covalent bond is with hydrogen and chlorine. Chlorine contains seven atoms in its outermost electron shell, and so forming a covalent bond with hydrogen allows it to share an electron with hydrogen and effectively have a complete outer electron shell.

Ionic bonding is based on the electronic forces between ions of atoms. Ions are atoms that have more or fewer electrons than protons, causing them to have a negative or positive electronic charge. Elements tend to form ions so that they have a closed outer shell. For example, sodium has 11 electrons – two in its inner shell, eight in its second shell, and one in the outermost shell. Therefore, it typically forms an ion with one less electron, a +1 ion. Chlorine has 17 total electrons – two in its inner shell, eight in its second shell, and seven in its outermost shell. Therefore it typically forms an ion with one more electron (because it "wants" eight electrons in its outer shell), a –1 ion. This makes sodium and chlorine atoms likely to ionically bond with each other and produces sodium chloride (NaCl), everyday table salt.

Ionic and covalent bonded materials tend to have different properties. Ionic compound materials tend to have higher melting and boiling points than covalent compounds. Because of this, ionic compounds are more often solids at room temperature than covalent compounds. However, ionic and covalent represent extremes on a spectrum of bonds that form compounds. The interatomic bonding in most compounds is some combination of covalent and ionic. A weaker form of bond is caused by the "van der Waals force" between two atoms. This force arises from the polarization of molecules or atoms into dipoles (i.e., where the positive and negative charges are separated and no longer spatially cancel each other out). This is sometimes caused by the orbiting of the negatively charged electrons around the positively charged nucleus of the atom. Though this is a weak bond, it can occur between all substances and is thus very important.

There are other types of bonds than those between atoms that are important in nanotechnology. The temporary bonding of a molecule onto the surface of a material can alter its conducting properties. When engineered in the right way, this bonding can be used to detect specific chemical species. As we discuss some applications it will be apparent where bonding is vital to understanding the workings and design of the technology.

Crystal structure

In most materials, atoms arrange themselves in a specific orientation that typically repeats itself in all three dimensions across the entire bulk of the material. This repeating arrangement of atoms is known as a crystal structure. A wide range of specific crystal structures exist and depend upon the bonding between the atoms of the structure as well as many other factors (for example, the way in which the material solidifies). The periodicity of the crystal is described by a mathematical lattice. In 1845, French physicist Auguste Bravais proved that there are only 14 unique lattices which can repeat (and fill up) three-dimensional space. These 14 lattices are now known as Bravais lattices in his honor and are: simple cubic, face-centered cubic (a simple cubic arrangement with an atom in the center of each face of the cube), body-centered cubic (a simple cubic structure with an atom in the center of the body of the cube), simple tetragonal, body-centered tetragonal, simple orthorhombic, body-centered orthorhombic, base-centered orthorhombic (an orthorhombic arrangement with atoms in the center of only two opposing faces), face-centered orthorhombic, rhombohedral, simple monoclinic, base-centered monoclinic, triclinic, and hexagonal. A crystal structure is made up of one or more atoms repeated at each lattice point.

The crystal structure is very important to all of a material's properties, from its mechanical strength to its electronic and magnetic properties. For example, both diamond and graphite are chemically of the same ilk: they contain only carbon atoms. The crystal structure of a diamond is a face-centered cubic Bravais lattice with a single carbon atom at each lattice point. In this form, diamond is the hardest natural material known to humans. However, graphite is a different arrangement of the same type of atom. In graphite, the carbon atoms form hexagonal sheets, which then are connected to each other via a van der Waals force. Carbon in this form is therefore not as strong in all directions as diamond (e.g., it rubs off from the "lead" of a pencil). Though the individual sheets of carbon are strong, they easily separate from each other.

So why do particular materials form particular types of crystal structure? While the specific answer is different for each type of material, the general answer is that atoms arrange themselves in the most energetically favorable way on formation of the crystal. Just as balls roll to the lowest point on a hill, atoms arrange themselves in the lowest energy state possible.

The arrangement of atoms is particularly powerful with nanoscale materials. In larger materials, different "grains" will form in which the same crystal structure is oriented in different directions within the same material. Also, defects in the crystal structure can occur. Both of these can cause the weakening of the strength of the material or the degradation of its electrical properties. However, in nanoscale materials these "grains" and defects are far less common.[15] As we discuss more nanoscale materials and their properties, it will become clear how the crystal structure is vital in the properties of the material.

These tools and theories do not represent the entire toolkit that scientists use in developing this new technology and understanding the scientific phenomena at the nanoscale. Any such treatment would require far more space than we can afford here. Instead, this treatment is intended to give an introduction to the types of tools and thought processes that scientists use in order to understand and manipulate design on the nanoscale.

---

[15] Nanoscale materials can contain certain types of defects, but most types of defects and grains increase a nanoscale material's energy too much to remain within the material during formation.

3

Nanomaterials

Because of the scale at which nanotechnology-based devices are formed and operate, the makeup of the material that each device utilizes is central to its functionality. In fact, in nanotechnology, material, device, and function are nearly inseparable. As such, nanotechnology as it is being researched today has a heavy focus on the discovery, characterization, and utilization of nanomaterials. This chapter provides an introduction to many of the materials that are enabling nanotechnology.

This chapter is heavily footnoted with scientific research papers. The purpose of this is twofold. First, it allows the reader to read the research as it is written if he or she is particularly interested in a certain aspect of the ongoing research at universities and other research institutions. Second, it highlights the amount of exploration in nanomaterials and the timeliness of this research. The research is recent and ongoing, and is a very exciting field in the scientific literature.

3.1 Formation of Materials

To have very small materials, it is necessary to have a way to form materials in the architectures and the morphology that you desire. Traditional methods of forming materials into specific shapes and devices include cutting, chipping, pounding, extruding, and other such bulk procedures. Nanotechnology is different. Whereas bulk procedures begin from a larger structure and form it into smaller structures, nanotechnology procedures can also include beginning at the atomic level and building up into larger, nanoscale structures. Where formation of smaller materials from larger is known as "top-down" technology, the formation of materials from atomic or molecular structures is known as "bottom-up."

Top-down methods can create nanoscale technology and they are used extensively today in the electronics industry. For example, faster computer chips require smaller parts to fit on the same size package at higher density; these parts are pieced together with more and more complex architectures and designs. Typically, top-down formation of nanoscale structures uses the same basic tools as those employed today to form smaller structures from larger. The technique known as photolithography is one such process. Photolithography is the most common technique used to create patterns on semiconductors and is capable of forming sub-100 nm patterns, although around 30 nm this becomes extremely difficult and costly. With photolithography in the computer industry, a silicon wafer is coated with a light-sensitive film called a photoresist that will harden when light is shone on it. A photomask with the design for each layer of the circuit is created and the photoresist-covered circuit board is then exposed to light (of a specific wavelength) through the mask, so that predetermined areas of the photoresist harden. The wafer is then exposed to an acid bath, hot ions, or other sloughing material so that the unhardened areas are removed. In this way, the features of a processor are built up layer by layer.

Bottom-up methods take the basic building blocks of nature and build up nanoscale materials from them. Typical methods of forming materials from the bottom up include vapor deposition (both physical and chemical), molecular beam epitaxy, and what is known as self-assembly. It was this type of material formation that Richard Feynman spoke of:

> What could we do with layered structures with just the right layers? What would the properties of materials be if we could really arrange the atoms the way we want them? They would be very interesting to investigate theoretically. I can't see exactly what would happen, but I can hardly doubt that when we have some control of the arrangement of things on a small scale we will get an enormously greater range of possible properties that substances can have, and of different things that we can do.[1]

We can divide the materials that are being developed into two basic categories: organic (i.e., carbon based) and inorganic (i.e., not carbon based).

3.2   Carbon Nanomaterials

Carbon is one of the most abundant elements on earth and carbon-based compounds form at temperatures typically found on earth. These qualities make carbon and its compounds the basis of all known life. As such, it is

---

[1]   Richard P. Feynman, *The Pleasure of Finding Things Out: The Best Short Works of Richard P. Feynman* (New York: Perseus Books Group, 2002), p. 135.

one of the most studied elements in the periodic table and its versatility of reacting with many other elements makes it the focus of an entire branch of chemistry known as organic chemistry. Carbon compounds show a large variety and range of applications because of their unique properties. Many products, from paints to pharmaceutical drugs, involve organic compounds.

Most of this versatility is due to the bonding ability of carbon. Because it has six electrons – with four electrons in its outer shell – carbon is very versatile in bonding with other elements. It can share electrons with a large combination of other elements. For example, in carbon dioxide (CO_2), two oxygen atoms bond with one carbon atom. Carbon shares two of its electrons with each oxygen atom (oxygen atoms have eight electrons in total and six in their outer shells). Because each of the oxygen atoms also shares two electrons with the carbon atom, the bond in carbon dioxide shares four electrons and is known as a double bond. At times, the bonding can be very strong. Another example is in the compound known as cyanogen chloride. This consists of three different elements: carbon, nitrogen, and chlorine. Chlorine normally has a total of 17 electrons and seven in its outer shell. So it shares one electron with carbon, making a single bond: two electrons form the bond, one from chlorine and one from carbon. Nitrogen normally has a total of seven electrons and five in its outer shell. So it shares three electrons with carbon, making a triple bond with carbon: six electrons form the bond, three from nitrogen and three from carbon.

Also, because carbon has four atoms in its outer shell, it can form a bond with itself very easily. If two carbon atoms were together, each would share all of its atoms with the other, theoretically forming a quadruple bond. However, this is a very high-energy bond (i.e., it is not very likely), and instead carbon tends to share its bonds with more than one carbon atom. It is possible to imagine a wide array of purely carbon alignments. In a chain of carbon atoms, each would have a double bond with the one prior to it and the one subsequent to it in the chain. Such a chain is the basis for forming polymers. At points along the chain of carbon atoms, one of the bonds can be broken and bond with another atom, thus forming functional aspects of a polymer. In fact, many different *allotropes*[2] of carbon do exist. Some of the allotropes of carbon include diamond, with a cubic type of crystal structure; graphite, with sheets of hexagonally arranged carbon atoms; Lonsdaleite, another crystalline form of carbon; buckyball, with 60 carbon atoms; C540, a spherical arrangement of 540 carbon atoms; C70, with 70 carbon atoms; amorphous carbon, which has no specific arrangement of the carbon atoms; and the nanotube.

In §2.2, the different properties that graphite and diamond have were used to illustrate the importance of the allotrope type in determining the

---

[2] Allotropy is the ability of the same compound to exist in different forms. In each of the forms, the atoms are bonded together in a different manner.

properties of materials. After receiving the Nobel Prize for Chemistry in 1996 – along with Robert Curl and Harold Kroto – for his part in discovering a new type of carbon, Richard Smalley discussed some of the beauty in the different allotropes of carbon:

> Nearly all of us have long been familiar with the earlier known forms of pure carbon: diamond and graphite. Diamond, for all its great beauty, is not nearly as interesting as the hexagonal plane of graphite. It is not nearly as interesting because we live in a three-dimensional space, and in diamond each atom is surrounded in all three directions in space by a full coordination. Consequently, it is very difficult for an atom inside the diamond lattice to be confronted with anything else in this 3D world because all directions are already taken up. In contrast, the carbon atoms in a single hexagonal sheet of graphite (a "graphene" sheet – in graphite, the carbon atoms form a sheet of hexagons) are completely naked above and below. In a 3D world this is not easy. I do not think we ever really thought enough about how special this is. Here you have one atom in the periodic table, which can be so satisfied with just three nearest neighbors in two dimensions, that it is largely immune to further bonding. Even if you offer it another atom to bond with from above the sheet . . . the only result is a mild chemisorption that with a little heat is easily undone, leaving the graphene sheet intact. Carbon has this genius of making a chemically stable two-dimensional, one-atom-thick membrane in a three-dimensional world. And that, I believe, is going to be very important in the future of chemistry and technology in general.[3]

How true this has turned out to be! In nanotechnology the two most studied and interesting carbon allotropes are the buckyball and the carbon nanotube.

The buckyball

Curl, Kroto, and Smalley's Nobel Prize was specifically for the discovery of the buckyball (officially named a fullerene). According to the Nobel Committee's press release:

> During an intense working week in the autumn of 1985, Robert Curl, Harold Kroto and Richard Smalley made the completely unexpected discovery that the element carbon can also exist in the form of very stable spheres. They termed these new carbon balls fullerenes.[4]

The spheres consist of 60 carbon atoms; for this reason, they are also designated C60. The geometry that they form consists of 12 pentagons and

---

[3]   Richard Smalley, "Discovering the Fullerenes," *Review of Modern Physics* 69 (1997): 723–30.
[4]   The Nobel Prize in Chemistry 1996. Available at http://nobelprize.org/nobel_prizes/ chemistry/laureates/1996/illpres/discovery.html (accessed October 12, 2008).

20 hexagons with carbon atoms at the corners. Each pentagon is surrounded on all sides by 20 hexagons. Incidentally, this is the same form as a soccer ball. The form was also used at the 1967 World Exhibition in Montreal by the architect Buckminster Fuller (and hence the name 'buckyball' or 'fullerene') in a spherical building. This shape was also known to the Greeks as a truncated icosahedron. Note that all of the carbon atoms in the structure bond with three other carbon atoms so that one of the bonds is a double bond. This double bond is always the shorter bond, which lies between two of the hexagons. The single bonds separate each hexagon from the pentagon that it is touching.

The properties of fullerenes have been studied significantly recently. It was realized very early on that it should be possible to enclose metal atoms in the fullerene cages. This would completely alter the properties of the metal. This was first successfully done with the rare earth metal lanthanum. Buckyballs can also act like a single atom and accept electrons, forming an "ion." These can then form a superconducting crystalline material such as K_3C_{60}. Buckyballs have also been shown to have possible uses in armor and for potential medicinal uses.

Buckyballs were initially formed when graphite was evaporated in an inert environment. This formed a vapor of carbon atoms. The vapor then reformed as the solid spherical formation. This vapor phase growth (vapor deposition) is common in forming novel nanostructures and it can exist in many specific forms. Vapor phase growth is highly utilized in materials development. In fact, this technique can be used to form many other allotropes of carbon. Specifically, by introducing a small amount of impurity into the vapor, the result is a very important cousin of the buckyball, the carbon nanotube.

Carbon nanotubes

The carbon nanotube consists of a one-atom-thick graphite sheet – the same sheet whose beauty Smalley discussed – rolled up into a seamless cylinder. If the carbon nanotube consists of only a single carbon sheet that meets end on end, it is referred to as a single-wall nanotube (SWNT). However, if the nanotube consists of multiple sheets rolled up coaxially or somewhat spirally, then it is referred to as a multi-wall nanotube (MWNT).

The other important characteristic of the nanotube is its chirality, which refers to the direction the nanotubes are rolled. The chirality has a large impact on the physical properties.[5] There are three distinct types of nanotube based on their chirality: chiral, armchair, and zig-zag. The difference between the three can be understood using the idea of chiral vector and

---

[5] Teri Wang Odom et al., "Atomic Structure and Electronic Properties of Single-Walled Carbon Nanotubes," *Nature* 391.1 (1998): 62–4; Jeroen W.G. Wilder et al., "Electronic Structure of Atomically Resolved Carbon Nanotubes," *Nature* 391.1 (1998): 59–62.

angle. The chiral vector and angle are modeled on a two-dimensional hexagonal lattice. Essentially, the chiral angle is a measure of the angle with which the graphite sheet is rolled up. To understand this more clearly, imagine rolling up a sheet of paper. It could be rolled directly with the length of the sheet as the axis: this would have a chiral angle of zero. It could also be rolled directly along the width of the sheet as the axis: this would have a chiral angle of 90°.

The chiral vector is determined as $C_h = na_1 + ma_2$, where a_1 and a_2 are unit vectors and n and m are integers. The chiral angle θ is measured relative to a_1. Nanotubes with chiral angles between 0° and 30° are known as chiral nanotubes. Armchair nanotubes are formed when $n = m$ and the chiral angle is 30°: they are so called because of the pattern the carbon lattice forms when it is rolled this way. Zig-zag nanotubes are formed when either n or m is zero and the chiral angle is 0°. Chirality is directly related to the electrical conductivity of carbon nanotubes. Armchair nanotubes have an electronic conduction closely resembling that of a metal. In fact, the standard rule is that for a given (n, m) nanotube, if $2n + m = 3q$ (where q is an integer), then the nanotube is metallic. In theory, metallic nanotubes can have an electrical current density more than 1,000 times that of metals like silver and copper. Alternatively, zig-zag nanotubes tend to have the same electronic properties as a semiconductor, where electrons must overcome a band gap in order to enter the conduction band. It is this wide range of electronic properties that makes nanotubes interesting in the field of electronics.

One interesting phenomenon associated with metallic conducting nanotubes is ballistic conduction. Ballistic conduction allows electrons to flow through the nanotube without collisions.[6] Therefore, it has quantized conduction and no energy dissipation, which means that it generates no heat when conducting electrons.[7] And having electronic components that have quantized conduction and generate no heat would be significant for electronic packaging applications. The electronic packaging industry is rapidly approaching the limits of current technology and heat management as miniaturization packs more and more components with increasingly fine feature sizes into ever-shrinking devices. Carbon nanotubes have the potential to provide a major breakthrough in thermal management with computer technology.

Carbon nanotubes also have novel physical properties. Nanotubes have been shown to have a stiffness (what engineers call a "Young's modulus") and a breaking stress (or "tensile strength") far in excess of those of diamond, which means that they are far more resistant to breaking when

[6] M.F. Lin and Kenneth W.K. Shung, "Magnetoconductance of Carbon Nanotubes," *Physical Review B* 51.12 (1995): 7592–7.
[7] Leonor Chico et al., "Quantum Conductance of Carbon Nanotubes with Defects," *Physical Review B* 54.4 (1996): 2600–6.

stretched.[8] In fact, in terms of these values, carbon nanotubes are one of the strongest known materials. This strength results from the covalent sp^2 bonds, a type of carbon bond in which each of two carbon atoms has four valence electrons, formed between the individual carbon atoms in the nanotube. This strong bond occurs only between carbon atoms within the same graphite sheet. The sheets themselves are loosely bonded together, making them easy to peel off a single sheet. This is why it is easy to write with the graphite in pencils. However, the intra-sheet bonds are very strong. MWNTs have been tested to have a tensile strength of 63 GPa,[9] far eclipsing high-carbon steel with a tensile strength of approximately 1.2 GPa. Further, steel has a Young's modulus of around 200 GPa, while the Young's modulus of nanotubes has been measured on the order of 1 TPa.[10] Since steel has a density of around 8 g/cm^3 and carbon nanotubes have a density of 1.3–1.4 g/cm^3, the latter's strength is staggeringly high. One of the main causes of these high values of the strength of nanomaterials is the lack of defects in the carbon nanotubes. This is probably also the reason why smaller nanotubes have higher values. The smaller a material is, the less thermodynamically stable it is for dislocations and line defects to exist within it. Defect concentration has not been directly linked to strength in carbon nanotubes, but it has in other materials.

It is important to note that the measurement of these properties is very difficult. This is true for a lot of measurements of nanomaterials. Typical measurement techniques are unrealistic for nanoscale materials. To solve this problem of dimensionality, researchers have developed a variety of new techniques for isolating and measuring the properties of a single nanostructure; often these are *in situ* transmission electron microscope (TEM) experiments.[11] An example of such a novel technique is where a TEM stage holder was created to allow an individual nanotube to be electrically excited into vibration (or resonance).[12] The resonating nanotube is modeled as a cantilever beam in resonance, governed by the theories of classical dynamics. As the resonant frequency and material constants of the nanotube are known, the

[8]   The Young's (or elastic) modulus is a number by which the strength of different materials can be compared. It is determined by measuring the resistance of a material to breaking and elongation by stretching.

[9]   Min-Feng Yu et al., "Strength and Breaking Mechanism of Multiwalled Carbon Nanotubes under Tensile Load," *Science* 287.5453 (2000): 637–40.

[10]   Jean Paul Salvetat et al., "Mechanical Properties of Carbon Nanotubes," *Applied Physics A: Materials Science & Processing* 69.3 (1999): 255–60.

[11]   Min-Feng Yu et al., "Strength and Breaking Mechanism"; Eric W. Wong, Paul E. Sheehan, and Charles M. Lieber, "Nanobeam Mechanics: Elasticity, Strength, and Toughness of Nanorods and Nanotubes," *Science* 277.5334 (1997): 1971–5; Nobuyuki Osakabe et al., "Time-Resolved Observation of Thermal Oscillations by Transmission Electron Microscopy," *Applied Physics Letters* 70.8 (1997): 940–2.

[12]   Philippe Poncharal et al., "Electrostatic Deflections and Electromechanical Resonances of Carbon Nanotubes," *Science* 283 (1999): 1513.

bending modulus of an individual nanotube can be calculated. This study revealed that there is a relationship between the radius of the nanotube and the bending modulus. The smaller tube diameters (8 nm) have the larger bending modulus (1 TPa) whereas the larger tube diameters (40 nm) have the smaller bending modulus (0.1 TPa). The bending modulus is related to the Young's modulus.

The strength and flexibility of carbon nanotubes make them of wide potential use in nanoapplications – particularly in controlling other nanoscale structures – which will give them an important role in engineering nano-technologies. Carbon nanotubes have been used in a number of other strengthening applications. They have already been used as composite fibers in polymers and concrete to improve the mechanical, thermal, and electrical properties. Also, it has been found that adding nanotubes to polyethylene increases the polymer's elastic modulus by 30 percent. Carbon nanotubes have been shown to be able to bend to extreme angles without fracture.[13] In concrete, they have been shown to increase the tensile strength and halt crack propagation.

Though carbon nanotubes have some properties that make them very appealing for applications, there are some drawbacks to their use commercially. It was stated above that the electronic properties of carbon nanotubes depend mainly on their chirality. Therefore, if nanotubes are to be used in commercial electronic applications, scientists need to be able to control the chirality of the nanotubes synthesized. However, in all the current techniques of nanotube synthesis, this control is not available. In current synthesis, metallic and semiconducting nanotubes are formed side by side. This forces the development of attempts to separate the types of nanotube post-synthesis. Recently, a technique was reported that successfully separated the types of nanotube that had been suspended in a solution.[14] This method took advantage of the difference of the relative dielectric constants of the two types of nanotube with respect to the solvent. Using alternating current dielectrophoresis, the metallic tubes were attracted to the microelectrode array faster, leaving the semiconducting tubes in the solvent. However, in the study, only 100 picograms (pg: 10^{-12} grams) of metallic nanotubes were recovered from 100 nanograms (ng: 10^{-9} grams) of starting tubes. This is a yield of only 1/1000. Though this is too small a quantity to be effective, efforts are ongoing to scale up the process, perhaps by using microfluidic dielectrophoretic separation cells, a common technique in biology. Another issue with the procedure is that during the processing, the nanotubes would bundle to form nanotube fibers. If these

[13] Henk W.C. Postma, Allard Sellmeijer, and C. Dekker, "Manipulation and Imaging of Individual Single-Walled Carbon Nanotubes with an Atomic Force Microscope," *Advanced Materials* 12.17 (2000): 1299.
[14] Ralph Krupke et al., "Separation of Metallic from Semiconducting Single-Walled Carbon Nanotubes," *Science* 301.5631 (2003): 344–7.

fibers contain a combination of metallic and semiconducting nanotubes, the process is largely inhibited.

These difficulties in nanotubes have caused research to explore many other types of one-dimensional nanostructures. Due to both the promise and the shortcomings of carbon nanotubes, interest in nanoscience has grown significantly since their discovery in 1991. Some research has been performed on inorganic nanotubes, particularly based on layered compounds such as molybdenum disulfide. They have excellent lubrication properties, resistance to shockwave impact, catalytic reactivity, and a high capacity for hydrogen and lithium storage. This implies a wide range of applications for carbon nanotubes, including energy storage (for hydrogen applications).

3.3   Inorganic Nanomaterials

Non-carbon-based materials are, in contrast to carbon-based materials, dubbed *inorganic* materials. Though related, they are often given separate treatment from carbon-based materials. Inorganic nanomaterials are made in a wide variety of shapes and chemistries. Electron microscopes allow for a proper classification of nanomaterials so that they can be discussed in an intelligible and useful manner. Because the main feature of nanomaterials is their small size in some dimensions, nanomaterials have come to be classified by the number of dimensions in which they are confined to the nanoscale. By convention, the names have been reduced and in fact refer to the number of spatial dimensions, which are *not* confined to the nanoscale – such is the oddity of scientific naming. As such, *two-dimensional nanomaterials* are confined to the nanoscale in only one dimension. *One-dimensional nanomaterials* have been confined to the nanoscale in two dimensions. Finally, *zero-dimensional nanomaterials* are confined to the nanoscale in all three dimensions. The physical confinement of the different nanostructures affects some of their applications – as will be shown in the remainder of this section – and so classifying them in this manner is useful in helping understand which materials are needed for certain applications.

Zero-dimensional nanostructures

One of the early nanomaterials to be extensively studied was the quantum dot (QD). QDs are zero-dimensional nanostructures that are semiconductors; they were first theorized in the 1970s and initially synthesized in the early 1980s. If semiconductor particles are made small enough, quantum effects begin to assert themselves. These effects limit the energies at which electrons and holes can exist in the particles, and photons are released. Because energy is related to the wavelength of the photons, this means that the optical properties of the particle can be finely tuned depending on its size.

Quantum mechanics determines that only certain discrete energy levels are allowed within a single atom. If two identical atoms are held at large distances from each other, electrons within each will have the exact same energy. As those two atoms are brought closer to one another, they interact, and no two electrons with the same spin can have the same energy; this governing principle is called the Pauli exclusion principle. When a large number of atoms are brought together to form a solid, the discrete allowed energy levels of the individual atoms become a continuous energy band. It is this energy band that differentiates conducting, semiconducting, and insulating materials. In a metal, the energy bands are completely filled, effectively having no band gap. With insulators, electrons are filled up to the level of the band gap, but no electrons are above the band gap. In semiconductors, some electrons at room temperature are above the band gap, the energy from the heat providing enough energy for these electrons to "jump" over this gap. This band structure directly impacts electronic and optical properties. In bulk structures, these properties can only be altered by adding constituents to create defects in the material. The impressive phenomenon in QDs is that the optical and electronic properties can be precisely tuned by changing the size of the dots in addition to adding dopants (i.e. designed chemical impurities). The electronic and optical properties that were thought to be inherent in a material are transformed when the material is formed in a small enough (nanoscale) size.

As the dimensions of a material decrease, quantum confinement effects begin to occur. Quantum confinement size is when a material is on the order of or smaller than the Bohr radius of its constituent compound (i.e., the effective size of that compound).[15] This restricts each atom's movement, resulting in the above mentioned discrete energy levels and differences in material properties. With QDs, quantum confinement effects have a significant impact on the optical properties of the material. One popular quantum dot material is cadmium selenide (CdSe). The optical properties of the CdSe QDs are very size dependent. As the particle size gets smaller, the color of the quantum dot solutions goes from blue to red, with the wavelength getting longer. The result is dramatic.

Scientists at Bell Labs were some of the first to determine the direct relationship between quantum confinement in zero-dimensional CdSe quantum dots and the induced higher energy shift in the electronic band structure.[16]

[15] Moungi G. Bawendi et al., "Luminescence Properties of CdSe Quantum Crystallites: Resonance between Interior and Surface Localized States," *The Journal of Chemical Physics* 96.2 (1992): 946–54.

[16] Paul A. Alivisatos et al., "Electronic States of Semiconductor Clusters: Homogeneous and Inhomogeneous Broadening of the Optical Spectrum," *The Journal of Chemical Physics* 89.7 (1988): 4001–11; Louis E. Brus, "Electron–Electron and Electron–Hole Interactions in Small Semiconductor Crystallites: The Size Dependence of the Lowest Excited Electronic State," *The Journal of Chemical Physics* 80.9 (1984): 4403–9.

It was demonstrated that when CdSe was reduced in one, two, and three dimensions to sizes that are on the nanoscale, the energy bands reconfigure to a band structure that resembles individual atoms.[17] That is, on a small enough scale, with a small enough number, groups of atoms act as a single atom. For this reason, QDs are often referred to as artificial atoms.

Because of their unique properties, QDs have some promising potential applications. The more prolific research involves using them for biological imaging[18] and computing[19] applications. Biological labeling exploits the luminescent properties of the QDs and requires attaching a functional group to the surface of the QD. The functional group preferentially binds itself to a specific organism, cell, or protein once the system is injected into a biological system. The QDs are then caused to emit light through luminescence or fluorescence, allowing the detection and tracing of the biological targets inside the body.[20] The advantage of using quantum dots for this application as opposed to organic dyes, which are currently used, is that QDs are brighter and more resistant to photobleaching.[21]

Zinc sulfide (ZnS) also plays an important role in quantum-dot-based nanosensors. CdSe is typically an ideal quantum dot material. However, under certain conditions quantum dots with a CdSe core have been found to be acutely toxic to biological cells. In *in vitro* applications, quantum dots used for labeling have been found to be toxic. This is due to cadmium release into the system. Capping the CdSe quantum dots with a ZnS shell seems to help the toxicity. This coating eliminates all cytotoxicity following exposure of the core-shell quantum dots to air, but UV light still induces some toxicity.[22]

---

[17]   C.B. Murray, D.J. Norris, and Moungi G. Bawendi, "Synthesis and Characterization of Nearly Monodisperse CdE (E = Sulfur, Selenium, Tellurium) Semiconductor Nanocrystallites," *Journal of the American Chemical Society* 115.19 (1993): 8706–15.

[18]   Warren C.W. Chan and Shuming Nie, "Quantum Dot Bioconjugates for Ultrasensitive Nonisotopic Detection," *Science* 281.5385 (1998): 2016–18; Marcel Bruchez Jr et al., "Semiconductor Nanocrystals as Fluorescent Biological Labels," *Science* 281.5385 (1998): 2013–16.

[19]   Xiaoqin Li et al., "An All-Optical Quantum Gate in a Semiconductor Quantum Dot," *Science* 301.5634 (2003): 809–11.

[20]   P.T. Tran et al., "Use of Luminescent CdSe-ZnS Nanocrystal Bioconjugates in Quantum-Dot-Based Nanosensors," *Physica Status Solidi B* 229.1 (2002): 427–32; Xiaohu Gao et al., "*In Vivo* Cancer Targeting and Imaging with Semiconductor Quantum Dots," *Nature Biotechnology* 22.8 (2004): 969–76.

[21]   Andrew M. Smith, Xiaohu Gao, and Shuming Nie, "Quantum Dot Nanocrystals for *In Vivo* Molecular and Cellular Imaging," *Photochemistry and Photobiology* 80.3 (2004): 377–85.

[22]   Wolfgang J. Parak, Teresa Pellegrino, and Christian Plank, "Labelling of Cells with Quantum Dots," *Nanotechnology* 16.2 (2005): R9–R25; Sathyajith Ravindran et al., "Quantum Dots as Bio-Labels for the Localization of a Small Plant Adhesion Protein," *Nanotechnology* 16.1 (2005): 1–4; Austin M. Derfus, Warren C.W. Chan, and Sangeeta N. Bhatia, "Probing the Cytotoxicity of Semiconductor Quantum Dots," *Nano Letters* 4.1 (2004): 11–18.

Quantum computing makes use of the electronic properties of quantum dots. Jeong et al. have linked two quantum dots, allowing for semiconductor-based quantum computers.[23] Computers now work by representing information as a series of bits (either on or off, ones or zero). This code is related by transistors. Quantum computers would take advantage of the formation of quantum bits, or qubits, that exist in both the on and off states simultaneously, making it possible for them to process information much faster than conventional computers. This is because a string of quantum bits would be able to calculate every possible on–off combination simultaneously, increasing the computer's power and memory drastically. Because quantum dots can act as an artificial atom, each dot can exist in this quantum state and can exhibit the spin state (up or down) that quantum computers would rely on.

One-dimensional nanostructures

The wire- or rod-like shape of one-dimensional nanostructures has caused them to be the source of somewhat intensifying research in recent years. In particular, their novel electrical and mechanical properties are of great interest to scientists. The category of one-dimensional nanostructures consists of a wide variety of morphologies, including: whiskers, nanowires, nanorods, fibers, nanotubules, nanocables, nanotubes, and others. Often, the distinguishing features between these are a little arbitrary. Whiskers and nanorods are essentially shorter versions of fibers and nanowires. These one-dimensional structures have been studied for some time. One-dimensional structures with diameters ranging from several nanometers to several hundred microns have been referred to as whiskers and fibers in early literature, whereas nanowires and nanorods are more recent and refer to one-dimensional nanostructures whose width does not exceed 100 nm.

One particular nanostructure has been the focus of intense research. The progression shifted from the one-dimensional nanostructure nanotubes to other one-dimensional nanostructures, such as nanowires. The term "nanowire" is widely used to represent one-dimensional nanostructures that have a specific axial direction while their side surfaces are less well defined.[24] Typically, nanowires have a radius that is negligible in comparison to their length. There are two distinct types of nanowire. Ultrafine wires or linear arrays of dots are formed by self-assembly. Nanowires have been

[23] Heejun Jeong, Albert M. Chang, and Michael R. Melloch, "The Kondo Effect in an Artificial Quantum Dot Molecule," *Science* 293.5538 (2001): 2221–3.
[24] John Westwater et al., "Growth of Silicon Nanowires via Gold/Silane Vapor–Liquid–Solid Reaction," *Journal of Vacuum Science Technology B* 15.3 (1997): 554–7; Y. Zhang et al., "Bulk-Quantity Si Nanowires Synthesized by SiO Sublimation," *Journal of Crystal Growth* 212 (2000): 115.

successfully synthesized out of a wide range of materials, including titanium oxide, indium oxide, indium-tin oxide, aluminum, and tungsten oxide.[25] In a similar way to QDs and nanotubes, one particular nanowire material – silicon – has been the focus of a large amount of research. The main synthesis technique that has been used to make silicon nanowires is physical vapor deposition (PVD). PVD is a process of transferring growth species from a source to a deposition substrate to form a structure. Several methods have been developed for vaporizing the source species and these will be explored in more detail later. In the case of silicon nanowires, the source material is typically high-purity silicon or silicon dioxide.[26]

One major benefit of silicon nanowires over carbon nanotubes is the electronic structure. Whereas nanotubes are either metallic or semiconducting depending on the chirality, silicon nanowires are always semiconducting.[27] This characteristic makes silicon nanowires immediately useful, bypassing the hurdle that carbon nanotubes faced as there is no need for separation of different electronic types of nanowires. However, a major drawback to silicon nanowires is that silicon forms an unavoidable passivation layer of silicon dioxide (SiO_2) – also known as silica – when silicon is in an oxidizing atmosphere such as air, just as with bulk silicon. As soon as it is exposed to an oxygen atmosphere, a silicon nanowire becomes encased in a sheath of amorphous SiO_2.[28] The problem with this is that SiO_2 is electrically insulating and this can affect any semiconducting properties that the silicon nanowire might have.

Silicon nanowires have other issues facing them. TEM studies have revealed that silicon nanowires are sometimes made of polycrystalline cores and can have dislocations and defects incorporated into the nanostructures.[29] This hinders electron and hole transport through a material;

[25] Changhao Liang et al., "Catalytic Growth of Semiconducting In$_2$O$_3$ Nanofibers," *Advanced Materials* 13.17 (2001): 1330–3; Pho Nguyen et al., "Epitaxial Directional Growth of Indium-Doped Tin Oxide Nanowire Arrays," *Nano Letters* 3.7 (2003): 925–8; Tomoya Ono, Shigeru Tsukamoto, and Kikuji Hirose, "Magnetic Orderings in A1 Nanowires Suspended between Electrodes," *Applied Physics Letters* 82.25 (2003): 4570–2; Hang Qi, Cuiying Wang, and Jie Liu, "A Simple Method for the Synthesis of Highly Oriented Potassium-Doped Tungsten Oxide Nanowires," *Advanced Materials* 15.5 (2003): 411–14.

[26] D.P. Yu et al., "Nanoscale Silicon Wires Synthesized Using Simple Physical Evaporation," *Applied Physics Letters* 72 (1998): 3458–60; erratum in *Applied Physics Letters* 85 (2004): 5104. See also D.P. Yu et al., "Amorphous Silica Nanowires: Intensive Blue Light Emitters," *Applied Physics Letters* 73.21 (1998): 3076–8.

[27] Yi Cui and Charles M. Lieber, "Functional Nanoscale Electronic Devices Assembled Using Silicon Nanowire Building Blocks," *Science* 291.5505 (2001): 851–3.

[28] Xin-Yi Zhang et al., "Synthesis of Ordered Single Crystal Silicon Nanowire Arrays," *Advanced Materials* 13.16 (2001): 1238–41.

[29] Mahendra K. Sunkara et al., "Bulk Synthesis of Silicon Nanowires Using a Low-Temperature Vapor–Liquid–Solid Method," *Applied Physics Letters* 79.10 (2001): 1546–8.

ideally, a material should be single crystal and defect-free for electronic applications. The grain boundaries in polycrystalline materials, as well as any dislocations, serve as scattering points in a material. These scattering points not only hinder the electrons from moving through the material, but also can ruin a material's thermal transport properties. Thermal transport properties are very important in nanoscale electronics.

Because of the problems created by the passivating layer, alternative methods of synthesizing silicon nanowires have been developed. Instead of employing a PVD technique, some researchers have used a chemical vapor deposition (CVD) technique, where a chemical reaction in the vapor phase is used to create the deposited nanostructure. In the case of silicon nanowires, a silane (SiH_4) gas is used as a source. This dissociates into silicon and hydrogen and the silicon reacts with gold on the deposition substrate to nucleate silicon nanowire growth. This type of synthesis leads to significantly reduced dislocation concentration in the nanowires and consistently produces single-crystal nanowires.[30]

However, the issue of the SiO_2 passivation layer still exists with this method and a post-synthesis treatment is necessary to remove it. The silicon nanowires must be treated with hydrofluoric acid in order to etch away the SiO_2 layer. This etching process leaves a chemically inert hydrogen-terminated surface on the silicon nanowires. The non-reactivity of this surface prevents a SiO_2 layer from reforming.[31]

One of the more distinguishing features of nanowires is their very large length with respect to their width, making the sides essentially non-faceted. There is a type of one-dimensional nanostructure that combines the length of nanowires with a faceted side structure, called a nanobelt. Nanobelts, also referred to as nanoribbons, have two of their dimensions confined to the nanoscale, with the third dimension being relatively very long. However, unlike nanowires, they exhibit faceted side surfaces, so that their cross-section is rectangular.

Pan et al. first reported transparent semiconducting oxides synthesized in a belt-like manner in 2001. The reported materials used to synthesize these nanobelts were zinc oxide (ZnO), cadmium oxide (CdO), indium oxide (In_2O_3), gallium oxide (Ga_2O_3), and tin oxide (SnO_2). These materials are all transitional metal oxides ranging over three different elemental groups (II–VI, III–VI, and IV–VI) and at least five types of crystallographic structures. Since this first report, nanobelts have also been synthesized in non-oxide

[30] Yi Cui et al., "Diameter-Controlled Synthesis of Single-Crystal Silicon Nanowires," *Applied Physics Letters* 78.15 (2001): 2214–16; Charles M. Lieber, "One-Dimensional Nanostructures: Chemistry, Physics and Applications," *Solid State Communications* 107.11 (1998): 607–16.
[31] D.D.D. Ma et al., "Small-Diameter Silicon Nanowire Surfaces," *Science* 299 (2003): 1874–7.

semiconductors such as zinc sulfide (ZnS), cadmium sulfide (CdS), CdSe, and zinc selenide (ZnSe).[32]

Nanobelts have several unique properties that make them amenable to study and for technological applications. In general, nanobelts can be synthesized as single crystals that are relatively long (about 1–2 mm). They have a rectangular cross-section that is generally uniform through the length of the belt. The width of the nanobelts can range from as much as 100 nanometers to as little as 6 nm. The aspect ratio has also been measured and ranges from about 5 : 1 to 10 : 1 (width-to-thickness ratio). Though stacking faults – where the planes of the atoms have not quite stacked perfectly – may exist, nanobelts are also essentially dislocation and defect free. So even if a stacking fault is present, it typically does not terminate within the nanobelt structure and provide a location for scattering processes to occur. Because of the extremely high surface-area-to-volume ratio, the presence of a dislocation is not energetically feasible. Nanobelts also have well-defined crystallographic planes. This means that not only is the fastest growth direction well defined, but also the top and bottom (along the width) planes and the side (along the thickness) planes are well-defined crystallographic planes and are synthesized along well-defined directions. This allows for tuning of properties and of catalytic surfaces and can have a profound effect on the structure and properties of the synthesized material.

It is important to note some of the ways in which nanomaterials differ from their bulk counterparts. One example in nanobelts has to do with their extreme flexibility. Many nanobelts are made of ceramic materials, notorious for their rigidity. However, at the nanoscale nanobelts are very flexible, enduring great strain without breaking. This strain is also fairly reversible because of the lack of dislocations, and the nanobelts should be extremely resistant to fatigue and failure. TEM images of ZnO nanobelts are used to reveal their geometrical shape. ZnO is usually a very hard ceramic material, but as a nanobelt it is highly flexible.

In some respects, nanobelts are not all that different from some of their more heavily researched cousins in one-dimensional nanomaterials

[32] Zheng Wei Pan, Zu Rong Dai, and Zhong Lin Wang, "Nanobelts of Semiconducting Oxides," *Science* 291 (2001): 1947–9; Yang Jiang et al., "Hydrogen-Assisted Thermal Evaporation Synthesis of ZnS Nanoribbons on a Large Scale," *Advanced Materials* 15 (2003): 323–7; Quan Li and Chunrui Wang, "Fabrication of Wurtzite ZnS Nanobelts via Simple Thermal Evaporation," *Applied Physics Letters* 83 (2003): 359–61; Christopher Ma et al., "Nanobelts, Nanocombs, and Nano-Windmills of Wurtzite ZnS," *Advanced Materials* 15 (2003): 228–31; Daniel Moore et al., "Wurtzite ZnS Nanosaws Produced by Polar Surfaces," *Chemical Physics Letters* 385 (2004): 8–11; Lifeng Dong et al., "Catalytic Growth of CdS Nanobelts and Nanowires on Tungsten Substrates," *Chemical Physics Letters* 376 (2003): 653–8; Christopher Ma et al., "Single-Crystal CdSe Nanosaws," *Journal of the American Chemical Society* 126 (2004): 708–9; Christopher Ma and Zhong Lin Wang, "Road Map for the Controlled Synthesis of CdSe Nanowires, Nanobelts, and Nanosaws: A Step Towards Nanomanufacturing," *Advanced Materials* 2005.17 (2005): 1–6; Yang Jiang et al., "Zinc Selenide Nanoribbons and Nanowires," *Journal of Physical Chemistry B.* 108.9 (2004): 2784–7.

– nanowires and nanotubes. For example, all three can be synthesized to varying lengths. All three have been developed in varying thicknesses (or diameters). Nanowires and nanobelts can controllably be synthesized along a single crystallographic growth direction. Reliably, nanowires and nanobelts can be synthesized with a precise electronic band structure. However, the differences between the nanostructures are significant. Notably, nanowires often contain defects and dislocations. Mechanically, the presence of these contributes to the fatigue of the material as a whole. Electronically, the presence of defects and dislocations in a material can decrease electron transport due to the increase in possible scattering sites. This decreases both the speed of response and the sensitivity of devices made with the nanowires. This also decreases their usefulness in lasing applications. Nanobelts have been shown to have significant use in such applications, even exhibiting a degree of tunability over certain wavelengths.

Of course, one-dimensional nanostructures are of interest for fundamental materials studies. They provide a physical manifestation of materials at extremes, being quasi-infinite lines or strings. Researchers can now test theoretical predictions of material surfaces and gain new insight into what properties really are material specific and which rely on size.

By contacting semiconducting one-dimensional nanostructures to metallic electrodes, field effect transistors (FETs) can be produced, thus allowing the exploration of the electrical properties of the structures. FETs have been fabricated with SnO_2 and ZnO nanobelts and with various nanowires using varying techniques.[33] Methods to create FETs with nanobelts include depositing a nanobelt onto a substrate and then using e-beam lithography to make electrodes or directly applying ethanol-dispersed nanobelts on predefined gold electrode arrays. In both cases a gold layer is used as the back gate electrode and the current–voltage (I–V) characteristics of the FET can be revealed. Nanobelt-based FETs have also been shown to be controlled by ultraviolet light illumination. When a UV light with a wavelength shorter than the band gap of the nanobelt is introduced, an immediate increase in the source-drain current can be observed. This indicates a switch from "off" status to "on." Nanobelt FETs have been shown to have switching ratios as large as six orders of magnitude, conductivities as high as 15 (ohm-cm)$^{-1}$, and electron mobilities as large as 125 cm^2/V·s.

When these one-dimensional nanostructure-based FETs are placed in different environments, they are shown to have high-quality applications as nanoscale sensors. The high surface-to-volume ratio of one-dimensional nanostructures makes them far more useful than thin films. This allows for higher sensitivity as sensors because the faces are more exposed and the small size is likely to produce a complete depletion of carriers into the nanobelt, which typically changes the electrical properties. Some nanobelts

---

[33] Michael S. Arnold et al., "Field-Effect Transistors Based on Single Semiconducting Oxide Nanobelts," *Journal of Physical Chemistry B* 107 (2003): 659–63.

have already been successfully shown to be gas sensors by being placed on platinum electrodes. Prompt current changes are observed when certain gases are introduced to the system. These results demonstrate the usefulness for fabricating highly sensitive nanoscale gas sensors using single one-dimensional nanostructures.

Zinc sulfide (ZnS) is an important phosphor host material, used in thin film, electroluminescent displays (ELDs), and many other phosphor applications. A typical electroluminescent display device has a very basic structure. There are at least six layers to the device. The first layer is a baseplate and it is usually a rigid insulator-like glass. The second layer is a conductor. The third layer is an insulator. The fourth layer is the phosphor material. The fifth layer is an insulator. Finally, the sixth layer is another conductor. Of course, at least one of the conductors must be transparent so that the light can escape the device. Essentially, ELDs are somewhat "lossy" capacitors that become electrically charged like a capacitor and then lose their energy in the form of light. The insulator layers are necessary to prevent arcing between the two conductive layers. An alternating current (AC) is generally used to drive an ELD because the light generated by the current decays when a constant direct current (DC) is applied.[34] Recently, the development of ZnS one-dimensional nanostructures has driven research into smaller, more precise ELD devices. ZnS serves as the host material and, by doping it with specific materials which serve as emission centers for the light, different color electroluminescence can be achieved. A significant amount of research has gone into doping ZnS nanostructures with manganese in order to develop their electroluminescence.[35]

Because of their size, nanomaterials are especially suited to interface with biological specimens. This is because many cells and cell components are sized in the nanometer to micrometer range. Typical biological entities are very amenable to use with and by nanoscale technologies. Cells are on the order of $10-100$ μm wide; viruses are approximately 100 nm wide; proteins are 10 nm in scale; and genes are $5-8$ nm long. Cells are much larger than the nanoscale. Notice that most nanomaterials contain at least one dimension that is smaller than or of the same order as most of the biomolecules. Nanoparticles can approach close to cells or components and may be coated or functionalized with biomolecules to change their surface chemistry, allowing tagging and manipulation of biologic targets. Magnetic nanostructures offer increased functionality for biomedical applications since they allow action from a distance through optically opaque living tissues via external

[34] Hideomi Ohnishi, "Electroluminescent Display Materials," *Annual Review of Materials Science* 19 (1989): 83–101.

[35] Steven C. Erwin et al., "Doping Semiconductor Nanocrystals," *Nature* 436 (2005): 91–4; Y.Q. Li et al., "Manganese Doping and Optical Properties of ZnS Nanoribbons by Postannealing," *Applied Physics Letters* 88 (2006): 013115; Anna W. Topol et al., "Chemical Vapor Deposition of ZnS : Mn for Thin-Film Electroluminescent Display Applications," *Journal of Materials Research* 19.3 (2004): 697–706.

magnetic field manipulation. They may also resonate in response to a time-varying field, transferring energy from the magnetic field to the particles themselves. Current biomedical applications of various types of nanoparticles include magnetic resonance imaging (MRI) contrast enhancement, cell separation, hyperthermic cancer treatment, drug targeting, genetic screening, disease detection, and others;[36] see Chapter 11 for more discussion.

Most recently, interest in carbon nanotube biocompatibility has arisen out of an observed need to study new biomedical uses of nanomaterials. Recent publications cite work to establish cytotoxicity, blood clearance, solubility, protein binding, and new carbon-nanotube-based architectures for bioprosthetics. Although some early work found unmodified carbon nanotubes highly toxic, later studies have found the materials to be well tolerated, especially after functionalization with a variety of active biomolecules. Carbon nanotubes are appealing in that they have been previously well characterized chemically and physically, and they often exhibit interesting electronic and mechanical properties.

Two-dimensional nanostructures

Two-dimensional nanostructures have traditionally been studied and categorized as "thin films" because of their confinement to the nanoscale only in one dimension. They have been developed for use for some time in fields as diverse as electronic devices and photovoltaic applications. For example, in the silicon integrated circuit industry, many devices rely on thin films for their operation, and control of film thicknesses approaching the atomic level is necessary. Monolayers are also used significantly in chemistry. The formation and properties of these layers are reasonably well understood from the atomic level upward, even in the quite complex layers that are used in lubricants. Significant research is being done in precisely controlling the composition and the smoothness of growth of these films.

Photovoltaic (PV) solar panels consist of several connected direct current PV cells, which are made out of a semiconducting material between two metallic electrodes. Typically this material is a silicon-based p–n junction[37] (i.e., a junction between a semiconducting material that has had positive charges added to it and a semiconducting material that has had negative

[36] Quentin A. Pankhurst et al., "Applications of Magnetic Nanoparticles in Biomedicine," *Journal of Physics D: Applied Physics* 36.13 (2003): R167–R181; Prashant K. Jain et al., "Calculated Absorption and Scattering Properties of Gold Nanoparticles of Different Size, Shape, and Composition: Applications in Biological Imaging and Biomedicine," *Journal of Physical Chemistry B* 110.14 (2006): 7238–48; Mads G. Johnsen et al., "Chitosan-Based Nanoparticles for Biomedicine," *Journal of Biotechnology* 118 (2005): S34; Jose A. Rojas-Chapana and Michael Giersig, "Multi-Walled Carbon Nanotubes and Metallic Nanoparticles and Their Application in Biomedicine," *Journal of Nanoscience and Nanotechnology* 6.2 (2006): 316–21.

[37] Martin A. Green, "Photovoltaics: Technology Overview," *Energy Policy* 28.14 (2000): 989–98.

charges added to it). But other such junction materials are also used, most notably indium gallium phosphide/gallium arsenide and cadmium telluride/cadmium sulfide.[38] The cells are encapsulated behind glass to waterproof them. Most PV systems today use mono- or multicrystalline silicon as the semiconducting material necessary for converting sunlight into electricity. However, a major drawback to using silicon is the high loss of materials during production of the wafers, and the efficiency achieved is relatively low. Thin films are one of the main alternatives to standard PV solar cells. Amorphous silicon is one of the most developed thin film materials. Also, newer materials such as copper indium diselenide (CIS) are being developed. Cell efficiencies of almost 20 percent have been achieved with CIS. Another thin film material that has been tested for use in PV cells is cadmium telluride (CdTe). This is a promising material because of the low cost of production of CdTe thin films, which uses techniques that include electrodeposition and high-rate evaporation.

Thin film nanostructures are good for highly efficient conversion of light to electrical power in photovoltaic cell devices due to their large surface area, on which photoelectrochemical processes take place. A large amount of research has investigated synthesizing titanium dioxide (TiO_2) electrodes to improve the PV structure for more efficient electron transport and good stability. Chemical vapor deposition (CVD) of Ti_3O_5 has been used to deposit layered crystalline anatase TiO_2 thin films that are optically responsive and stable.[39] Also, compression of TiO_2 powder has been used as a technique to form thin films. However, TiO_2 films have trouble achieving an efficiency of over 10 percent. So efforts have focused on wide brand gap semiconducting oxide materials such as ZnO and SnO_2.[40] In addition, combination

[38] Arvind Shah et al., "Photovoltaic Technology: The Case for Thin-Film Solar Cells," *Science* 285.5428 (1999): 692–8; Steven A. Ringel et al., "Single-Junction InGaP/GaAs Solar Cells Grown on Si Substrates with SiGe Buffer Layers," *Progress in Photovoltaics* 10.6 (2002): 417–26; Derk Leander Batzner et al., "Stability Aspects in CdTe/CdS Solar Cells," *Thin Solid Films* 451–2 (2004): 536–43; Alessandro Romeo et al., "Recrystallization in CdTe/CdS," *Thin Solid Films* 361 (2000): 420–5; Ayodhya N. Tiwari et al., "CdTe Solar Cell in a Novel Configuration," *Progress in Photovoltaics* 12.1 (2004): 33–8.

[39] Mukundan Thelakkat, Christoph Schmitz, and Hans-Werner Schmidt, "Fully Vapor-Deposited Thin-Layer Titanium Dioxide Solar Cells," *Advanced Materials* 14.8 (2002): 577–81.

[40] Karin Keis et al., "Nanostructured ZnO Electrodes for Dye-Sensitized Solar Cell Applications," *Journal of Photochemistry and Photobiology A: Chemistry* 148.1–3 (2002): 57–64; Subbian Karuppuchamy et al., "Cathodic Electrodeposition of Oxide Semiconductor Thin Films and Their Application to Dye-Sensitized Solar Cells," *Solid State Ionics* 151.1–4 (2002): 19–27; Kirti Tennakone et al., "Dye-Sensitized Composite Semiconductor Nanostructures," *Physica E* 14.1–2 (2002): 190–6; Shlomit Chappel, Si-Guang Chen, and Arie Zaban, "TiO_2-Coated Nanoporous SnO_2 Electrodes for Dye-Sensitized Solar Cells," *Langmuir* 18.8 (2002): 3336–42; Shlomit Chappel and Arie Zaban, "Nanoporous SnO_2 Electrodes for Dye-Sensitized Solar Cells: Improved Cell Performance by the Synthesis of 18 nm SnO_2 Colloids," *Solar Energy Materials and Solar Cells* 71.2 (2002): 141–52; Si-Guang Chen et al., "Preparation of Nb_2O_5 Coated TiO_2 Nanoporous Electrodes and Their Application in Dye-Sensitized Solar Cells," *Chemistry of Materials* 13.12 (2001): 4629–34.

structures comprised of semiconducting oxide film and polymeric layers for solid-state solar cell devices have been examined in hopes of increasing the overall efficiency. So far, these devices have increased efficiency up to 5 percent for ZnO devices, 1 percent for SnO_2 devices, and up to 2 percent for hybrid devices.[41]

The semiconductor industry, specifically the manufacture of computer processors and related industries, has for a long time driven functional materials development – in particular, driving materials cost down and shrinking the size of those materials. As we noted in §3.1, photolithography is the most common technique used to create patterns on silicon. Photolithography is capable of forming sub-100 nm patterns. It becomes more difficult and costly as the size of the features on the pattern gets smaller, limited by the wavelength of light and the resist used. Even though interesting techniques such as double patterning have been developed to achieve smaller features with current resist and photolithography tools, new materials and techniques are being explored to enable processors to achieve higher and higher processing speeds in the technology nodes to come.[42]

As an established industry with fierce competition, striving for better, cheaper, and more energy-efficient devices, the semiconductor industry has pushed the edge of technologies forward in the past and continues to do so as we move into the development of nanomaterials. The economic and business need is driving intense investment into nanomaterials such as carbon nanotubes and silicon nanowires. It seems likely that this industry will remain a driver for the development of nanomaterials in the near future as well.

[41] Wendy U. Huynh, Xiaogang Peng, and A. Paul Alivisatos, "CdSe Nanocrystal Rods/Poly(3-hexylthiophene) Composite Photovoltaic Devices," *Advanced Materials* 11.11 (1999): 923–7.

[42] Semiconductor industry developments are guided by the International Technology Roadmap for Semiconductors (ITRS), which defines technology nodes based on feature sizes on chips and determines specifications for the various enabling technologies and standards that are needed so that the entire industry can work in concert. Each different generation of technology is defined in a "node" typically named by the smallest feature size on the computer chip: for example, the *65 nm node* or the *45 nm node*.

4

Applied Nanotechnology

With such a wide array of novel materials and properties being developed and studied in laboratories at universities, national labs, and companies around the world, the challenge becomes to develop useful applications for them. Already, products that use the functionality of nanoscale materials to produce new, innovative designs and new uses have made their way onto the market. These products mark just the beginning of the development of nanotechnology, but they also provide insight into the pervasiveness of the technology that will mark nanotechnology's introduction.

This chapter looks at various current applications of nanotechnology and examines how they work, why nanotechnology is vital to their function, and what insight these applications offer into the future of nanotechnology. Some of these applications may at first seem trivial. However, they imply other, more revolutionary applications – and also more serious issues related to their social and ethical implications, as later chapters will discuss. Such revolution aims at changing the way we interact with technology.

4.1   Using Nanomaterials

The properties of nanomaterials challenge the extremes of physical law. The strength of materials is a very simple example of this. The theoretical limit of the strength of a material is directly related to the energy it takes to break the atomic bonds connecting the atoms in that material. However, no bulk material ever comes close to approaching this theoretical limit. This is because the bonds in a material do not need to break all at once in order to fracture the material. Instead, a single bond can break and this break can propagate through the material, thus limiting the material's strength; this is called a 'dislocation.' A dislocation is a type of defect in the crystal structure of the material, and, in general, it can be visualized by the ending of a plane of atoms in the middle of the material. In this case, the planes around it are not straight and do not follow the perfect crystal structure of the material. This introduces stress into the material. Nanoscale materials, in contrast to bulk materials, typically cannot retain the presence of this defect in the material and it is rejected from the crystal or never allowed to form. As such, nanomaterials usually do not contain defects such as dislocations and their strength approaches the theoretical limit.

A similar story can be told when discussing other internal properties of materials, such as electrical conduction and a material's interaction with light (be it reflection or transmission): nanotechnology allows for a wide array of applications. Nanomaterials can achieve better results with less material, thus making that material lighter, smaller, and so forth. Another way in which nanomaterials serve newer applications concerns their surface properties. The relative amount of the material that makes up the surface – i.e., the surface-to-volume ratio – is a much greater percentage in nanomaterials than in larger bulk materials. We can consider the example of a bulk sphere in relation to a nanoscale sphere. Mathematically, the surface of a sphere is $A = 4\pi r^2$ (where r is the radius of the sphere) and the volume of a sphere is $V = 4\pi r^3/3$. This gives a surface-to-volume ratio of $3/r$.

Notice that the greater the radius of the sphere (the bigger the particle being considered), the less is the surface-to-volume ratio. Why is this important? The surface of materials is where a lot of specific processes take place. These include adsorption, chemical reactions, and processes in which the material interacts with its environment. The greater the surface-to-volume ratio, the faster these processes can take place or the greater the number of processes that occur. As a result, nanoscale materials are far more reactive and functional than their bulk counterparts.

We already see some applications that take advantage of some of the other properties of nanoscale materials. Nanoscale materials are used in electronic, biomedical, pharmaceutical, energy, and cosmetic applications, along with many others. Most computer hard drives contain nanoscale thin films of material allowing for huge increases in storage capacity. Further innovative applications use many of the properties of nanoscale materials. One of the most well-marketed applications of nanotechnology is stain-resistant, wrinkle-free cloth, best known as stain-free pants.

Resisting stains is typically a matter of repelling liquids so that the material does not absorb them. Instead, the liquids bead up and roll off the surface of the material. Such stain-resistant material has a property that prevents water absorption. When a material or a molecule has this property, it is called "hydrophobic." Hydrophobicity takes advantage of some unique properties of water.

Water molecules have fairly remarkable chemical properties. Many substances can dissolve in water – so many that it is generally referred to as the universal solvent. It is also one of the most common pure substances that are found naturally in all three states of matter. A very important property of water is that it is electrically polar, which means that its positive and negative centers of charge are separated; this creates an electrical dipole. In bonding, the electrons of the hydrogen atoms are attracted to the oxygen atom. (Oxygen has eight total electrons and six on its outer shell, so receiving two electrons – one from each of the hydrogen atoms – completes its outer shell.) The hydrogen molecules also are not symmetrically aligned on the oxygen molecule. In fact, the bonding angle is 104.45°, well off the 180° necessary for the charges to balance out. Because of this charge difference, oxygen is known as a polar material. This causes it to be attracted to other polar molecules, including other water molecules;[1] this polarity is also what makes water such a good solvent. When an ionic or other polar compound is mixed with water, hydration occurs in which the molecules of the compound are surrounded by water molecules. The negatively charged dipole ends of the water molecule are attracted to the positively charged ends of the compound and vice versa.

However, not all molecules have dipoles, and non-polar molecules actually reject water and do not dissolve in a water solution. An example of this is mixing oil in water. The two compounds do not mix and the oil will remain clumped together in a ball within the water. The reverse also occurs: if water is dropped into a solution of oil, it will remain a ball within the oil. Because non-polar molecules have this property and seem to reject mixing with water, they are known as hydrophobic molecules. An interesting application of hydrophobicity in nature occurs with the cell membranes in biology. Our cell membranes are made up of a dual layer of lipids (i.e., fats with hydrophilic phosphate heads and hydrophobic tails). Because a biological body is a very aqueous environment, the cell walls arrange in an interesting configuration, thus allowing for effective protection of the cell from the outside environment. This configuration of the cell walls consists of hydrophobic tails inside the wall, and hydrophilic heads on the outside of the cell and the inner side of the wall.

This property of water is taken advantage of in creating stain-resistant fabrics. Though it would be possible to create pants with a macroscale coating of hydrophobic material, this would not be practical since the pants would be uncomfortable and probably very expensive. However, by including hydrophobic nanoscale particles (or wrapping nanoscale fibers) with the regular cotton textile of the pants, the rejection of water (and therefore stains) can be achieved while maintaining the everyday functionality of the garment. Furthermore, making the cloth less polar achieves additional

---

[1] In fact, it works in much the same way that two magnets are attracted to each other; the positive charge of one water molecule is attracted to the negative charge of another water molecule and vice versa.

effects, such as that the material becomes resistant to the buildup of static electricity. Also, because the fabric keeps liquids on the surface, they evaporate faster, allowing the fabric to breathe better than regular fabrics. This application may sound trivial and meaningless, but even these sorts of nanotechnologies can have greater importance and impact than appear at first glance. Soldiers in battle situations often go days without being able to change their clothes. As the uniform repels water, not only is it kept clean and the soldier more comfortable, but the wear and tear on the uniform and the weight carried by the soldier are reduced.

Another application of nanomaterials that has already reached the market is the introduction of tennis balls that keep their bounce longer. Regular tennis balls left out in the air become flat and unplayable after about 14 days, sometimes less. However, by utilizing nanoparticles properly, tennis balls can be made that last twice as long while not changing the elastic properties of the tennis ball. How does this work? The natural rubber that is used in tennis balls – necessary to having the right bounce on the ball – has a lot of very small holes in it, thus making it very permeable. Through these holes, air escapes from the tennis ball, deflating it over time. If the tennis balls are kept in an enclosed or a pressurized environment – such as the unopened can that they come in – then they can last for a very long time without the air escaping. However, once the tennis balls are placed in an open environment, they begin to deflate.

Enter nanoparticles and nanocomposites. Nanocomposites (i.e., composites of dissimilar nanoscale materials) can be used to coat the inside of the rubber of the tennis balls. These nanocomposites have significantly fewer holes in them than standard rubber and, therefore, adding them to tennis balls significantly reduces the amount of air that leaks out. While this is possible using a larger bulk technology coating, the nanoscale of the coating helps the balls more easily retain their elastic properties. Tennis balls made in this fashion are already gaining wide acceptance and are the official ball of the Davis Cup. Again, this application may sound somewhat trivial and frivolous. However, a similar use for this technology is coating the inside of tires so that they retain their air longer and become more resistant to puncture. The point is that nanotechnology is going to impact even the smallest aspects of manufacturing and technology. Some of its impact will not be widely heralded, even though the scientific advances used in the technology are extremely creative.

It is also important to note that just because a technology is possible does not mean that it is productively valuable. One of the major issues with utilizing nanotechnology is organizing the various parts into place in a reliable, cost-efficient manner. The structures (e.g., nanowires, nanotubes, quantum dots, etc.) all exist, but methods must be developed to utilize them effectively and efficiently. This means we need to be able to place the structures in specific places, forming specific architectures, and do this fairly quickly. In bulk technologies, much of the placement of parts is done by simply moving the pieces together and attaching them. A crane moves the beams of steel into place in building a new skyscraper. Robots in an assembly line help place the parts of automobiles together and attach them. For a variety of reasons, similar processes are not considered practical for nanotechnology.

First, they would be too slow in building up structures and devices. A similar system of moving atoms (or molecules) into precise locations has been demonstrated, perhaps most famously in having the IBM logo spelled out with atoms.[2] More recently, assembly of carbon nanotubes through nano-robotic manipulations has been accomplished.[3] Having to move molecules into precise place can take minutes per placement as it requires finding the object to be moved, picking up or somehow gaining control over the object, moving the object to the desired location, releasing the object from control, and then measuring to confirm that the placement was successful. For the large number of objects that are included in complex structures, the time that this takes hinders the ability to make enough devices and structures for production.

Second, there is the problem of "sticky fingers," which we briefly mentioned in §1.4. The problem arises because it is difficult for a nanoscale crane (or other such system) to let go of a molecule it has picked up once a bond has formed between two nanostructures. Think about the trouble you may have had throwing away a piece of tape. Though there have been proposed mechanisms to avoid this problem in directed assembly of nano-structures, there remains a significant debate on the respective merits of these mechanisms and the extent to which it is possible to overcome this problem.[4]

However, both of these problems and more can be overcome using an assembly technique known as "self-assembly." Self-assembly is a generic term referring to a process through which many molecular-scale objects arrange themselves without directed assembly. This is typically accomplished by utilizing chemical bonding and linking molecules attached to nanostructures. One structure that is studied extensively for self-assembly purposes is DNA.

DNA (deoxyribonucleic acid) consists of two strands of interlocking nucleotides. The nucleotides that make up DNA are adenine, cytosine, guanine, and thymine. They are arranged in a specific order on each strand so that when put together, adenine in one strand bonds to thymine in the

---

[2] Donald M. Eigler and Erhard K. Schweizer, "Positioning Single Atoms with a Scanning Tunneling Microscope," *Nature* 344 (1990): 524–6.

[3] Toshio Fukada, Fumihito Arai, and Lixin Dong, "Assembly of Nanodevices with Carbon Nanotubes through Nanorobotic Manipulations," *Proceedings of the IEEE* 91 (2003): 1803–18.

[4] Eric Drexler and Richard Smalley, "Point–Counterpoint: Nanotechnology," *Chemical and Engineering News* 81 (2003): 37–42.

other strand, and guanine in one strand bonds to cytosine in the other. This molecular recognition allows two complementary single strands of DNA to come together in a very specific way. It also can provide the building blocks for creating arbitrary nano-sized shapes using DNA. Through a process known as DNA origami,[5] very complex shapes have been designed and created using only DNA. This process utilizes the molecular recognition of DNA to cause a single large strand of DNA to fold by using smaller strands to pinch the long strand of DNA, bending it in very specific ways. Shapes that have been made include a map of North America, a five-pointed star, smiley faces, and even written words. This process can be extended further. By using linker molecules, nanostructures such as carbon nanotubes or silicon nanowires can be attached to DNA strands in specific places on a long strand.[6] The DNA folding would then organize the various nanostructures into predesigned architectures.

This is just one example of self-assembly that could help move complex nanotechnologies from the research lab to full-scale production. Other examples exist, but the concept of self-assembly is generally the same: utilizing chemical bonds and linking molecules to arrange nanostructures into complex patterns. These self-assembly techniques are vital to the development of nanotechnology into products, particularly computing devices.

Because of the large number of industries and companies involved in developing and utilizing nanotechnology and the extensive network of enabling technologies needed, such as high-resolution microscopes, characterization tools, and computer simulations, a roadmap similar to the International Technology Roadmap for Semiconductors (ITRS)[7] is valuable. The ITRS allows the disparate companies that supply, develop, and use semiconductor technology to cooperate and to be able to predict innovation and coordinate development without sacrificing company secrets and intellectual property. With this, the semiconductor industry is able to plan out the next 15 years. In a similar vein, the Foresight Nanotech Institute, a nanotechnology thinktank, and Battelle, a research and development organization, have developed the Technology Roadmap for Productive Nanosystems.[8] The roadmap is "a call to action that provides a vision for atomically precise manufacturing technologies and productive nanosystems."[9] The roadmap provides general direction and a specific roadmap and, though it is not as

---

[5]   Paul Rothemund, "Folding DNA to Create Nanoscale Shapes and Patterns," *Nature* 440 (2006): 297–302.

[6]   Kinneret Keren et al., "DNA-Templated Carbon Nanotube Field-Effect Transistor," *Science* 302 (2003): 1380–2.

[7]   For more information, see International Technology Roadmap for Semiconductors (ITRS). Available at http://www.itrs.net/ (accessed November 7, 2008).

[8]   "Productive Nanosystems: A Technology Roadmap," Battelle Memorial Institute and Foresight Nanotech Institute, 2007. Available at http://www.foresight.org/Roadmaps/ Nanotech_Roadmap_2007_main.pdf (accessed November 8, 2008).

[9]   Ibid., p. vii.

widely followed as the ITRS, it represents an attempt to formalize the steps that need to be taken for nanotechnology to be developed. By following a roadmap, companies developing nanotechnology can set milestones and develop with some idea about what other developments are on the horizon. Even this small level of predictability is essential for allowing industry to soundly invest in product development. In short, it aims to bridge the gap between the promise of nanotechnology and the current state of technology. Whether it is effective and valuable remains to be seen, but it represents the kind of thinking that can motivate extensive industrial investment and development and move nanotechnology from the university and national research labs into corporate development and products.

4.2   Nanotechnology Computing and Robotics

Nanotechnology is about the future: according to its proponents at least, it represents the next revolution in technology, on a par with the Industrial Revolution and the Information Technology Revolution. Some of the claims of what nanotechnology will provide us in the future seem like mere science fiction and fantasy. And, frankly, we have been down the science-fiction-as-future path before, with obvious results: "Every general futurist gets the future wrong to a significant extent."[10] By the turn of the millennium, mankind has failed to achieve commonplace space travel, household humanoid robots, and other marvels that were predicted in the mid twentieth century.[11] Often the problem is not the scientific or technological abilities involved in predicting future technology, but rather how that technology will be used by the public. Technological development – much as nature – abhors a vacuum and does not occur simply because it can. Technologies develop because they fulfill some need or popular desire to achieve a certain goal. However, this does not mean that we should not analyze and look into the science of the future. As we try to develop science and technology for a specific goal, other technologies are developed along the way even if the final technology is not achieved (at least in its imagined form). Indeed, it is important to understand the direction of scientific developments.

In §1.4, we mentioned several components of this future nanotechnology. These components include the ideas of exponential growth, nanoscale

---

[10]   J. Storrs Hall, *Nanofuture: What's Next for Nanotechnology?* (New York: Prometheus Books, 2005), p. 28.

[11]   However, computers, the Internet, robot vacuum cleaners, and other such technologies have been developed. While the space travel in *2001: A Space Odyssey* has not come true yet, the ubiquitous use of computers pictured by Arthur C. Clarke has occurred. Also, mankind has recently figured out how to create a humanoid robot that can balance itself as it walks up the stairs very slowly.

robotics (nanobots), and molecular manufacturing. These three concepts represent very important portions of the future conception of nanotechnology. Exponential growth refers to any type of growth that doubles over some specific time period. A classic example is the tale of the gold on the chessboard. As the story goes, a Persian king was given a beautiful chessboard by a courtier, wanted to reciprocate, and asked the courtier what he desired. The courtier asked for one piece of gold on the first square of a chessboard, two pieces on the second, four pieces on the third, and so on, until the end of the chessboard was reached. A chessboard has 64 squares (8×8), and by the end of the first row the amount of gold on the eighth square was only $2^7 = 128$ pieces. By the end of the second row, it was 32,768 pieces. At the end of the chessboard, 4.6×10^{18} (2^{63}) pieces of gold were needed to cover the 64th square, and around 9.2×10^{18} ($2^{64} - 1$) pieces were needed to cover the entire board. There was not enough gold in the world to do this.[12]

Another example of accelerating growth demonstrates how suddenly it has an impact. A hypothetical colony of ants lives in an ant farm. They reproduce at such a rate that they double their number every day – imagine that there are no predators in this farm and food is abundant – and they will fill up the colony in 30 days. At first, they have the entire world of the ant farm at their disposal. They barely fill up a minute amount of the ant farm. Then the colony slowly fills up the farm, but not very much. Even one day before they run out of room, they have twice as much as they have ever needed in 29 generations. Then, suddenly, they are out of room the very next day.

In a strict sense, exponential growth is not necessarily fast. For example, if something grows at 5 percent per year it will take over 14 years for it to double. However, exponential growth is typically used to describe something growing extremely fast. It is that sort of growth that is predicted by nanotechnology futurists. Exponential growth means that "things" will grow and advance always at a faster rate than they did before. Consider:

> As exponential growth continues to accelerate into the first half of the twenty-first century, it will appear to explode into infinity, at least from the limited and linear perspective of contemporary humans. The progress will ultimately become so fast that it will rupture our ability to follow it. It will literally get out of our control. The illusion that we have our hand "on the plug" will be dispelled.[13]

Moore's law is a perfect example of accelerating returns and exponential growth. Computing devices, be they mechanical, vacuum tube, or integrated

[12] This parable is sometimes told with rice instead of gold.
[13] Ray Kurzweil, "The Law of Accelerating Returns," March 7, 2001. Available at http://www.kurzweilai.net/articles/art0134.html (accessed April 17, 2007).

circuit based, have always provided accelerating price–performance growth. It is probable that, just as new technologies replaced old to maintain this accelerating growth, some type of new technology will replace current computing technologies and this growth will continue. Nanotechnology provides the next pathway to continue this type of growth.

It provides this pathway in a couple of different ways that we will consider here. The first, and simplest, is that it allows electronics to go beyond the limits of photolithography (i.e., the current methods by which the small features of computing devices are fabricated; see §3.1 for more discussion). The way to increase computing power at greater and greater rates is to make the features on chips smaller and smaller so that more can be placed onto an area and the costs will be driven down as the speed rises. Photolithography accomplishes patterning these features by using light to transmit a pattern onto a surface using a mask (literally) to block certain areas of the surface from being exposed to the light.

The problem is that we are limited in the same way as with optical microscopes. The wavelength of light can only take us so far (can only produce features that are so small) before it becomes not cost-effective to have light sources capable of allowing us to produce smaller features. Current state-of-the-art technology uses deep ultraviolet light with wavelengths of 193 nm to produce minimum feature sizes down to 50 nm. This minimum feature size is restricted by two main equations. The first is

$$F = k\frac{\lambda}{N_A}$$

where F is the feature size, k includes process-related factors (it is usually around $\frac{1}{2}$), λ is the wavelength of light, and N_A is the aperture of the lens used to focus the light (as seen from the surface). The second equation is

$$D_F = 0.6\frac{\lambda}{N_A^2}$$

where D_F is the depth of field (the range of the depth that is focused). What is desired is a large depth of field and a small feature size; these are competing values. According to the first equation, making a larger and larger lens and moving it closer and closer to the surface will increase the feature size that is needed. However, just as looking at something up close blurs your vision of farther away objects, the depth of field will suffer by doing this. Because of this, it is generally believed that current photolithography techniques will not be worth the cost of feature sizes below about 30 nm. Of course, the death of photolithography has been predicted before and photolithography may, once again, prove more resilient than expected.

Nanotechnology techniques provide a way out of the entire paradigm. By assembling structures from the ground up, molecule by molecule, or atom by atom, the resolution and depth-of-field limitations can be eliminated entirely. Assembly of nanoscale structures can be achieved by using macro-molecules to link them together in complex architectures. Using techniques known collectively as self-assembly, disordered components form larger structures and devices without specific placement or "outside" manipulation; they do this by utilizing their chemical attributes and linking molecules adsorbed onto the surface of the component.[14] This sort of assembly can be used for structures with features well below the limits of lithography and is seen as a very potent successor to replacing the current method for producing electronic components, as smaller and smaller parts are required.

The second pathway that nanotechnology provides to continue exponential growth in computing is through the properties of the materials themselves. As we discussed earlier, nanomaterials have properties that far exceed their bulk counterparts. In carbon nanotubes, to reiterate, conduction of electrons occurs without hindrance by defects in the material and therefore at speeds far exceeding carbon on a larger scale. Nanoscale copper has a conductance that exceeds that of bulk copper. Furthermore, by using quantum properties of nanoscale materials, nanotechnology can help enable quantum computing, which is an entirely different method of computing. Contemporary computing consists of memory made in bits, where each can hold one of two values (traditionally, "1" or "0"). However, a quantum computer consists of memory made (fittingly) in qubits. These qubits can hold one of the two values, or a superposition of these values (i.e., any addition of the values together). The qubit essentially simultaneously possesses both values. (Recall, from §2.2, Schrödinger's cat, which was simultaneously alive and dead.) By manipulating these, quantum computers can have exponentially greater speed than contemporary computers.

For example, let us consider a contemporary computer that operates with two bits. These two bits can place the entire computer into any one of four states, such as 11, 01, 11, 00; each bit has a 1 or a 0 representing which of its two states it is in. In a quantum computer, the qubits can be a superposition of all of the allowed states of the entire system. That is, the state of the system is described by the following wavefunction:

$$|\psi> = a|00> + b|01> + c|10> + d|11>$$

where a, b, c, and d are complex numbers (i.e., they contain the square root of negative one).

---

[14] Of course, oftentimes outside manipulation is used, but because the process uses very little outside control (beyond the original parameters), it will still be considered self-assembly.

These coefficients represent the probability that the system will be in the given state (e.g., $|a|^2$ is the probability that the system is in the state 00). Recording the state of a quantum computer requires an exponential number of these coefficients: the two-qubit system above required $2^2 = 4$ numbers, a four-qubit system would require $2^4 = 16$, a 10-qubit system would require $2^{10} = 1,024$ coefficients, and so on. This provides an exponential growth in computing power. Nanotechnology enters into quantum computing through quantum dots (see §3.3). Because quantum dots begin to exhibit and control the quantum properties of materials, they represent nanotechnological boxes that contain a distinct number of electrons. By utilizing the quantum properties of these dots, a powerful quantum computer could be developed. Quantum dots are typically formed by augmenting existing technology. This means that when one qubit is fabricated, the process can be scaled to make many more.

Computing is not the only area in which nanotechnology has the potential to provide incredible growth and development. To see how nanotechnology can contribute to other areas, it is important to consider nanobots, which we met in §1.4. Nanobots would utilize nanoscale components in such a way that they could perform complex tasks in a variety of environments, ranging from extreme conditions such as space to the conditions within our bloodstream. The size of nanobots allows for access to many areas that are not available to current technologies. One such example is inside the cell. Access to this environment could provide untold possibilities for curing disease and enhancing our basic human capabilities. As Robert Freitas Jr writes:

> Humanity is poised at the brink of completion of one of its greatest and most noble enterprises. Early in the 21st century, our growing abilities to swiftly repair most traumatic physical injuries, eliminate pathogens, and alleviate suffering using molecular tools will begin to coalesce in a new medical paradigm called nanomedicine. Nanomedicine may be broadly defined as the comprehensive monitoring, control, construction, repair, defense, and improvement of all human biological systems, working from the molecular level, using engineered nanodevices and nanostructures . . .
>
> The very earliest nanotechnology-based biomedical systems may be used to help resolve many difficult scientific questions that remain. They may also be employed to assist in the brute-force analysis of the most difficult three-dimensional structures among the 100,000-odd proteins of which the human body is comprised, or to help ascertain the precise function of each such protein. But much of this effort should be complete within the next 20–30 years because the reference human body has a finite parts list, and these parts are already being sequenced, geometered and archived at an ever-increasing pace. Once these parts are known, then the reference human being as a biological system is at least physically specified to completeness at the molecular level. Thereafter, nanotechnology-based discovery will consist principally of examining a particular sick or injured patient to determine how he or she deviates from molecular reference structures, with the physician then interpreting these

deviations in light of their possible contribution to, or detraction from, the general health and the explicit preferences of the patient.[15]

It is possible in theory to make nanobots that are self-replicating (that is, they can create a copy of themselves out of the materials in their environment). The exponential growth of these nanobots was first described in Eric Drexler's book *Engines of Creation*: "[T]he first replicator assembles a copy in one thousand seconds, the two replicators then build two more in the next thousand seconds, the four build another four, and the eight build another eight. At the end of ten hours, there are not thirty-six new replicators, but over 68 billion."[16] This type of (exponential) growth allows for the rapid creation of many nanoscale robots that, working together and programmed properly, could build highly advanced devices and structures in parallel and allow for a revolution in manufacturing.

4.3 Predicting the Future of Technology

We have explored many of the fundamental scientific principles and engineering challenges that are forging applications in nanotechnology. We sought in the preceding chapters to provide a basic look into some of the fundamental theories and basic technologies that contribute to nanotechnology and, hopefully, this allows us to have a realistic discussion about nanotechnology and its potential impact on society. In part, this involves anticipating the future of technology, which is always a difficult but an important task. As Douglas Adams wrote, "trying to predict the future is a mug's game. But increasingly it's a game we all have to play because the world is changing so fast and we need to have some sort of idea of what the future's actually going to be like because we are going to have to live there, probably next week."[17]

Predicting the future has a rich history. However, as we quoted John Storrs Hall in the previous section, "every general futurist gets the future wrong to a significant extent."[18] The problem, as Hall explains it, is not that the futurist predicts changes that are too fantastical. The problem is that futurists tend to *underpredict* technological changes. To be sure, futurists may have predicted robotic servants in our homes, but they did not predict

[15] Robert A. Freitas Jr, *Nanomedicine, Volume I: Basic Capabilities* (Georgetown, TX: Landes Bioscience, 1999), §1.1, §1.2.1.13. Available at http://www.nanomedicine.com/NMI.htm (accessed November 23, 2008).

[16] Eric Drexler, *Engines of Creation: The Coming Era of Nanotechnology* (New York: Broadway Books, 1987), p. 58.

[17] Douglas Adams, *The Salmon of Doubt: Hitchhiking the Galaxy One Last Time* (New York: Macmillan Books, 2002).

[18] Hall, *Nanofuture*, p. 28.

near-universal cellular phones, personal MP3 players, and devices ubi-
quitously connecting individuals to a worldwide network without wires.
Futurists also tend to predict incorrectly the use of technology in the future.
Again, consider Hall: "It is . . . a truly remarkable fact that on the very
brink of an economic-technological revolution unparalleled in history no
one foresaw the universal motor car and all that it was soon to imply."[19]
The futurists were wrong not in the actual technology (the motorized car-
riage), but in the specific use of the technology. There is a rich history of
this sort of misprediction. In 1916, Charlie Chaplin declared, "The cinema
is little more than a fad."[20] In 1939, Winston Churchill predicted that atomic
energy "might be as good as our present-day explosives, but it is unlikely to
produce anything very much more dangerous."[21] In 1946, Darryl Zanuck,
a producer at 20th Century Fox, declared of television, "People will soon
get tired of staring at a plywood box every night."[22] And, as recently as
1977, Ken Olson, the founder of Digital Equipment Corporation, stated,
"There is no reason anyone would want a computer in the home."[23]

People have also been wrong about what is technologically possible and
the impact that technology will have, or they have developed predictions
from bad models or inputs. In 1972, the Club of Rome used modeling
simulations to study the growing world population and finite resources. The
predictions suffered from bad inputs to the model – including too many
unknown factors – and predicted catastrophe within 20 years. Though the
authors admitted that their estimates of the availability of resources were
poor, the predictions took hold.[24] On the other hand, science and technology
have also led to successful predictions. Mathematical models and theories predicted the discovery of many celestial bodies, from black holes to
the planet Neptune.[25] Technologists – and President John F. Kennedy –
predicted that man would travel into outer space and walk on the moon.
After developing space flight and devoting resources, this forecast was realized. In other fields prediction also has a mixed record. The problem with
bad predictions is that we, as a society, often invest large sums of money
and resources, through government policy and/or private enterprise, to react
to and prepare for a future that does not actualize. Of course, the flip side

[19]   Ibid.
[20]   David Robinson, *Chaplin: The Mirror of Opinion* (Bloomington: Indiana University Press, 1983), p. 20.
[21]   Winston S. Churchill, *The Gathering Storm* (New York: Mariner Books, 1986), p. 345.
[22]   Robert I. Sutton (ed.), *Weird Ideas That Work: 11½ Practices for Promoting, Managing and Sustaining Innovation* (New York: Free Press, 2001), p. 105.
[23]   BBC News, "The Tech Lab: Gordon Frazer," September 7, 2007. Available at http://news.bbc.co.uk/1/hi/technology/6981704.stm (accessed November 13, 2008).
[24]   Donella H. Meadows, *The Limits to Growth* (New York: University Books, 1974).
[25]   John Couch Adams, "Explanation of the Observed Irregularities in the Motion of Uranus, on the Hypothesis of Disturbance by a More Distant Planet," *Monthly Notices of the Royal Astronomical Society* 7 (1846): 149.

is also true: if we ignore predictions, then we fail to prepare for a future that may indeed happen. It might seem that predictions get more and more incorrect the further ahead we look. However, this is not necessarily true, partly because we have different needs met by predictions as the time frame changes. Weather predictions are a certain case in point. We need fairly accurate predictions of the weather for very short time frames: whether or not it rains tomorrow is important. We are also able to predict relatively accurately the weather for tomorrow. But we do not have the ability to predict as accurately the weather forecast 200 years from now. We have averages and general forecasts, aided by models, but accurately determining the rainfall on August 23, 2250 is beyond our predictive ability. Perhaps it is true that "the best way to predict the future is to invent it."[26]

None of this means that prediction is useless. Though it is wrought with danger and potential error, understanding what the future will provide is important for several reasons. First, it inspires us to action, whether it is true speculation or not. Predictions from science fiction of humans traveling to faraway planets inspired legions of engineers and scientists to develop space flight, rocketry, and many other technologies. The "House of the Future" at Disneyland inspired and showcased hands-free phones, very large televisions, electric razors, and the wide number of uses of plastic in the home. Second, predicting the future is important in helping to prepare society for the changes that new technologies and developments will bring. This understanding can help us monitor and be prepared for any adverse effects, and it can help us disseminate effectively the positive impacts, whether through laws or some other societal mechanism. Bill Joy wrote in his (in)famous essay, "Why The Future Doesn't Need Us," of this need for planning for the future:

> The experiences of the atomic scientists clearly show the need to take personal responsibility, the danger that things will move too fast, and the way in which a process can take on a life of its own. We can, as they did, create insurmountable problems in almost no time flat. We must do more thinking up front if we are not to be similarly surprised and shocked by the consequences of our inventions.[27]

The need for prior understanding of new technologies is even more important now than it was with atomic technology. This is because the rate of technological change has sped up and will most likely continue to accelerate. Ray Kurzweil recently expressed this in his book, *The Singularity Is Near*.[28] If the long-term pattern of accelerating change continues, then new

[26]   Alan Kay, "Predicting the Future," *Stanford Engineering* 1.1 (1989): 1–6.
[27]   Bill Joy, "Why the Future Doesn't Need Us," *Wired Magazine* 8.04 (2000).
[28]   Ray Kurzweil, *The Singularity Is Near: When Humans Transcend Biology* (New York: Viking, 2005).

technologies and technological revolutions will be introduced to society at a rate much faster than societal institutions (both governmental and non-governmental) move. Getting out in front of these changes allows for the necessary development of these institutions.

Because of its breadth of application, nanotechnology represents a technological revolution that will impact many elements of human life, changing the way things are manufactured and the way we interact with technology, as well as removing many barriers that have traditionally stood in the way of technological adaptation. It is important, then, to consider the societal impact and implications of nanotechnology. The remainder of this book aims at that charge: building a framework for thinking about the impacts that these technologies will have and how those impacts can be navigated with ethical and social responsibility.

Unit II

Risk, Regulation, and Fairness

5

Risk and Precaution

In the first unit of this book, we became familiar with the nanoscale and some of nanotechnology's basic features. The third unit of the book is concerned with applications of nanotechnology in specific areas, as well as the ethical and social implications that derive from those applications; in particular, we consider environmental, privacy, military, medical, and enhancement applications. This middle unit is transitional, and in several ways. We develop here a general framework that will accommodate these later discussions. Many of the ethical and social implications we consider are mixed: nanotechnology has much promise, though that promise carries corresponding hazards. Furthermore, there are cases where we aren't sure just what the hazards are; as with any new technology, there are likely to be unanticipated effects. This chapter, then, offers discussion of some of the relevant concepts and principles that are appropriate given the promise and (potentially unknown) hazards of nanotechnology. Thinking about how to move forward with this technology requires us to think carefully about this promise and these hazards, and the fact that some of the hazards are potentially unknown further complicates the deliberative framework insofar as we have to make decisions under uncertainty. Much of this discussion abstracts away from nanotechnology in particular; the point is to afford ourselves the conceptual apparatus that will allow us to address the specific applications we consider in the remainder of the book. The next chapter, in a similar vein, considers issues of regulation. Once we have the apparatus developed in this chapter alongside discussion of central regulatory issues, we can move on to particular applications of nanotechnology.

5.1   Risk

As we will see in the third unit of the book, various applications of nanotechnology carry risks. This chapter will clarify how to conceptualize risk, as well as how to move forward in light of the risk involved in various courses of action (e.g., with regards to emerging technologies).[1] This discussion complements the one offered in the following chapter about regulation. First, though, let us try to think more abstractly about what risk *is*, as this will help us be more precise both in this chapter and beyond. We will follow the work of Sven Ove Hansson, who has written extensively on risk and its philosophical implications.

In technical discussions, there are various different senses in which 'risk' is used; here, we will discuss four.[2] First, risk can be some unwanted event which may or may not occur. In this sense, environmental impacts are one of the risks of nanotechnology; this is to say that nanotechnology may or may not have these impacts and, furthermore, that the impacts would be negative. Note that both of these features are important, in this context, for attribution of risk. If nanotechnology definitely had some specific impact, then we would more appropriately call it a consequence rather than a risk: uncertainty is one of the features of risk. Additionally, the impact needs to be a bad thing for it to be a risk; otherwise we would call it something else, such as a (potential) benefit. These points might be obvious, but they help us to conceptualize risk. Second, risk can be the *cause* of an event which may or may not occur. If we say that nanotechnology carries environmental risks, what we mean is that nanotechnology causes – or, more accurately, tends to cause – negative environmental impacts. This postulation of a causal mechanism is more committed than the first conception of risk.

The third and fourth conceptions of risk are quantitative, as opposed to qualitative. The third conception holds that risk is the probability of an unwanted event which may or may not occur. So imagine that someone asks what the risk is of nanotechnology having a certain environmental impact. An appropriate answer here might be, for example, 10 percent. The first sense of risk treated the environmental impact itself as the risk, whereas the second treated nanotechnology as the risk (i.e., that which caused the impact). This third conception, though, tells us how likely it is that some impact will be realized. Fourth, and similarly, we could talk about the *expected outcome* of unwanted events. So imagine that there are 100 fish in some river that we are going to purify using nanoparticles. Further imagine that those nanoparticles are toxic to the fish population, and that some of the fish will die through the purification. We do not know which fish will die, but, given various epidemiological studies, we might reasonably say that 20 percent of the fish will die. The risk, then, is 20 fish in the sense that this is how many fish we expect to lose. On the third sense, we are given the likelihood that something will happen (e.g., as a percentage), whereas the fourth sense gives us an expected outcome (e.g., in terms of

---

[1]   This chapter is adapted from Fritz Allhoff, "Risk, Precaution, and Emerging Technologies," *Studies in Ethics, Law, and Society* 3.2 (2009): 1–27.

[2]   Sven Ove Hansson, "Philosophical Perspectives on Risk," *Techné* 8.1 (2004), p. 10.

some number of units lost). This fourth conception is the standard use of 'risk' in professional risk analysis. In particular, " 'risk' often denotes a numerical representation of severity, that is obtained by multiplying the probability of an unwanted event with a measure of its disvalue."[3]

Henceforth, it is this fourth conception that we shall be most interested in, though some of the other conceptions will also recur. There are various reasons that we shall focus on this fourth conception; as already mentioned, it is the standard use in risk analysis. One advantage it has is that it allows us to assess risks quantitatively, which helps make them commensurable with benefits. For example, if we can say that some remediation will lead to an expected loss of 20 fish, this loss can then be compared somehow to the benefits of the remediation, such as more long-term benefits for fish, cleaner water for a local township, and so on. On the first conception, the risk would just be "the loss of fish." The second conception acknowledges that the remediation will *cause* the loss of fish, and the third conception tells us the likelihood. The fourth conception, though, ties all of these things together, telling us what we can *expect* to happen, given the remediation. And this is why it is the most useful for risk analysis, even if we can speak of risk in the other three senses.

Now that we have various conceptions of risk, as well as the one that we will follow, we look at how decision making relates to risk. Of course, one of the hallmarks of risk is that we do not know for sure what will happen, given some course of action. It is this lack of epistemic certainty that makes decision making under risk philosophically and practically interesting. If we knew that some course of action had a set of determinate consequences, some of which were good and some of which were bad, then decision making would be a lot easier. To be sure, we might disagree about how to weight those good and bad consequences, such as if some application of nanotechnology had a positive economic upshot while having a negative environmental impact. Some might think that the environmental consequences were worth it, while others might not; this problem will recur in any society with pluralistic values. In such a situation, we would have to think about how to render positive and negative consequences commensurable, and we would further need to establish some democratic (or other) process for adjudicating disagreement. But in cases of epistemic uncertainty, this problem is exacerbated by the epistemological one, which is to say that we not only have to deal with a plurality of values, but we also do not even have epistemic certainty what the consequences will be. The values problem, then, is common to either scenario and is therefore not endemic to our discussion of risk.

There are, logically, four epistemic situations that we can be in with regards to risk.[4] The first of these is that we know the probability of some

---

[3] Ibid.
[4] Sven Ove Hansson, "What Is Philosophy of Risk?," *Theoria* 62 (1996), p. 170.

negative outcome. Imagine, for example, a case of Russian roulette in which a bullet is placed in one of six chambers of a revolver. Here we know the probability of a bullet being discharged, which is 1/6. Call this decision making under known probabilities: someone makes a decision whether or not to fire the gun, knowing what the probability is that a bullet will fire. Contrast decision making under known probability with decision making under uncertainty, wherein we know the probability only with insufficient precision. Return to the gun scenario and imagine that, last week, I put either two or three bullets in the chambers, but I have forgotten how many. If I choose to fire this gun, then I am doing so without known probabilities for the risks.

Finally, think of an extreme case of decision making under uncertainty: decision making under ignorance.[5] The ignorance, though, could be of two different sources, either of which would compromise our ability to determine some expected outcome. First, we might have little to no information about some specific outcome. Again, return to the gun example. Imagine that I pick up someone else's gun, not having any information about how many bullets are in the chamber, and then contemplate firing it. Assuming that I do not look in the chamber (and cannot otherwise tell anything by weight), I then have no information about the probability of a bullet discharging: that probability could be anywhere from zero if all chambers are empty to one if all chambers are loaded. Second, though, we might not even know what the outcomes *are*, much less how certain they are. Consider asbestos, for example, which became increasingly popular in the late nineteenth century as insulation. Despite the fact that even the Ancient Greeks observed lung damage in the slaves who wove it into cloth, the proclivity of asbestos to cause lung damage was not widely noted until the 1920s.[6] When asbestos became prevalent, the adverse health effects were largely unknown altogether, not just the probabilities that those effects would occur. It was not just that certain specific effects (e.g., mesothelioma and asbestosis) were unknown, but it was not even common knowledge that anything bad would happen to those who inhaled asbestos.

Technically, these three situations are all instances of decision making under uncertainty, though it is useful to think about the different variants involved in each case. Putting all four together, then, here are the epistemic situations we can have in relation to risk:

[5] See Sven Ove Hansson, "Decision Making under Great Uncertainty," *Philosophy of the Social Sciences* 26.3 (1996): 369–86.

[6] See, for example, W.E. Cooke, "Fibrosis of the Lungs Due to the Inhalation of Asbestos Dust," *British Medical Journal* 2 (1924): 147–50. A post-mortem in 1899 by a London doctor, H. Montague Murray, connected the death of a factory worker to asbestos inhalation. The Cooke paper, though, as well as a report that came out shortly thereafter, were what established widespread recognition of the link. For the report, see E.R.A. Merewether and C.W. Price, *Report on Effects of Asbestos Dust on the Lung*, HM Stationery Office (1930).

1 decision making with full knowledge of outcomes and probabilities;
2 decision making with full knowledge of outcomes and some, though not all, knowledge of probabilities;
3 decision making with full knowledge of outcomes and no knowledge of probabilities; and
4 decision making with incomplete knowledge of outcomes (as well as their associative probabilities).

Again, 2, 3, and 4 are all instances of decision making under uncertainty, and 3 and 4 are both instances of decision making under ignorance; what separates these from each other is not their formal relationship, but rather the degree (or type) of uncertainty.[7]

When dealing with technology in general, or even nanotechnology in particular, in which epistemic situation are we likely to find ourselves? A ready observation is that it almost certainly is not the first one. The only time that we have full knowledge of outcomes and probabilities is likely to be in sorts of idealized cases, such as when we are talking about rolling dice or flipping coins that are known to be fair.[8] In reality, epistemic uncertainty is sure to abound. When trying to decide whether to pursue some course of action – especially more complex ones, like policy decisions – there will almost certainly be some negative consequences that may or may not follow, and it is very unlikely that we will have epistemic certainty of either the relevant probabilities or even the relevant consequences. Of course, we at least hope to know the latter, and we also hope to know the former within some reasonable range of error. In subsequent chapters, we will explore whether this is the case with applications of nanotechnology; this theoretical framework will be useful for that discussion.

So what do we do with the uncertainty that we almost certainly face? Hansson, following Charles Sanders Peirce, offers an account of "uncertainty reduction" (cf. Peirce's "fixation of belief").[9] Hansson proposes that we reduce decision making under unknown probabilities to decision making under known probabilities, or even to decision making under certainty. For example, imagine that we are trying to figure out whether it will rain tomorrow. Various meteorologists get together, and they all come up with estimates as to the likelihood of rain. Just for simplicity, suppose that three camps converge on reasonably close estimates: 70, 80, and 90 percent chance of rain. We must now make a decision about our day that hangs on whether it will rain (e.g., whether to plan a picnic). Further suppose that this is an

---

[7] Technically, everything could be classified as decision under uncertainty, so long as zero and one were allowed as the probabilities that some consequence would attain.
[8] Hansson, "What Is Philosophy of Risk?," p. 171.
[9] Ibid., p. 172. See also Charles Sanders Peirce, "The Fixation of Belief," in Charles Hartshorne and Paul Weiss (eds), *Collected Papers of Charles Peirce* (Cambridge, MA: Harvard University Press, 1934), pp. 223–47.

instance of situation 2 above; we know what the outcomes are, but we do not have epistemic certainty as to the probabilities since the meteorologists disagree. What we will probably do is look at the testimony and aggregate it in some manner – thus, psychologically, abrogating the uncertainty. So, for example, we might take the testimony on board and then take the likelihood of rain to be 80 percent, effecting something like an average of the reports. There are other things we could do, such as taking the median, picking our favorite meteorologist, excluding our least favorite meteorologist, and so on. But, pragmatically, we are certainly going to look to a way to reduce the uncertainty.

This reduction, seemingly, improves our epistemic status from 2 to 1. Of course, our actual epistemic status has hardly changed at all: we have not gained any more information, but have rather just adopted some strategy to convince ourselves that we know more than we do. Hansson thinks that we often take it a step further, moving ourselves toward known probabilities and then toward certainty. If we collapse the different testimonies to an 80 percent aggregation of rain, do we go on our picnic? Probably not: 80 percent is high enough that we convince ourselves that it *will* rain (i.e., that the chance of rain is 100 percent). And now our epistemic status is even better than 1 since we have full knowledge of the outcomes. Or so we would like to think; obviously we are still, actually, in 2.

What are we supposed to do with all this uncertainty? Some approaches, like Bayesianism, would have us assign probabilities to everything.[10] If we have no information at all, then maybe we just assign probabilities of 0.5; those prior probabilities will thereafter get revised as we start to garner evidence. In the long run, maybe these sorts of approaches will get it right, though they are not terribly practical, and they otherwise face short-term limitations. For example, imagine that some technological application may have some disastrous consequence, but we really have no idea whether it will. Should we proceed with the application? We could take the consequences, multiply them by 0.5, and then derive some expected cost; this expected cost can be compared to the expected benefits. But if, unbeknownst to us, the objective probability of the negative consequence is 0.9 (rather than our subjective 0.5), we could be really far off with our risk assessment. Of course, we could be off with it in the other direction, too, thus overestimating the risks rather than underestimating them. However, we might think that there is some sort of *asymmetry* between these sorts of errors: it is worse to be insufficiently cautious than it is to be overcautious. This sort of attitude gives rise to precautionary approaches, which will be presented in §5.3 and critically evaluated in §5.4. In the next section, though,

[10] For an accessible introduction to Bayesianism, see Peter Godfrey-Smith, *Theory and Reality: An Introduction to the Philosophy of Science* (Chicago: University of Chicago Press, 2003), ch. 14. For a more technical discussion, see John Earman, *Bayes or Bust: A Critical Examination of Bayesian Confirmation Theory* (Cambridge, MA: MIT Press, 1992).

let us take a step back and talk about cost–benefit analysis in general; the relationship between cost–benefit analysis and precautionary approaches will receive further discussion in subsequent sections.

5.2   Cost–Benefit Analysis

Section 5.1 has two upshots. First, we wanted to conceptualize risk: we offered various conceptions and then proposed to focus on the expected outcome conception. Second, we wanted to highlight the central role that uncertainty plays in risk, including the various guises under which it can appear. Now we will consider how cost–benefit analysis can be applied to decision making under risk, with particular emphasis on how it looks under conditions of uncertainty.[11] This emphasis will be used to motivate precautionary approaches, though we will then return to the relationship those approaches bear to cost–benefit analysis.

Imagine that we are considering whether to perform some action, say φ. If we knew that φ had good consequences G and bad consequences B, then we could just think about whether the net effect was positive or negative (i.e., whether $G - B > 0$). There are a lot of challenges here: the consequences need to be commensurable, they probably need to be (at least somewhat) quantifiable, people might disagree on how to weight them, and so on.[12] But we can imagine stripped-down examples that elide all of these interesting features. Imagine that we are running a business and are considering some marketing plan for the new carbon nanotubes that our company has just developed; furthermore, imagine that the marketing plan would cost $10,000 to execute and would increase our sales by $20,000. Furthermore, imagine that there are no other marketing plans under consideration and that, given our fiscal cycles, the decision has to be made immediately (i.e., before any other marketing plans could be developed). In this case, it seems straightforward that we should effect the plan since the benefits outweigh

---

[11]   For our purposes, various nuances and conceptions of cost–benefit analysis are largely unimportant, though there is an important literature in this regard. One of the most ardent defenders of cost–benefit analysis is Richard Posner; see, for example, his *Catastrophe: Risk and Response* (New York: Oxford University Press, 2004). Cass Sunstein has written extensively on this topic; see, especially, his *The Cost–Benefit State* (Washington, DC: American Bar Association, 2002) and *Risk and Reason: Safety, Law, and the Environment* (Cambridge: Cambridge University Press, 2004). Frank Ackerman and Lisa Heinzerling critique cost–benefit analysis in *Priceless: On Knowing the Price of Everything and the Value of Nothing* (New York: New Press, 2003). Sunstein offers a review essay of contemporary scholarship, including Posner, *Catastrophe*, and Ackerman and Heinzerling, *Priceless*, in "Cost–Benefit Analysis and the Environment," *Ethics* 115 (2005): 351–85. See also Kristen Shrader-Frechette, *Taking Action, Saving Lives: Our Duties to Protect Environmental and Public Health* (New York: Oxford University Press, 2007).

[12]   See, for example, W. Kip Viscusi, *Fatal Tradeoffs* (New York: Oxford University Press, 1993). See also Ackerman and Heinzerling, *Priceless*.

the costs, there are no other alternatives to consider, there are none of the messy complexities mentioned above, and so on.

Now imagine that, unlike the epistemic certainties of that case, there is the sort of epistemic uncertainty postulated in situation 1 in §5.1: we have known probabilities, but not certainties. The marketing plan still costs $10,000 to execute, but there is a 40 percent chance that it will fail, thus eliciting no increased sales. There is a 60 percent chance that it will succeed, thus eliciting the $20,000 in increased sales. All other details are the same. What do we do now? We already know the costs with certainty (namely $10,000), but there is uncertainty about the benefits. We therefore calculate the expected benefits, which are:

$$0.4 \times \$0 + 0.6 \times \$20,000 = \$12,000$$

The $12,000 in expected benefits is greater than the $10,000 in actual costs, so we are still justified in pursuing the marketing plan, even given the possibility of its complete failure. Cost–benefit analysis, then, works not only when we have certainty regarding outcomes, but also when we have uncertainty but known probabilities.

Again, there are numerous other complexities to the cost–benefit approach; some were mentioned above. Returning to our previous example of the river purification project, imagine that those 100 fish will be killed, but the local township will have cleaner drinking water. These sorts of assessments have myriad complexities. Some of them are empirical: how *much* cleaner would the drinking water be? Would this *matter* in any significant way, such as health outcomes? How much? And then come the issues of commensurability and values: imagine that the purification, while killing 100 fish, will lead to a 10 percent decrease in the local incidence of a certain water-borne disease, giardiasis, while having no other demonstrable effects. Is this worth it? There are not general answers to these sorts of questions, though we will return to them in subsequent chapters when discussing particular applications of nanotechnology; for now, we just want to acknowledge some of these complexities.[13]

But, for present purposes, let us press on with our discussion of uncertainty. As shown above in the marketing plan cases, cost–benefit analysis seems promising when dealing with either known outcomes or known probabilities. Known outcomes, though, are not instances of risk at all, and so are not germane to that discussion. Known probabilities, as mentioned in §5.1, are only likely to occur in idealized cases, such as those involving fair dice and coins. While known outcomes or probabilities constitute positive epistemic statuses, these are not the epistemic statuses in which we are likely to find ourselves. Rather, we are more likely to find ourselves in

---

[13] Sunstein, *The Cost–Benefit State*, pp. 153–90 offers more discussion in a chapter called "The Arithmetic of Arsenic."

situations 2, 3, and/or 4: uncertain probabilistic knowledge, no probabilistic knowledge, and/or incomplete knowledge about outcomes. What guidance can cost–benefit analysis offer us now?

Return to the marketing example and make the parameters as follows: the plan still costs $10,000 to execute, and either it will increase our sales or it will not (known outcomes). Imagine there to be a 40–80 percent chance that the plan will succeed in increasing revenues and a 20–60 percent chance that it will not (i.e., such that the chances of success and failure sum to 100 percent); our experts just cannot agree on the proper assessments. As before, sales go up by $20,000 if the plan is successful. Do we implement it? It is hard to figure out what to say. We could try to effect the sort of uncertainty reduction discussed above: maybe we act as if the probabilities are in the middle of the ranges, thus there being a 30 percent chance that the plan will fail and a 60 percent chance that it will succeed. The expected outcome, then, is $12,000, which means that we should execute the plan. But there is something overly simplistic about this approach. For example, even though there was a 20–60 percent chance of failure, it hardly follows that the *actual* chance of failure is 30 percent; all we really know is that the probability falls somewhere within that range. The same is true with the probability of success. Maybe the actual chance of failure is 60 percent and the actual chance of success is 40 percent. In that case, the expected increase to sales is $8,000, which is less than the cost of the marketing plan, so it should not be pursued. So, unlike when we know the probabilities in which our privileged epistemic status leads to infallibility, we could make the *wrong decision* by applying cost–benefit analysis (in the above way, at least) when the probabilities are uncertain.

It is even worse when we move from limited knowledge of probabilities to no knowledge of probabilities. So imagine that we are considering the marketing plan, but we just have *no information* whether it will succeed or fail; maybe the CEO calls in looking for an immediate decision while all the relevant advisors are indisposed. Should we pursue the plan? As mentioned above, we could just give arbitrary assignments of probability to each outcome, 0.5 being the most plausible in cases of full ignorance. So there is a 50 percent chance that it will succeed and a 50 percent chance that it will fail, with an expected outcome of $10,000. Since this is how much the plan cost, we are neither any better off nor any worse off by pursuing it or not. But this is almost certainly the (objectively) wrong answer since *any* other probability assignment would give a deterministic answer about our course of action. It is worse yet again if we do not even know what the outcomes are. Imagine that, unbeknownst to us, the marketing plan infringes on copyrights held by another company, thus exposing us to legal liability. If there is a 60 percent chance of success and a 40 percent chance of failure, then we might put our expected outcome on $12,000, thus meaning that we should pursue the plan. But this would actually be a disaster because, once we release it, we get sued for $50,000. Obviously, if we do not have all the information regarding outcomes, our abilities to make good decisions can be compromised.

These epistemic situations – 2, 3, and 4 from §5.1 – are the ones in which we are most likely to find ourselves, and we see how cost–benefit analysis can get the wrong answer in these cases. This is not to say that we should not use cost–benefit analysis; indeed, as we will see in §5.4, it is not obvious that there is even an alternative. Rather, the point is to show how uncertainty challenges cost–benefit analysis. This should not be surprising as uncertainty challenges *any* decision making approach but, in these contexts, we might think more about how and when to move forward on decision making. Return to the above example where the CEO needs a decision on whether to effect the marketing plan, and we do not have any information on its prospects. One thing we might consider is to delay the decision until we have more information. Or, in the case where there is some unknown lawsuit waiting in the wings should we pursue the marketing plan, maybe we should not pursue the plan until we have reasonably convinced ourselves that no lawsuits are likely, or that there are no other negative externalities. We shall critically evaluate these possibilities in §5.4, but now let us consider the sort of approach that they suggest: precaution.

5.3   Precautionary Principles

Cost–benefit analysis under uncertainty poses risks, namely the risk of making the wrong decision. If we could somehow reduce the uncertainty, then we would occupy an improved epistemic status and be correspondingly more likely to make the right decision. The most obvious way to get rid of uncertainty is to hold off making the decision until we have better knowledge regarding probabilities and outcomes. For example, if there is uncertainty regarding the probability of some outcome, then we could do more research and try to reduce the uncertainty. If there are unknown outcomes, then we could take more time and try to make sure that we have uncovered all of them. Particularly when we risk substantial and negative consequences, we should be wary of making hasty decisions. To wit, we might adopt something like a "precautionary principle."[14] Part of the challenge with the precautionary principle approach is clarifying exactly what such a principle says, and various formulations abound. Charitably, there definitely seems to be merit to a principle that says we should not act hastily given the potential for substantial and negative consequences. When we try to pin down the details, though, it gets somewhat more complicated.

---

[14]   Note that much of the literature refers to *the* precautionary principle, though we shall talk about *a* precautionary principle or else precautionary principle*s*. The reason is that there is hardly any sort of definitive statement of "the" precautionary principle; rather there are many different formulations that bear various relations to each other.

Precautionary principles have been offered in various contexts, with environmental applications being the most common; the reasons for this emphasis will become clearer below.

For starters, let us look at some actual precautionary principles in order to understand their key features. There are many different formulations, as codified in national laws or international treaties.[15] For present purposes, however, we can take the context of issuance, as well as details regarding the issuing bodies, to be largely irrelevant. Consider the following three examples, which are representative. The first comes from the 1992 Rio Declaration of the UN Conference on Environment and Development (Principle 15):

> In order to protect the environment, the precautionary approach shall be widely applied by States according to their capabilities. Where there are threats of serious or irreversible damage, lack of full scientific certainty shall not be used as a reason for postponing cost-effective measures to prevent environmental degradation.[16]

Another formulation is the 1998 Wingspread Consensus Statement on the Precautionary Principle, which holds that "When an activity raises threats of harm to human health or the environment, precautionary measures should be taken even if some cause and effect relationships are not fully established scientifically."[17] And, finally, consider the European Union's Communication on the Precautionary Principle (2000):

> The Communication underlines that the precautionary principle forms part of a structured approach to the analysis of risk, as well as being relevant to risk management. It covers cases where scientific evidence is insufficient, inconclusive or uncertain and preliminary scientific evaluation indicates that there are reasonable grounds for concern that the potentially dangerous effects on the environment, human, animal or plant health may be inconsistent with the high level of protection chosen by the EU.[18]

These formulations are so varied that it is not even immediately obvious what they all have in common. We are told of a precautionary approach,

---

[15] Sunstein, "Cost–Benefit Analysis," p. 351 argues that Europe has been more sympathetic to precautionary approaches whereas the US has defended cost–benefit analysis. We are more interested in the philosophical underpinnings of the approaches than their applications, but this phenomenon bears notice. See also Sunstein, *The Cost–Benefit State*; Arie Trouwborst, *Evolution and Status of the Precautionary Principle in International Law* (London: Kluwer Law International, 2002); and Poul Harremoes et al. (eds), *The Precautionary Principle in the 20th Century: Late Lessons from Early Warnings* (London: Earthscan, 2002).

[16] Available at http://www.un.org/documents/ga/conf151/aconf15126-1annex1.htm (accessed June 23, 2008).

[17] Available at http://www.sehn.org/wing.html (accessed June 23, 2008).

[18] Available at http://www.gdrc.org/u-gov/precaution-4.html (accessed June 23, 2008).

precautionary measures, and a precautionary principle, though it hardly seems clear what any of these entails; reading the full documents, rather than just these excerpts, is not much more help. Risk (or threat) resounds throughout the different formulations, as does lack of evidence or certainty. But how do these pieces fit together in any meaningful sort of way? And, furthermore, how can we use those pieces to yield a *generalized* precautionary principle that abstracts away from the particular language used in these cases? In other words, what is the *logical structure* of precautionary principles? Is there one that they all share?

In trying to answer these questions, Neil Manson's work will help.[19] Manson develops an account of precautionary principles by first looking to see what sorts of generic features they share; he then considers what relationship those features have to each other. This does not presuppose that there is a single and general precautionary approach, as those features might have different relationships (or even be different) in various formulations; nevertheless, he thinks that there is at least something that the different formulations must have in common in order for them to be plausibly considered precautionary principles.

Before moving forward, it is worth acknowledging that much of the discussion regarding precautionary principles takes place in environmental contexts. This is because the environment is an especially complex system and such complexity gives rise to a lot of uncertainty regarding risks. For example, consider the introduction of rabbits into Australia.[20] Rabbits first came to Australia in the late 1780s, though the population explosion is thought to date to 1859. That fall a landowner living near Melbourne released 24 rabbits into the wild to simulate British hunts; other landowners then followed suit. Within 10 years there were literally millions of rabbits in the wild, and as many as 600 million nationwide by the mid 1900s; there are myriad ecological reasons for this population explosion, including mild winters and widespread farming (i.e., availability of food). The effects on Australia's environment have been disastrous, from species loss to erosion. Various countermeasures have been employed, such as shooting, poisoning, and fencing. Most dramatic (and effective) was the intentional introduction of a myxomatosis, a disease designed to kill rabbits; the disease caused the rabbit population to fall to approximately 100 million, though resistance eventually spread. Australia again has in excess of 300 million rabbits, despite the introduction of calicivirus – another biological measure – in 1996.

---

[19] Neil A. Manson, "Formulating the Precautionary Principle," *Environmental Ethics* 24 (2002): 263–72. See also Per Sandin, "Dimensions of the Precautionary Principle," *Human and Ecological Risk Assessment* 5.5 (1999): 889–907; and Carl F. Cranor, "Toward Understanding Aspects of the Precautionary Principle," *Journal of Medicine and Philosophy* 29.3 (2004): 259–79.
[20] For a further discussion of this account, see Tim Low, *Feral Future: The Untold Story of Australia's Exotic Invaders* (Chicago: University of Chicago Press, 2002). For a more general theoretical account of invasive species, see Julie Lockwood, Martha Hoopes, and Michael Marchetti, *Invasion Ecology* (Hoboken, NJ: Wiley-Blackwell, 2006).

The upshot of this example is that apparently trivial and benign acts can have catastrophic consequences: the release of 24 rabbits led to a rabbit population of over 600 million with dire economic and environmental impacts. Furthermore, those consequences could not have been predicted given the best scientific and other theories available. In addition to the consequences being negative and substantial, they can also be *irreversible*.[21] The release of those first rabbits began the inexorable march to present circumstances. This is not to say that things could not possibly have been any other way than exactly as they are today (e.g., the first round of hunters could have caught all of their prey, myxomatosis could have had a slightly different epidemiological trajectory, etc.); but rather that the impacts on the relevant environmental system have been so substantial that any sort of complete remediation of the problem is almost certainly now impossible. And, pragmatically, aside from the hunters catching all/most of the first rabbits, some roughly similar cascade of events would probably already have been prefigured.

More generally, any intervention into a well-functioning and complex system can have profound (and often negative) consequences. Complex systems, by definition, have many parts, which fit together in complicated ways. Affecting either the parts or the relationships among them can have implications for the other parts and their interactions. Furthermore, feedback cycles can multiply these effects. And, finally, complex systems are the most epistemically intractable. In such systems, we are almost certain to have limited knowledge about their proper functioning and, therefore, limited knowledge of how some intervention will affect that functionality. As in the case of the Australian rabbits, small perturbations can be disastrous. The environment is obviously one such system, but there are others. Perhaps most analogous is the human body, which we revisit in Chapters 11, on nanomedicine and nanotechnology, and 12, on human enhancement. (In some senses, this will also turn up in Chapter 9 on nanotechnology and the military.) But even social systems can manifest many of these features, an issue discussed in Chapter 10 on nanotechnology and privacy. The present point, though, is to establish the particular risks that complex systems offer, given our limited knowledge about them. As intimated and exemplified above, many precautionary approaches derive in environmental contexts for exactly this reason. Now we have seen the motivation for precautionary principles: to recognize the potential for dramatic and irreversible damage

---

[21] The concept of irreversibility is hardly transparent, though we shall not pursue further discussion here. For some of the conceptual complications, see Cass R. Sunstein, "Two Conceptions of Irreversible Environmental Harm," *University of Chicago Law & Economics, Olin Working Paper No. 407*, May 2008. Available at SSRN: http://ssrn.com/abstract= 1133164 (accessed October 2008). See also Neil A. Manson, "The Concept of Irreversibility: Its Use in the Sustainable Development and Precautionary Principle Literatures," *Electronic Journal of Sustainable Development* 1.1 (2007): 1–15.

in complex systems and to appreciate the limited epistemic situations in which we are likely to find ourselves in regard to those systems. With this in mind, we return to our discussion of the logical structure of precautionary approaches.

Manson argues that given some *activity*, which may have some *effect* on the environment, a precautionary principle must indicate some *remedy*.[22] We think this sounds right, though a couple of comments are worth making. Note the use of 'may,' which bears emphasis. Central to all precautionary approaches is the notion of uncertainty: if we knew what the consequences were, then we could just see whether the net effect was positive or negative. Even if we had known probabilities for the consequences, we could formulate an expected outcome, as we worked through in §5.2. But, if we have unknown probabilities or unknown outcomes, everything becomes more complicated; we used these unknowns to motivate the idea of precaution in the first place. So it is critical to precautionary approaches that there be unknowns, as is reflected by 'may' above; other weak modal language, like 'possible,' would also be appropriate, though see further discussion below. Second, Manson frames his discussion explicitly in terms of the environment, but we think that it generalizes beyond that context; as mentioned above, there are other contexts in which we have the same salient features, and our discussion will apply to those contexts as well. In what follows, we will converse at this more general level.

In addition to this acknowledgment of activities, effects, and remedies, Manson argues that all precautionary principles must share a three-part structure. The first part is the *damage condition*, which specifies some characteristics of the effect in virtue of which the precautionary approach is warranted. The second part is the *knowledge condition*, which specifies the state of scientific knowledge regarding the relationship between the activity and the effect. Finally, the third part specifies the *remedy*, which is the course of action that decision-makers should take *vis-à-vis* the activity. Putting this all together, all precautionary principles must share this structure: if the activity meets the damage condition and if the link between the activity and the effect meets the knowledge condition, then decision-makers ought to effect the remedy. This is a very general structure, leaving many possibilities for particular precautionary principles. For example, consider the damage conditions, which could characterize the relevant effects in any of the following ways, among others: serious, harmful, catastrophic, irreversible, destructive of something irreplaceable, reducing or eliminating biodiversity, violating the rights of future generations, and so on. Knowledge conditions could invoke parameters like: possible, suspected, indicated by precedent, reasonable to think, not certainly ruled out, not reasonably ruled out, etc. And remedies could be: bans, moratoria, postponements, research into

---

[22] Manson, "Formulating," p. 265.

alternatives, attempts to reduce uncertainty, attempts to mitigate the damage conditions, and so on.[23]

So, for example, we could say that if some effect is serious and possible given some activity, then we ought not to perform that activity. The damage conditions do not always scale in a simple way (i.e., in terms of increasing damage) but, to the extent that they do, as the damages become greater, then we might require improved epistemic status before avoiding the remedy. For example, imagine that the damages could be either "serious" or "catastrophic," the latter obviously being worse. And then imagine that knowledge conditions could be "possible" or "not certainly ruled out." If catastrophic harms are possible, then we might trigger the remedy more readily than we would were the harms to be merely serious. Since "not certainly ruled out" carries a higher epistemic threshold than "possible" (i.e., it requires us to have greater knowledge), we should apply that knowledge condition more readily to the catastrophic damages than to the serious ones. Or, less abstractly, it should just be the case that we require greater certainty regarding the relationship between some activity and its effects as those effects become more negative in terms of being immune from the remedies, all else being equal.

But all else does not have to be equal: rather than adjusting the knowledge condition as the effects become more negative, we could also adjust the remedy. Keeping the knowledge condition the same, then, again think of whether the effects proffer serious or catastrophic harms. As the harms become more substantial, then the remedy can simply become more restrictive. For example, we could say that, if the harms are serious, then the activity should be postponed. Alternatively, we could say that, if the harms are catastrophic, then the activity should be banned. This is not to say that in either case would the harms necessarily be realized, because of the uncertain relationship between the activities and the effects; it is just that, all else being equal, the remedy should be sensitive to the damage condition. All this is to say (and somewhat contrary to Manson's presentation) that these three conditions are not completely interchangeable, but rather should be interrelated to each other such that the above comparative desiderata attain. Given two different harms, should the knowledge condition be adjusted or the remedy? It depends. In some cases, our epistemic situation might be fairly hard to improve, so we might then adjust the remedies as the harms look more severe. In others, it might be the case that the remedies are hard to move (e.g., for legislative reasons), so we might then adjust the knowledge condition.

There are other things worth discussing, though many of them take us too far afield. Let us nevertheless make a few more observations before moving on. First, note that the knowledge condition effectively amounts to

---

[23] All of these lists are adapted from Manson, "Formulating," p. 267.

a burden of proof issue between the would-be practitioners of the activity and its opponents. As this condition becomes more stringent, proponents of the activity have more work to do in terms of ruling out some negative effect of their activity. For example, and perhaps counterintuitively, "possible" is more stringent than "likely" in the sense that it is easier to rule out some effect being likely than its being possible; we might be able to show that it is not likely that our carbon nanotubes will have some negative effect on the environment without being able to show that such an effect is impossible. Whether we are willing to proceed with some activity given a possible effect rather than a likely effect, as suggested above, probably has to do with what that effect is, as well as whatever other recourses are available to us *vis-à-vis* remedies. Second, the remedies postulated by precautionary principles are quite commonly bans on the corresponding activities, though this hardly need be the case; above, we saw a wide range of other available remedies. Bans might make particular sense as the effects become worse, but it bears emphasis that the precautionary approach is not committed in this way.

Having gone through much of this abstract and theoretical discussion, let us return to the examples of precautionary principles presented above in order to see how well this theoretical account holds up against actual principles. Consider again Principle 15 from the 1992 Rio Declaration of the UN Conference on Environment and Development:

> In order to protect the environment, the precautionary approach shall be widely applied by States according to their capabilities. Where there are threats of serious or irreversible damage, lack of full scientific certainty shall not be used as a reason for postponing cost-effective measures to prevent environmental degradation.

The damage condition here is explicit: the specified damages are ones that are "serious or irreversible." The knowledge condition is also apparent: lack of full scientific certainty. The remedy is the non-postponement of measures to prevent environmental degradation. Putting it all together and simplifying some of the language: if the damages are serious or irreversible, and if we lack full scientific certainty that those damages will occur, then we should not postpone measures to prevent environmental degradation. These statements are hardly transparent, though recognition of the underlying logical structure is definitely useful. To make this even less abstract, return to our example about river purification with nanoparticles. If we cannot rule out (cf. lack of scientific certainty) the possibility of those nanoparticles destroying the biodiversity of the river (cf. serious and irreversible), then we should not postpone measures that would prevent those harms. Those measures could include, for example, preventing the use of the nanoparticles at all. Now that we have a well-formed conception of the precautionary principle, let us subject it to critical evaluation.

5.4   Evaluating the Precautionary Principle

In evaluating the precautionary principle, it will be useful to have a particular conception in mind. The account developed above gives us the logical structure of precautionary principles, and there is nothing inherently problematic with a formal proposal that, given some potential for damage and given some epistemic status regarding the causal links between some activity and that damage, we should effect some remedy. Rather, it is when we start specifying the damage condition, knowledge condition, and remedy that substantive critiques are possible. The hazard of picking a specific precautionary principle, though, is that criticisms of it will not necessarily apply to other variants; those variants could have different features that immunize them from the criticisms. Aware of this hazard, we nevertheless propose to focus on a particular conception, while offering discussion of alternatives as we proceed.

The specific principle we will consider is one that is most commonly discussed in the literature: the catastrophe principle.[24] This principle specifies the damage condition as catastrophic, as opposed to lesser damages, such as harmful or serious ones. Its knowledge condition specifies possibility, which is comparatively permissive: a lot of effects are possible even if they are not, for example, likely. And, finally, the remedy is a ban. As mentioned above, bans are common remedies offered by precautionary principles even if they are not, strictly speaking, required by such principles. So we specify the catastrophe principle as follows: if, given some activity, some catastrophic effect is possible, then we should ban the activity. This formulation is substantive enough to be evaluated (i.e., the three parts are specified) while still being general enough that the following discussion cuts across various ways it could be further specified *vis-à-vis* the particular activities or effects. There are three broad sorts of criticisms that have been lodged against this formulation; we shall consider them in turn.[25]

The first criticism is of the knowledge condition, particularly its (extremely) weak modal operator: possibility. On the catastrophe principle, mere possibility of some catastrophe is enough to produce a ban against some activity. It is possible, in at least some sense, that nanotechnology could destroy the world. Surely that is catastrophic; ergo, no nanotechnology. But what is the sense of 'possibility' that matters? It has to be something stronger than mere logical possibility: it is (logically) possible, for example, that our nanotechnology could lead, tomorrow, to some catastrophe on some inhabited planet in the deepest recesses of some other galaxy. But

---

[24]   See also Manson, "Formulating," pp. 270–4. Note that Posner, *Catastrophe*, explicitly defends cost–benefit analysis even under prospective catastrophe.

[25]   See also John Weckert and James Moore, "The Precautionary Principle in Nanotechnology," *International Journal of Applied Philosophy* 2.2 (2006): 191–204.

surely this is not physically possible, if for no other reason than it could not get there fast enough. Rather, what we need is some sort of physical possibility or, even better, empirical possibility: things may be physically possible that are nevertheless not likely to happen (e.g., decreased entropy in some complex system). This sort of possibility at least forestalls straw man objections against the catastrophe principle.[26]

Still, though, empirical possibility is extremely weak: a *lot* of things are empirically possible. For example, consider the notion that self-replicating nanobots will somehow cause human extinction. Is that empirically possible? In some sense, yes: these nanobots could replicate to the extent that they take over whatever environments humans would otherwise occupy. Is this likely? No. Does any reasonable scientific evidence suggest that it would happen? No. But is it (empirically) *possible*? Yes. So, on the catastrophe principle, it would seem that we cannot have whatever technology might give rise to the nanobots. This seems like the wrong answer, though, particularly given the (extreme) unlikelihood that these negative consequences would be realized.

Proponents of such an approach, however, could point out that the *magnitude* of the catastrophe justified the triggering of the remedy (e.g., a ban) *despite* the low probability of the catastrophe. And there has to be at least something right in this sentiment. Consider, for example, two cases. In the first, something extremely bad is going to happen with a 1 percent probability; and in the second, something somewhat bad is going to happen with a 50 percent probability. Which scenario is better? It has to matter what the magnitudes of the bad effects are. Imagine that we could render them financially, just to make the conceptualization simple. The first case has a 1 percent chance of having US$1 billion in damage. The second has a 50 percent chance of having US$1 million in damage. Even though the probability is lower in the first case, the expected damages are 20 times higher. Therefore, we cannot just look at the (low) probability and say that we should proceed regardless. But what if the probabilities are really low and the consequences really bad (e.g., self-replicating nanobots destroying the world)? From an expected outcome approach, it does not matter; these would just "cancel out," thus giving results commensurable with more moderate values.

This gives rise to a second worry about the precautionary principle, which is to identify its relationship to traditional cost–benefit analysis. We think that this relationship has been poorly understood, particularly insofar as the precautionary approach is sometimes characterized as an "alternative" to cost–benefit analysis.[27] To motivate this part of the dialectic, consider

[26] For more discussion, see David B. Resnik, "Is the Precautionary Principle Unscientific?," *Studies in the History and Philosophy of Biological and Biomedical Sciences* 34 (2003): 329–44.
[27] See, for example, Manson, "Formulating," p. 264; Weckert and Moore, "The Precautionary Principle," p. 191. Sunstein, "Cost–Benefit Analysis," alleges a "tension" between precautionary and cost–benefit approaches (p. 352) though then goes on to suggest that the views are "complementary" (p. 355). These certainly look like different claims, but we are ultimately sympathetic to the latter, as will be expressed below.

that the precautionary approach either is something new (*vis-à-vis* cost–benefit analysis) or else it is not. On the former, it is supposed to be problematic and, on the latter, it is not even interesting. Starting with the latter, remember that the defender of the catastrophe principle owes us some account of 'possible,' in terms of both what it means and why it matters. Following the above discussion, let us assume that it means something like "empirically possible" and it matters because, despite the low probabilities, the potential effects are catastrophic. This sounds perfectly plausible, but then it just says the same thing as cost–benefit analysis; cost–benefit analysis can certainly accommodate low probabilities of catastrophes in terms of formulating expected outcomes.

Another way to go is to say that the precautionary principle really is saying something different. For example, the defender of the precautionary approach might deny that the environment is or has some singular value, which is commensurable among other values. Given that there is some catastrophic risk to the environment – however unlikely – that risk just trumps all other considerations. This sort of line is different from cost–benefit analysis in the sense that the latter would allow us to consider the *benefits* of some activity, rather than merely having to stop at an identification of the risks. But, for a number of reasons, this has to be wrong. First, it allows extremely low probabilities to derail entire activities. (Again, one could point to the magnitude of the consequences, but then this just brings us back to the first horn of the dilemma.) Second, these low probabilities – which nevertheless establish *possibility* – could be effectively impossible to reduce to zero. Imagine we can show a 1 percent chance of some effect, or a 0.1 percent chance, or a 0.01 percent chance: in no case have we shown that it is impossible. If the precautionary approach is meant to do something different from cost–benefit analysis, then it becomes paralyzing. Third, this is simply irrational. Imagine that, if we φ'd, there was an X percent of some cost C; further imagine that C is really bad. Should we φ? It is impossible to even conceptualize this question without knowing the benefits that would attain by φ'ing (as well as their associative probabilities). Imagine there is some evil deity who asks for a tithe, lest he destroy the planet. Furthermore, imagine that he might destroy the planet anyway, given the (remote) probability that he finds the tithe unacceptable. So, if we tithe, then it is possible that he will destroy the planet. The defender of the catastrophe principle therefore has to say that we cannot tithe, even if the deity will certainly destroy our planet if we do not. This does not make any sense: it is completely irrational to allow remote risks to entirely preclude our consideration of the associative benefits for some course of action.

The evil deity example gives rise to a third criticism of the catastrophe principle: this criticism holds not just that the principle is false, but rather that it is incoherent. Consider Cass Sunstein: "Because risks are on all sides of social situations, and because regulation itself increases risks of various sorts, the principle condemns the very steps that it seems to require."[28] So imagine that it is possible that some activity gives rise to some catastrophe. Therefore, we ban that activity. But surely it is possible that *the ban* risks a catastrophe as well. So we cannot ban the activity. Return to our example about water purification using nanoparticles: this practice could (even if not likely) have disastrous effects on the environment. But a failure to have clean water could (and probably more likely would) lead to disastrous effects, particularly *vis-à-vis* the world's poor who are increasingly without drinking water (see §7.3). The effects on them directly are bad enough, but there could be added effects in terms of political destabilization, global conflict, and so on. The catastrophe principle would say that we cannot purify the water and, similarly, that we cannot effect the ban against the purification. In other words, it says we cannot φ and we cannot $\sim\varphi$. This is logically impossible; therefore the principle is incoherent. The incoherence charge is a strong one and certainly one best avoided. For example, so long as one of the catastrophic effects is more likely than the other (e.g., as follows from φ or $\sim\varphi$), then maybe the advocate just guards against the more likely catastrophe. But this would require further emendation to the principle, and then risks some of the other criticisms presented above.[29]

Having seen various criticisms, let us now offer our own view.[30] We think that there are two fundamental issues with precautionary approaches. The first has to do with the knowledge condition. In the catastrophe formulation, mere possibility was enough to force the ban on some activity. Some people have wanted to say that this leads to bans too easily since negative effects will always be possible, even in our sense of empirical possibility. This does not worry us, though, because of the potential magnitude of those effects. If the probability of the effects is really low, but the negative consequences of the effects are really high, then we should take the risk seriously. Part of the problem is undoubtedly epistemic as we will not always know what the probabilities are, and we certainly cannot rule out that they

---

[28] Sunstein, "Cost–Benefit Analysis," p. 355; see also pp. 366–9. Sunstein means this criticism to apply to the precautionary principle more generally, rather than to the catastrophe formulation in particular. We disagree and think that the criticism, at best, attaches to catastrophe-like formulations because different knowledge conditions (e.g., ones requiring "likely" rather than "possible") are unaffected by the criticism. See also Cass R. Sunstein, "Beyond the Precautionary Principle," *Pennsylvania Law Review* 151 (2003): 1003–58; and Cass R. Sunstein, *Laws of Fear: Beyond the Precautionary Principle* (Cambridge: Cambridge University Press, 2005). For a detailed response to the incoherence objections, see Jonathan Hughes, "How Not to Criticize the Precautionary Principle," *Journal of Medicine and Philosophy* 31 (2006): 447–64.
[29] Another criticism, not presented here, is that precautionary approaches contribute to, and even promote, unfounded public fears. See Adam Burgess, *Cellular Phones, Public Fears, and a Culture of Precaution* (Cambridge: Cambridge University Press, 2004).
[30] For more detailed responses to some of these criticisms, see Stephen M. Gardiner, "A Core Precautionary Principle," *Journal of Political Philosophy* 14.1 (March 2006): 33–60. Gardiner defends a particular version of the precautionary principle, arguing that his formulation – different from the catastrophe principle – is immune to standard criticisms.

are zero (as the catastrophe approach would seemingly require). We will return to that below; but suffice it to say that unlikely but catastrophic risks should obviously play a part in our decision making. We hardly need a precautionary approach, though, to tell us that; no reasonable person would deny it.

One obvious way to make the precautionary approach more permissive is to relax the knowledge condition. For example, we might say that the negative effects need to be, not just possible, but likely. This project becomes more epistemically tractable in the sense that it is easier to establish likelihood than it is to rule out possibility; this is to say not that establishing likelihood is easy, but that ruling out possibility is extremely hard. Note that, pragmatically, this suggestion transfers the burden of proof from the proponent of some activity to its detractor. For example, we might not be able to rule out the possibility of self-replicating nanobots destroying the world, but can it be proven to be likely? Defenders of precautionary principles think that the burden of proof should be on the would-be facilitator of some catastrophe; opponents claim that the principles are too restrictive. Where should the burden of proof go? We do not think that this question or the corresponding conception is very useful. Rather, what matters is what the risks *are*. They might be hard to determine but, conceptually, the risks are what matter, not where the burdens of proof fall. From a procedural or regulatory perspective, burden of proof might be important, but there are ways of dealing with it (e.g., further research, independent commissions). But, philosophically, the focus should be on the risks themselves.

This, then, brings us back to the second fundamental issue with the precautionary approach, which is its relationship to cost–benefit analysis. As suggested above, we think that there has been a lot of confusion regarding this issue, particularly in claims that precautionary principles are alternatives to cost–benefit analysis. Cost–benefit analysis cannot possibly be wholesale wrong as an approach to decision making. In our everyday lives, we continually weigh costs and benefits (discounted by their perceived probabilities) and make decisions based on those assessments; such an approach is almost the paragon of rationality.[31] If you were facing a great but unlikely benefit versus a great and likely benefit for opposing courses of action, which would you pick? The answer is so trivial as to lead us to wonder what all this dialogue over the precautionary principle is supposed to contribute.

There seem to be two possibilities in this regard. The first is that there are certain domains in which the precautionary principle is supposed to supplant cost–benefit analysis. For example, consider environmental contexts

[31] A similar attitude is expressed by Posner, *Catastrophe*, p. 139 who argues that cost–benefit analysis "is an indispensable step in rational decision making," even under catastrophic risk; quoted in Sunstein, "Cost–Benefit Analysis," p. 363.

in which serious and irreversible harm is possible; this is the sort of context in which we often see precautionary principles surface. But why would cost–benefit analysis be ill-equipped to handle this situation? Certainly cost–benefit analysis can accommodate concepts like 'serious' and 'irreversible' since these have obvious upshots in terms of risk assessment. It cannot be those concepts that activate the precautionary approach as an alternative to cost–benefit analysis. What about the environmental context itself? Maybe we should exercise extra caution when dealing with the environment because of the sort of thing that it is or the moral features that it has. Even if this is true, though, cost–benefit analysis would still work: only the *weighting* of the relevant considerations would change once we properly appreciated environmental values. In other words, imagine that we rally behind the precautionary approach because we really decide that the environment is important. What have we gained? We could just do cost–benefit analysis and maintain that environmental costs are really bad and environmental benefits are really good. If precautionary approaches effectively increase the weighting of environmental considerations, we could afford similar weightings through cost–benefit analysis.

Whoever wanted to defend the "supplanting" model would now have to argue that the environment is not simply one value among many – or even an important value, as cost–benefit analysis could surely accommodate – but rather that it is patently incommensurable with other values. So, presumably, we cannot use cost–benefit analysis because the environment is special and cannot be compared to other values. To figure out whether to destroy the redwood forest, we hardly focus on the joy that would be derived from the proposed theme park, or on how much money people would be willing to pay to access those trees as against the alternative theme park. This joy and the associative economic preferences are morally relevant, but these are incommensurable with the value of the forest.[32] Surely the forest must be preserved regardless. Or so the dialectic might proceed. However, it cannot be right that environmental values are incommensurable with others. Imagine that terrorists will destroy the entire world unless we destroy a single tree. Save the tree? Forests and trees matter, but so do a lot of other things, and we have to have some complex value system that accommodates all of those values. For these reasons and those above, we therefore reject the idea that precautionary approaches are meaningful alternatives to cost–benefit analysis.

Rather, we think that precaution *supplements* cost–benefit analysis *given uncertainty*.[33] As we saw in §5.1, there are various epistemic situations in which we might find ourselves with regards to risk. If we know that some

---

[32] A classic on this issue is Mark Sagoff, "At the Shrine of Our Lady of Fatima; Or, Why All Political Questions Are Not All Economic," *Arizona Law Review* 23.4 (1981): 1283–98.
[33] Cf. Posner, *Catastrophe*.

act A has an X percent chance of realizing some benefit B while, at the same time, having a Y percent chance of realizing some cost C, then we just compare $X \times B + Y \times C$ with the alternatives to A and pick the best expected outcome. As we discussed in §5.2, this becomes more complicated when we do not know X or Y. It is even worse when we do not know B and C, either. Precaution is a risk-averse strategy for dealing with uncertainty.[34] If we know that there is a Y_1 to Y_2 percent chance of some cost C, precaution might, for example, tell us to act *as if* the probability were the higher value, Y_2 percent. And, if we were considering some uncertain benefit, we might act as if the probability were the lower value. But this then integrates quite well with cost–benefit analysis: it just requires us to be conservative in our assessments.

Whether we should be conservative does not depend on the (non-epistemic) values at stake or on their probabilities, which are treated straightforwardly through cost–benefit analysis. Rather, the conservativeness is dictated by the (epistemic) value of uncertainty and our predilections against it. The disvalue of uncertainty is hardly obvious; there are certainly contexts in which most of us prefer it (e.g., opening presents). When making decisions about applications of nanotechnology, we just have to think about how tolerable uncertainty is, particularly given the potential consequences. In later chapters, we shall explore these issues as they pertain to particular applications, though we do not think that there is a generalized answer to such concerns. Regardless, the point of this chapter is to develop conceptual apparatuses regarding risk, cost–benefit analysis, and precaution; we shall return to these discussions throughout the book. For Chapter 6 let us turn to issues in regulation, wherein we will see how some of these theoretical discussions play out in practice.

---

[34] Consider, for example, Ackerman and Heinzerling, *Priceless*, p. 227 who mean to be offering a critique of cost–benefit analysis and a defense of precaution. If, for example, our nation spends more than it needs to on regulatory protection, its "preference is to tilt toward over-investment in protecting ourselves and our descendants" (quoted in Sunstein, "Cost–Benefit Analysis," p. 359). But this "precaution" is just demonstrating a collective agreement that the prospective negative consequences are really bad and that we hardly want to countenance their actualization. (1) However, this is hardly antithetical to cost–benefit analysis, which has to be able to accommodate our preferences: what else would "cost" and "benefit" even *mean* as wholly independent of our preferences? There might be cases, like the forests, where we want to ascribe value independently of our preferences, but there are other cases, like whether to effect a tax increase to support a farm subsidy, that cannot be understood without thinking about how put off we would be by the tax increase, what impacts it would have on the food supply (and whether we would care about those), and so on. (2) So, when we "over"-invest, all we are doing is demonstrating that we take the cost to be worth the protection that it affords us against a negative outcome; this protection could be economic, psychological, moral, or symbolic. (Cf., for example, the war on terror, which almost certainly costs far more money than it could ever prevent in terms of economic damages.) This is not to say that we are infallible with all of our protective investments, though it is to say that we can rationally accommodate risk aversion under a cost–benefit framework.

6

Regulating Nanotechnology

In the preceding chapter, we provided some analysis of the twin concepts of risk and precaution. Why do they matter? Because there is a fierce and urgent debate today – a real dilemma that affects real lives – about whether nanomaterials and nanotechnology research ought to be regulated. No matter which way this debate goes, there is much at stake: both benefits and harms to jobs, worker safety, consumer health, animals, the environment, business, innovation, and global competitiveness.

The issue of regulating nanotechnology came to the forefront with a January 2006 report by the Woodrow Wilson International Center for Scholars (WWICS) that argued for stricter policies in the US.[1] What motivates the push for more or stronger regulation is not just the possible risks associated with nanotechnologies, but also very much our own uncertainty and/or ignorance about both the risks and the technologies themselves (see §5.1). This chapter will explain and explore the regulation debate, laying out the major positions as well as suggesting a simpler, more politically feasible alternative to stricter laws, at least in the meantime.[2,3]

As a preliminary note: while the aforementioned WWICS report dealt exclusively with US laws, the underlying debate is parallel to ongoing

[1] J. Clarence Davies, *Managing the Effects of Nanotechnology* (Washington, DC: Woodrow Wilson International Center for Scholars, January 2006).
[2] This chapter expands on Patrick Lin, "Nanotechnology Bound: Evaluating the Case for More Regulation," *NanoEthics: Ethics for Technologies that Converge at the Nanoscale* 2 (2007): 105–22.
[3] In this chapter, we will use "laws" and "regulations" interchangeably for the sake of simplicity, though we recognize a distinction between the two, which is not material to this discussion. Furthermore, while regulations may take a wide range of forms (e.g., labeling requirements, penalties and fees, market access, outright ban) and apply to many stages in the "supply chain" or product lifecycle (e.g., research, manufacturing, marketing, disposal), we will not discuss these nuances here for the sake of brevity and to stay within our goal to deliver a broad, conceptual evaluation of the debate.

investigations around the world and the concerns are the same. Because we need not refer to any particular law or regulation in this chapter, our discussion here can be applied equally well to those investigations outside the US. Indeed, we do not take up an analysis of any particular law or regulation, because they continue to evolve incrementally and because, as noted in the preface, this chapter (and book) is meant to provide not specific policy recommendations, but rather a conceptual analysis of the debate at large.

Also, the risks we will address here are primarily environmental, health, and safety (EHS) risks as opposed to, say, the risk that there is no democratic control over the technologies; this latter issue does not appear as urgent as EHS risks but is an entirely separate issue that nonetheless deserves its own investigation.[4] We do, however, address some of these non-EHS risks toward the end of this chapter as well as in the third unit of this book.

6.1 The Stricter-Law Argument

While we have not yet seen anyone or any organization clearly articulate or formalize the argument for stricter laws in nanotechnology, as opposed to implying the argument, it can be characterized as the following.

While more and more nanotechnology products are introduced into the marketplace, some studies have already suggested that engineered nanomaterials may be harmful, for instance, causing brain damage in animals.[5] As a specific example discussed earlier in this book (see §3.2), the carbon nanotube happens to resemble the whisker-like asbestos fiber. This is troubling because the shape of asbestos fibers is what makes them so difficult to dislodge from one's lungs. Further, nanoparticles are so small, by definition, that they might easily and undetectably slip into a person's body and cells with undetermined effects.[6]

There are also unknown environmental impacts of nanomaterials (see Chapter 8 for more discussion). Because they are created to be more durable than existing materials, it begs the question of how long they will persist in our landfills.[7] If nanoparticles can be absorbed by cells, as studies have

[4] See, for example, Sheila Jasanoff, *Designs on Nature: Science and Democracy in Europe and the United States* (Princeton: Princeton University Press, 2005).
[5] Eva Oberdörster, "Manufactured Nanomaterials (Fullerenes, C60) Induce Oxidative Stress in the Brain of Juvenile Largemouth Bass," *Environmental Health Perspectives* 112.10 (2004): 1058–62.
[6] Vicki Colvin and Mark Wiesner, "Environmental Implications of Nanotechnology: Progress in Developing Fundamental Science as a Basis for Assessment," a keynote presentation delivered at the US EPA's "Nanotechnology and the Environment: Applications and Implications STAR Review Progress Workshop" in Arlington, Virginia, August 28, 2002.
[7] Lawrence Gibbs and Mary Tang, "Nanotechnology: Safety and Risk Management Overview," NNIN Nanotechnology Safety Workshop at Georgia Institute of Technology, December 2, 2004.

shown, then they could slip into our food chain and eventually into humans. This conjures up the related scenario of food poisoning from shellfish that had fed on toxic algae and other lessons in bioaccumulation, e.g., involving the pesticide, DDT.[8] Again, the effects of nanoparticles on our biology are still unknown, so food poisoning may be the least of our worries, with genetic damage and death as other possibilities.[9]

So the problem is this: for many people, it seems to be commonsensical that if there are real questions about the EHS impact of nanotechnology products or any other product, then we should investigate them further before these products enter our marketplace. Society has learned this precaution from past lessons involving the introduction of such hazardous products as asbestos or lead paint or DDT into the public space.

Of course, there are laws and regulations already in place that, in theory, should prevent harmful products from ever reaching the marketplace, such as the Toxic Substance Control Act (TSCA)[10] and the Occupational Safety and Health Act (OSHAct)[11] in the US. But are these laws really equipped to handle nanotechnology? We will not attempt to make those determinations here, particularly given the comprehensive analysis offered by the WWICS report and subsequently by other organizations,[12] but there is good-faith reason to believe that current laws are not perfect, which suggests a real possibility that they cannot account for the nanomaterials in question.

Even if one does not know much about the relevant laws, it is understood that laws are created based on the available facts and circumstances of the time and foreseeable future. Those facts and circumstances continue to evolve, to be refined, and even to be repealed over time, as appropriate. And given how little we know about nanotechnology – though we do know nanomaterials have novel and unpredictable properties – it is difficult to see why we should expect current laws to *not* need updating (or an overhaul) as we learn more about nanotechnology. At any rate, it is better to be safe than sorry, i.e., to be open to the possibility that we need stricter laws rather than to risk damaging our health and environment – or so the stricter-law argument goes.

At this point, the stricter-law argument concludes that current laws, such as those in the US, which have not prevented such products from entering the marketplace, have failed. This is because if there are serious and continuing questions about a product's safety, then common sense – or some version of the precautionary principle (see §§5.3–5.4 and below) – requires

---

[8]   Ibid.
[9]   Ibid.
[10] Available at http://epw.senate.gov/tsca.pdf (accessed September 1, 2008).
[11] Available at http://www.osha.gov/pls/oshaweb/owasrch.search_form?p_doc_type=oshact (accessed September 1, 2008).
[12] See, for example, David Berube, "Regulating Nanoscience: A Proposal and a Response to J. Clarence Davies," *Nanotechnology Law & Business* 3.4 (2006): 485–506.

that product not be released into the marketplace until its safety is more convincingly established.

In the case of the 2006 WWICS report, the recommendation is to strengthen existing laws and regulations, as well as to enact new ones. The report also provides an analysis of the legislation relevant to nanotechnology, including not just TSCA and OSHAct, but also the Food, Drug, and Cosmetic Act[13] and the major environmental laws such as the Clean Air Act[14] and the Resource Conservation and Recovery Act. Again, in this chapter, we will not look at that analysis to determine whether or not these acts really are equipped to deal with nanotechnology, but instead focus more on the supporting reasons behind the (implied) arguments to evaluate their soundness and consistency.

Calling for stricter laws, of course, is not the only possible response to the alleged failure of current laws. Some have already proposed a moratorium or full ban on nanotechnology research and products until EHS risks are better understood and mitigated as needed.[15] We will also not investigate this particular position here, since if the argument for stricter laws cannot be defended, then it seems unlikely that an argument for a moratorium, which we take to be a significantly more extreme position, can also be defended. Furthermore, the stricter-law argument might reason that the lack of support for a moratorium, other than from the few groups that have proposed it, may indicate that this position is an over-reaction, all things considered.

Short of new or stronger regulations, other measures have been taken in response to these EHS issues to questionable effect. In January 2008, the US Environmental Protection Agency (EPA) implemented a program for the nanotechnology industry to *voluntarily* report their use of nanomaterials, expecting data from 180 companies; instead, it received submissions from only 19 companies, as of its July 2008 deadline.[16] Thus, half-measures such as this have not been effective enough to pacify the growing chorus of calls to regulate nanotechnology, though the EPA should be applauded for recognizing the concern and adjusting its policies where (politically) feasible.

---

[13] Available at http://www.fda.gov/opacom/laws/fdcact/fdctoc.htm (accessed September 1, 2008).

[14] Available at http://www.epa.gov/air/caa/ (accessed September 1, 2008).

[15] See, for example, Australian Green Party, "Call for Moratorium on Nanotechnology," Australian Associated Press article dated March 17, 2007; Friends of the Earth (FOE), "Nano-ingredients Pose Big Risks in Beauty Products: Friends of the Earth Report Highlights Unregulated Risks of Nanoparticles in Cosmetics and Sunscreens," press release dated May 16, 2006; and ETC Group, "No Small Matter II: The Case for a Global Moratorium," report dated April 2003.

[16] See US Environmental Protection Agency site, http://www.epa.gov/oppt/nano/stewardship.htm (accessed August 21, 2008); also, Rebecca Trager, "EPA Nanosafety Scheme Fails to Draw Industry," *Chemistry World*, August 5, 2008, available at www.rsc.org/chemistryworld/News/2008/August/05080801.asp (accessed August 21, 2008).

6.2   Learning from History

If the stricter-law argument sounds a bit far-fetched or paranoid, we can look at recent US history to see where our laws have failed us in protecting the public, including industry workers, from commercial hazards. There does not need to be an elaborate conspiracy theory that such laws have been poorly designed, perhaps as a result of misinformation, corporate influence, or political haggling (though invariably some laws have suffered from these and other more insidious manipulations). Rather, it is a plain fact that people are fallible, both scientists and legislators alike, especially when it comes to predicting the future. When it comes to nanotechnology, existing regulations may not be adequate to protect the public and they may not fully address the safety risk posed by new materials.

Asbestos, lead paint, and DDT are frequently cited case studies, as well as the diet drug "fen-phen" (a combination of fenfluramine and phentermine) and other pharmaceuticals. In each case, these materials made their way into the marketplace and into our homes, only for it to be discovered later that they are hazardous to our health and/or environment. However, one can object that these incidents, having occurred years or decades ago, are an unfair comparison to today's risk in nanotechnology. In defense of current regulations, we have since evolved our laws and our thinking to more rigorously test new products, ever mindful to prevent such incidents from occurring again.

But why should we believe that the evolution of our product and environmental safety laws have reached an end point now? Are we really so clever as to have finally created a system of safeguards to protect us from every conceivable EHS risk? Current events suggest otherwise. In July 2008, one of the most respected US cancer centers, the University of Pittsburgh's Cancer Institute, issued an internal warning that mobile phones may cause cancer.[17] In October 2006, the also-renowned Cleveland Clinic and others found that mobile phones appear to be linked with infertility in men.[18] And a steady stream of studies continues to point to other health risks with mobile

---

[17]   See University of Pittsburgh's Cancer Institute's site, http://www.upci.upmc.edu/news/ (accessed August 21, 2008); also, Seth Borenstein and Jennifer C. Yates, "Pittsburgh Cancer Center Warns of Cell Phone Risks," *Associated Press*, July 23, 2008, available at http://ap.google.com/article/ALeqM5hwzQ6Jsq3cSWa721yR84l99_pnlAD923R4DO1 (accessed August 21, 2008).

[18]   Ashok Agarwal, Fnu Deepinder, and Kartikeya Makker, "Cell Phones and Male Infertility: Dissecting the Relationship," *Reproductive BioMedicine Online* 15.3 (2007): 266–70, available at www.clevelandclinic.org/reproductiveresearchcenter/docs/agradoc250.pdf (accessed August 21, 2008).

phones.[19] Yet, mobile phones are one of the most ubiquitous devices we possess and use daily worldwide, and stronger precautionary regulations or measures have still not been taken even given these studies.[20] (Also, see the discussion below about the new risks discovered with Teflon.) This is not to say that mobile phones actually do cause us harm, but only to point out that nanotechnology is not alone in its tension with existing laws and regulations that are supposed to prevent harm to workers, consumers, and the environment.

Of course, it is perhaps unreasonable to expect that we can *guarantee* that commercial products, particularly emerging technologies, are completely safe. Undoubtedly, the use of mobile phones in the last 15 years has contributed immensely to business operations (and profit), emergency rescue, personal freedom, and more; it is still a key enabler of a mobile workforce, which makes new careers and business models possible. But it seems backward to release a product into the marketplace and then conduct the EHS testing needed to answer basic questions surrounding the product.[21]

This scenario of discovering too late the harms of a product and having to recall it from the global market is one we want to avoid. This is the point of the nanotechnology regulation debate. It would be a catastrophe on many levels if mobile phones are shown to be hazardous to our health, since we have already been using such devices for more than a decade. If they are hazardous, we would hope that risk would have been caught by EHS processes designed to protect against this contingency. But since real, basic questions are still open surrounding mobile phones, as just one product, many suspect that EHS regulations today need to be repaired, independently of any special risks posed by nanomaterials.

---

[19] Christian Nordqvist, "Extensive Cell Phone Use Linked to Brain Tumors, Swedish Study," *Medical News Today*, April 1, 2006, available at www.medicalnewstoday.com/articles/40764.php (accessed August 21, 2008); Editors, "Testing Cell Phone Radiation on Human Skin," Reuters News, March 3, 2006, available at www.sarshield.com/news/Testing%20cell%20phone%20radiation%20on%20human%20skin%20%20CNET%20News_com.htm (accessed August 21, 2008); M.T. Whitney, "Mobile Phones Boost Brain Tumor Risk by Up to 270 Percent on Side of Brain Where Phone is Held," *Natural News*, February 22, 2007, available at www.naturalnews.com/021634.html (accessed August 21, 2008).

[20] For skepticism, see Adam Burgess, *Cellular Phones, Public Fears, and a Culture of Precaution* (Cambridge: Cambridge University Press, 2004).

[21] There also may be non-EHS-related effects from cell phones. If they helped to create mobile workforces – the Internet is the other key enabler – then they seem to also contribute to the commoditization of houses in that individuals and families are quick to move to another city or state; not too long ago, houses were much more our homes that rooted families to their communities, possibly for generations. And the commoditization of houses, namely the desire to buy and "flip" a house for quick profit, seems to be a large factor in the current mortgage and financial market crisis.

Returning to nanotechnology, the "common sense" we referred to in constructing the argument above gives rise to the precautionary approaches that we discussed in the previous chapter.[22] There we focused on the so-called catastrophic formulation, which has been prevalent in the philosophical literature; this says that if, given some activity, some catastrophic effect is possible, then we should ban the activity. For regulative discussions, though, we might be equally concerned with scenarios under which weaker damage conditions apply, such as serious harm (as opposed to catastrophe). Such weaker conditions are more easily satisfied. And, of course, outright bans are among the most drastic of remedies. Less severely, we might countenance other remedies, such as postponement of the activity until the risk is mitigated. These features then give rise to a weaker precautionary approach such that any activity that might lead to serious harms should be postponed until those harms are mitigated.[23] Turning this formulation to nanotechnology, the issue is whether to implement stronger laws and regulations in order to keep nanotechnology off the marketplace, given that it might pose serious harms. And, given previous and ongoing research, it is a credible claim to say that, at best, we are uncertain whether nanomaterials are safe, and at worst, there is good evidence to show at least some are not.

6.3   Objections to the Stricter-Law Argument

If we take the preceding as an accurate characterization of the stricter-law argument, then we might (loosely) formalize the argument as the following set of premises P and conclusions C:

P1   Some *prima facie* evidence exists that some engineered nanomaterials may be harmful to EHS interests.

P2   Current laws may be inadequate in accounting for EHS risks in nanotechnology (as a general liability of any law that is now relevant to an area that did not previously exist or was not properly/fully considered during the legislative or regulatory process).

C1   Therefore, there is a possibility that nanotechnology, as it advances and absent stricter laws, may lead to EHS harms.

P3   EHS harms are an unacceptable consequence, especially since our laws and some governmental agencies exist specifically to protect us from those harms.

---

[22]   See §§5.3–5.4 for more discussion. See also John Weckert and James Moor, "The Precautionary Principle in Nanotechnology," *International Journal of Applied Philosophy*, 2.2 (2006): 191–204.

[23]   Again, this possibility must be empirically possible and not just logically possible.

P4 Common sense suggests we should adopt a precautionary principle in this case, which requires that: if an action might possibly lead to an unacceptable consequence, then we should postpone that action until that risk is mitigated.

C2 Therefore, we should postpone our development of nanotechnology (at least commercially) without taking some action to mitigate nanotechnology's EHS risk.

P5 We can sufficiently mitigate nanotechnology's EHS risk by either enacting a moratorium or implementing stricter laws (among a range of possible remedies).

P6 A moratorium on nanotechnology research or commercialization has limited support and may be an over-reaction, so it is not a viable or reasonable option.

C3 Therefore, the action we should take, if we want nanotechnology to proceed, is to implement stricter laws.

In other words, the stricter-law argument can be described as follows. New research suggests that some nanoparticles may be harmful or at least that they have not been shown to be safe. Many, if not most, laws relevant to our discussion here were designed prior to these new studies and without nanoparticles in mind; it is possible that our current laws do not adequately protect us from EHS risks. Given this uncertainty, we ought to create new laws and regulations to mitigate these risks, as a reasonable alternative to a more drastic remedy such as a ban on nanotechnology.

There has been much debate over the main conclusion of the stricter-law argument (C3). The objections, both actual and possible, to that argument include the following, in order of the weakest to strongest:

1 *Ordinary materials objection*: nanomaterials are not any more harmful than other materials, so they need no special regulations.

2 *Status quo objection*: current regulations are enough to safeguard the public from these harms.

3 *Precautionary principle objection*: the precautionary principle is too strict and should not be applied here, so the entire argument that rests on it is flawed.

4 *Self-regulation objection*: self-regulation, not more governmental regulation, is the answer.

5 *Other harms objection*: stricter regulation would stunt the growth of a nascent nanotechnology industry.

6 *Future harms objection*: more than the usual, near-term economic harms cited in 5, there may be more serious harms in the future if the nanotechnology industry were hindered now.

7 *Better-us-than-them objection*: increasing regulation only puts that nation at a disadvantage with respect to other nations that may then

develop and reap the benefits of nanotechnology first; so it is better that we (your respective nation) rush to develop nanotechnology first, unencumbered by extra regulation, even if that causes some harms.

Though some of these objections can be and have been combined, we will consider each separately in the following. The last few objections are the most compelling, so we will spend more time on those positions than others in our discussion.

Ordinary materials objection

This objection asserts that existing laws and regulations are adequate to account for nanotechnology, because nanomaterials are essentially the same kinds of substances that we have been using for decades. That is to say, a carbon nanotube is still only made up of carbon, and nano-sunblock is still only made up of zinc or titanium oxide; and these are materials that current regulations have proven sufficient to handle. In some cases, nanomaterials are simply much smaller versions of the familiar, bulkier materials; they are not novel substances. In other cases, they are the same material with a different molecular arrangement. Therefore, we do not need stricter laws to account for nanotechnology.

In fact, nanotechnology is something that has arguably existed since the beginning of the world, if not earlier: "Nanostructures – objects with nanometer scale features – are not new nor were they first created by man. There are many examples of nanostructures in nature in the way that plants and animals have evolved. Similarly there are many natural nanoscale materials . . . catalysts, porous materials, certain minerals, soot particles, etc. that have unique properties particularly because of the nanoscale features."[24] (For more extended discussion, see §1.5.)

To analyze this objection: in the formalized argument stated above, this objection disputes either premise P1 or P2, or both, and therefore rejects conclusion C1 as well as the dependent conclusions C2 and C3. However, the objection seems scientifically naive at best and contradictory at worst. First, it is precisely the different molecular arrangement of the same materials that creates different properties. In a certain arrangement, carbon can be made into pencil lead and useful for writing; in another arrangement, it is a diamond; in yet another, it is a carbon nanotube that, for example, is useful in building lighter cars or aircraft. This is not a strange phenomenon: we see water in its different forms every day, from a gas to a solid to a liquid. Therefore, the issue is not so much that nanotechnology works with common materials, but rather that by manipulating these materials at the

---

[24] US National Nanotechnology Infrastructure Network website, available at www.nnin.org/ nnin_what.html (accessed August 21, 2008).

nanoscale, we can create uncommon results – results that today's laws and regulations could not have anticipated.

Second, it is not just the molecular arrangement that gives nanomaterials their unique properties; it can also be their size (see §1.3). For instance, aluminum is often considered an everyday, safe element, e.g., soda cans are made from it. But if aluminum is ground up fine enough into dust, it can spontaneously combust in a highly energetic reaction with air. Further, nanoparticles are, by definition, so small that existing air and water filters would be unable to prevent their escape from manufacturing facilities, thus opening the possibility for toxic emissions that affect workers and the outside environment.[25]

Given their size, nanomaterials may also be able to slip by current methods of testing for safety and health risks. In other words, current regulations that require such testing may not be enough, if the testing methods they require cannot catch nano-sized materials. Recognizing this challenge, researchers continue to devise new approaches to nanotoxicology.[26]

Finally, the ordinary materials objection also seems to be a case of "wanting to have your cake and eat it too." The allure of nanotechnology in the first place is that the materials we are creating have novel and useful properties that we are still trying to understand and exploit. So it would be inconsistent to say that these nanomaterials are nothing special that we need to worry about, when the entire point is that they are extraordinarily special.

As something special and unpredictable, it would be reasonable to think that they might be more (or less) hazardous or toxic than ordinary materials of the same element or chemical, which is exactly the concern that is prompting calls for more regulation. The EPA, for instance, seems to have reached such a conclusion in its recent policy adjustment of requiring nanomaterials to be evaluated as novel substances (through TSCA), in contrast to its previous assumption that allows generalizations about the safety of materials from the bulk level to the nanoscale.[27]

Status quo objection

This objection to the stricter-law argument asserts that, as a matter of fact, current regulations are enough to safeguard the public from these harms. They have served us well over the years, and without definitive proof that

---

[25]   Sheldon K. Friedlander and David Y.H. Pui, "Emerging Issues in Nanoparticle Aerosol Science and Technology," report published by US National Science Foundation, November 2003.

[26]   See, for example, Ning Li et al., "Toxic Potential of Materials at the Nanolevel," *Science* 311.5761 (2006): 622–7.

[27]   James B. Gulliford, "Toxic Substances Control Act Inventory Status of Carbon Nanotubes" (Washington, DC: Environmental Protection Agency, October 27, 2008). Available at http://nanotech.lawbc.com/uploads/file/00037805.PDF (accessed April 21, 2009).

nanomaterials are actually harmful in consumer products or manufacturing, it is premature to subject the nanotechnology industry to more regulations. This is not to say that all regulatory decisions are correct or fully mitigate EHS risks, but only that current laws and regulations strike an appropriate balance between EHS risks and other interests, e.g., economic benefits from nanotechnology.

To analyze this objection: it disputes premise P2 in our formalized argument, thereby throwing into question all conclusions from C1 to C3. We have previously pointed out the obvious, that current laws and regulations are fallible – or might not achieve the right balance among interests or accurately perceive the risks in the first place – and probably do not fully protect us from EHS risks in all consumer products or their manufacturing. The following is a specific, recent example of that failure.

In 2006, the US Environmental Protection Agency urged companies, including DuPont and 3M, to phase out their use of a chemical (perfluorooctanoic acid, or PFOA) used to make Teflon®, the non-stick material found in everyday items such as cookware, carpeting, clothing, food packaging, and thousands of other products.[28] In use for more than 50 years, the chemical is linked to cancer and organ damage in laboratory animals. It is so pervasive in our environment that it is now found in the blood of nearly every American. One Teflon manufacturer has already paid more than $100 million to settle lawsuits from residents who live near its factory, including claims of birth defects.

"The science on [Teflon] is still coming in, but the concern is there, so acting now to minimize future releases of PFOA is the right thing to do for our environment and health," explained an EPA official.[29] Environmental watchdogs support the EPA move, explaining: "It would be hard to imagine a chemical that is more widespread in our environment. It is found everywhere from babies in the womb to whales in the ocean. And beyond that, it is indestructible in the environment. It lasts forever."[30]

So given this current and apparent failure of US regulations to discover or account for EHS risks posed by Teflon's manufacturing processes – never mind other continuing controversies, such as currently available pharmaceutical drugs that might unknowingly cause severe health problems – a similar failure with respect to nanotechnology is not just possible but highly plausible.

In the case of the Teflon-making chemical, even though studies have not definitively proven that it is harmful to humans in the amounts present in

---

[28] US Environmental Protection Agency, "EPA Seeking PFOA Reductions," press release, January 25, 2006, available at http://yosemite.epa.gov/opa/admpress.nsf/a543211f64e4d1998525735900404442/fd1cb3a075697aa485257101006afbb9!OpenDocument (accessed August 21, 2008).

[29] Ibid.

[30] Brian Ross, "Government Moves to Curb Use of Chemical in Teflon," *ABC News*, January 25, 2006, available at http://abcnews.go.com/WNT/story?id=1540964 (accessed August 21, 2008).

our everyday lives (otherwise a much stronger and immediate ban perhaps would have been proposed, we would hope), there are enough data to suggest a real risk to the environment and our health. So the EPA seems to be guided by a precautionary principle or something similar in concluding that companies should produce less of this chemical, in case its presence in consumer goods and in manufacturing emissions is truly harmful. In other words, the EPA decided that the best course of action is to err on the side of precaution, even though a $2 billion per year business is reputedly at stake.[31]

Further, even if current laws are adequate to account for nanomaterials in production today and in the near future, the industry is still learning about the science and working on new materials, and these materials may slip past existing laws. Either way, it is also prudent to believe that the processes we have established to regulate business in general are imperfect and will continue to be a work in progress, as long as businesses and research organizations continue to innovate. Therefore, we should be open to the possibility that current laws and regulations are not enough, particularly when the consequences of their failure may be catastrophic. This objection, therefore, does not appear to be defensible against the stricter-law argument.

Precautionary principle objection

Following the last discussion, if we believe that current laws are fallible and may not be precautionary enough, this objection maintains that the precautionary principle is not an obviously correct or common-sense rule that we should follow: it is *too* strict, and therefore the stricter-law argument becomes unreasonable since it depends critically on precautionary approaches. The most serious criticism we examine here is that precaution represents a risk-averse strategy that is too conservative, at least as it applies to the considered case of nanotechnology where the EHS risk is still unclear.[32]

Risk aversion, according to this argument, is not the only workable strategy in life, business, or politics. After all, if Americans never took unnecessary or perhaps unreasonable risks, then we never would have accomplished such things as expanding the country westward to California, inventing the airplane, and putting a man on the moon. In fact, America was built on the backs of explorers and frontiersmen, such as Christopher Columbus and Captain John Smith, who risked and sometimes lost everything. And many other nations can say similar things about their forebears, pioneers, and inventors.

Such may be the case with nanotechnology: it is a new frontier in science that, while admittedly containing unknown danger, also holds much promise. However, if we were to follow precautionary approaches, we might

---

[31] Ibid.
[32] Cass R. Sunstein, "Cost–Benefit Analysis and the Environment," *Ethics* 115.2 (2005): 351–85.

lose a great opportunity to develop a science that has been called "the Next Industrial Revolution."[33]

To analyze this objection: it attacks premise P4 in the formalized argument, evoking powerful emotions of national pride and adventure, so it may appeal to many. However, it is unfair to compare our current debate on strengthening laws relevant to nanotechnology with, say, the Wright brothers' debate on whether they should jump off a cliff on what amounts to a bicycle-powered deathtrap, or with any of the other situations cited above.

One reason is that the individuals associated with the above events, from Christopher Columbus' crew to Neil Armstrong, presumably had consented to such risks. Their decisions more or less directly affected only their own lives. But in our considered debate surrounding nanotechnology, countless people may be put at risk without their consent. Indeed, surveys have shown that most Americans are unaware of what nanotechnology is or have not even heard the word before, so it would be impossible for them to give informed consent anyway, even if asked.[34]

The issue of rights might be relevant here. Our basic human right to not be unjustifiably harmed plausibly entails a right to not have one's life unjustifiably endangered or otherwise put at risk of significant harm. That is, not only are we morally barred from harming others without just cause, but we should also not put others at risk of such harm or even cause theoretical or statistical harm. Without their consent to be subjected to such risk, failing to apply precaution in the case of nanotechnology may violate this right.

Of course, one possible reply to this is that by participating in a democratic system such as that in the US, we are in effect "consenting" to the outcomes of elections, ballot propositions, as well as any legal actions of the leaders we elect. So if an elected legislative body were to pass some measure or law that runs contrary to precautionary approaches, then it can be said that we had consented to such a decision by electing politicians prone to such aggressive policies. If current laws and regulations are allowed to stand as they are, that is a decision by which the public must abide (or seek to reverse through the established political channels).

However, political theorists have pointed out that we cannot consent to unjustifiably lay down our lives or submit to unreasonable harms, since that would defeat the very purpose of government in the first place.[35] So

---

[33] *National Nanotechnology Initiative: Leading to the Next Industrial Revolution* (Washington, DC: US National Science and Technology Council's Committee on Technology, February 2000).

[34] Jane Macoubrie, *Informed Public Perceptions of Nanotechnology and Trust in Government* (Washington, DC: Woodrow Wilson International Center for Scholars, September 2005).

[35] See, for example, Thomas Hobbes, *Leviathan* (New York: Penguin Group, 1982 [1651]), §14.8.

extending this line of reasoning to our discussion here, it may be argued that we also cannot consent to or countenance policies that lead to harm to our persons. And it can be argued that whatever (implicit) consent we have given might still allow for several forms of regulatory remedies; as previously noted, we will not delve into these here.

At any rate, it does not seem unreasonable to suggest that if the stakes are high enough – as apparently is the case with nanotechnology, where real human and animal lives as well as the environment are at risk – then minimizing risk should be a guiding principle rather than, say, the pursuit of profit, adventure, or glory. Individual actors may arrive at a different conclusion, depending on their tolerance for risk and what value they place on their own welfare. But if the decision involves risking the welfare of countless others, it may be irresponsible not to adopt something like a precautionary principle, in which case the stricter-law argument again survives.[36]

Self-regulation objection

This position opposes more governmental regulation as a way to mitigate EHS risks. Rather, it advocates self-regulation as an alternative, such that if any additional regulation is needed, it should be left up to the industry to decide what measures are appropriate, as well as what form (e.g., product labeling).

There are several reasons why this view is attractive to many.[37] First, it promotes a smaller governmental footprint in business and individual lives, so it instantly appeals to libertarians and some conservatives. Also, it may make more sense for the nanotechnology industry – that presumably knows its field better than lawmakers do and has a real stake in its work processes – to devise and implement regulations, rather than some distant bureaucracy whose edicts are inevitably born from political compromise. Self-monitoring and self-regulation foster a sense of responsibility within the industry. Further, self-regulation seems to succeed, as evidenced by any number of professional codes of ethics.

To analyze this objection: a persistent criticism of the idea of self-regulation is that it seems to let the proverbial fox guard the henhouse or, in other words, there is a sizable conflict of interest.[38] Can we trust an industry – any industry – to make its own rules when money is involved? Can they fairly create processes that protect EHS interests of the public, even at the expense of their own interests, financial or otherwise?

---

[36] Again, see §§5.3–5.4 for more detailed discussion in this regard. See also Weckert and Moor, "The Precautionary Principle."

[37] William J. Witteveen, "A Self-Regulation Paradox: Notes Towards the Social Logic of Regulation," *Electronic Journal of Comparative Law* 9.1 (2005), available at www.ejcl.org/91/art91-2.html (accessed August 21, 2008).

[38] Ibid.

Some have called it a pragmatic paradox to ask a person or organization to obey the law and, at the same time, *be* the law.[39] Because there is no real separation between those enforcing regulations and those subject to the regulations, the door seems to be open for self-imposed regulations to be selectively enforced and for potentially covering up illegal or unsafe practices.

Of course, an enlightened company might see that it is in their best interest to deliver only safe products, since harming one's own customers is counterproductive to one's reputation and business as well as opening the company to possible litigation. But will every company arrive at the same conclusion, ignore short-term gains for long-term interests, and follow the rules? For self-regulation to work, nearly every industry actor needs to comply, since all it takes is one clever company to sidestep industry-imposed regulations for possible catastrophe to occur, i.e., the EHS risks may still exist and are not sufficiently mitigated by self-regulation.

The diagnosis of why actors fail to cooperate even though it is in their collective interests to do so, also known as a Prisoners' Dilemma, is well covered in literature.[40] As has been shown by groups such as OPEC – whose members are notorious for ignoring their own self-imposed quotas for oil production, even though cooperation gives them a means to control oil prices – it is a real challenge to get organizations to do what they have committed to, even if breaking that commitment will make them worse off in the long run.

Further, if governmental regulations are believed to be imperfect because they contain political compromise, it is unclear why matters should be different with self-regulation. An industry coalition is merely composed of companies, research organizations, and individuals of varying influence and interests, collectively representing a government of sorts, albeit a smaller and more direct model, with the same tendencies and weaknesses.

In nanotechnology the problem is worse, since there is no single "industry" that encompasses all the possible or even current applications in nanotechnology. Unlike associations for architects, engineers, lawyers, or medical doctors, there is no such group for nanotechnologists, because their work and interests are so varied, cutting across myriad industries and companies from Applied Materials to BMW to L'Oreal to Merck to Zyvex and countless others. In fact, discussions about nanotechnology are usually prefaced with the disclaimer that 'nanotechnology' itself is a misnomer and properly should be 'nanotechnologies' to reflect the different lines of research and applications.

Given the above concerns, it is not apparent that self-regulation is a more viable or desirable alternative to governmental regulation. Further, we should note that this objection does not dispute the soundness of the stricter-law argument: it agrees that something should be done to mitigate risk associated

---

[39] Ibid.
[40] Robert Axelrod, *The Evolution of Cooperation* (New York: Basic Books, 1984).

with nanotechnology. That is to say, it does not dispute any of the premises or logic in the formalized argument. It does not even dispute any of the conclusions, including C3 that mandates stricter laws; rather, the objection merely prefers self-regulation to governmental regulation.

Even if the reasons given for self-regulation over governmental regulation are defensible, they appear irrelevant to attacking the stricter-law argument; therefore, the stricter-law argument survives this objection. At most, the objection might highlight the stricter-law argument, as formulated in this chapter, as incomplete: further argument is needed to show that the stricter laws need to be mandated by government agencies rather than an industry coalition, which would then be the point of attack for this objection.

Other harms objection

This objection, perhaps the most popular of the seven considered, suggests that if stricter laws were imposed, there would be unacceptable costs or harms to the nanotechnology industry now. Few objectors have specified these costs, but we can imagine what some might be. If tougher regulation makes it more difficult for a nanotechnology product to be delivered to market – e.g., due to extended product testing cycles or more comprehensive environmental impact reports – then a business can reasonably expect to generate less revenue over a given period, since they will not or will not as quickly have that product on the shelves.

This also means businesses might not be able to afford to keep the same number of researchers or other employees on staff, leading to a loss of jobs. Without as many active researchers – including those in academic or other non-business labs, to the extent that these new laws affect their work – nanotechnology will not advance as quickly as it might otherwise have. And if other nations do not have the same stringent restrictions that we do, the US may suffer a real competitive disadvantage globally. (We will discuss other potential costs later, but these seem to be the primary ones associated with this objection.)

Indeed, a recent report from Cientifica argues that today, even without the stronger regulations proposed, the pace of funding, research, and development in the US is not fast enough to sustain business efforts and compete with other nations.[41] The report warns that not enough government spending in nanotechnology is focused on areas of immediate commercial impact. And accessing this funding is a slow process, taking an average of two to three years before it even reaches the lab. The report also finds that, as a proportion of its gross domestic product, the Japanese government spends three times as much on the technology than the US does. As it applies to

---

[41] "Where Has My Money Gone? Government Nanotechnology Funding and the $18 Billion Pair of Pants" (London: Cientifica, January 2006), available at www.cientifica.eu/index.php?option=com_content&task=view&id=33&Itemid=63 (accessed August 21, 2008).

the other harms objection, this report would lend defense to the claim that the nanotechnology industry needs more support, not more hurdles that would slow it down further.

To analyze this objection: it disputes premise P3 in our formalized argument above. By itself, it does not deny that there may be EHS harms from nanotechnology (i.e., it does not dispute conclusion C1), but it asserts that EHS harms are *not* an unacceptable consequence if stricter laws would cause greater harm, thereby questioning conclusions C2 and C3.

The objection – that an action will have burdens on the business side – is a common response to nearly any proposal to introduce new tax or regulations. For instance, a higher minimum wage would mean that some businesses will need to spend more on payroll and perhaps pass along that expense to customers in higher prices. And considering that some of these companies might be barely profitable, it seems reasonable to predict that some may go out of business. They simply cannot afford to spend more, without somehow increasing revenue; and if they knew how to do that, they probably would have done it already. (Though even in hindsight, it is not clear how many, if any, legitimate businesses have closed as a direct result of paying a higher minimum wage.)

But a loss of jobs and revenue by itself is not necessarily a bad thing, if there are other redeeming results. For instance, even if nanotechnology products were allowed to reach the marketplace unhindered by new or strengthened laws, their success would inescapably cause other sectors and companies to lose jobs, just as word-processing software displaced workers in the typewriter industry. A nanopaint company whose products are more durable and scratch resistant than traditional paint may likewise displace competitors, so a loss of jobs and revenue by somebody may be unavoidable. This is an illustration of "economic Darwinism," presumably a desirable situation where new, better innovations and businesses replace older, less efficient, or less effective ones. In the nanotechnology regulation debate, the economic harms potentially caused by stricter nanotechnology laws may be offset by the lives, animal and human, that the stricter laws potentially save.

So while we can empathize with the other harms objection, it appears to be one-sided and ignores the fact that there must be a tradeoff. Every proposal has its costs, and there seems to be a necessary downside for somebody. But there is also an upside or benefits too; otherwise, the proposal would have not been worth making in the first place, even if there were no costs. The challenge is not just to identify these costs, but also to evaluate the costs and benefits to determine whether the proposal is worthwhile overall (see §5.2 for some difficulties to this approach).

For instance, we know that for every bridge or skyscraper that is built, a certain number of construction workers can be expected to die or be seriously injured on that job. This is not a trivial concern, but should it be enough to derail a bridge or skyscraper project? We are loath to put a price tag on a person's life, such as in making a cold, utilitarian calculation of lives lost versus economic benefits from the bridge. But the fact is that such calculations must happen in real-world projects.

As it applies to the issue at hand, the benefit of stricter regulations is that we reduce the risk that nanotechnology may pose to the public, industry workers, and the environment. Does reducing that risk justify the potential loss of profit, jobs, and competitive advantage in the nanotechnology industry?

Finding the answer to this question depends on additional considerations. How much would these new regulations reduce the risk: a substantial or an incremental amount? How much burden exactly would these regulations put on the industry? If we could quantify likely and worst-case scenarios, how many consumers might be harmed, and what kind of harm would they suffer, without new regulations, and how much would the industry lose in jobs and profits? These are questions that require more research to answer, although a common intuition might be that people should be valued more than profits, no matter how much is at stake for an industry – a point we will not take up in this chapter.

However, matters can quickly become more complicated without debating that point, when one considers the role of rights in this discussion. If we have a basic human right not to be unjustifiably harmed and the government has an obligation to protect its citizens (from internal and external threats, including unsafe commercial products), then it seems that stricter regulations are needed to protect this right and fulfill the government's obligation.

But on the business side, it is less clear what rights would be violated by the introduction of stronger laws. Do we have a right to the jobs that might be lost in nanotechnology? Do businesses have a right to develop products that are in compliance with existing laws (or is there a corporate moral responsibility to employees and customers, beyond what is required by law)?

Even if we answer in the affirmative to these questions and others, it may be useful to note the types of rights at stake. As some have made the distinction, our individual right to not be unjustifiably harmed is a "negative" right, meaning that it requires others *not* to interfere with us in certain ways.[42] But any business-related rights seem to be "positive" rights in that they require some good or service to be provided.[43] As other examples, our right to free speech is a negative right, since it requires that others refrain from preventing us from speaking our minds. In contrast, our right to education is a positive right, since it requires that we be provided with access to learning.

The relevance of this distinction is that negative rights, it has been argued, are stronger than positive rights (again, assuming that the distinction is

[42]   Leif Wenar, "Rights," *The Stanford Encyclopedia of Philosophy*, July 9, 2007, http://plato.stanford.edu/entries/rights/ (accessed August 21, 2008).
[43]   Ibid.

intelligible and useful).[44] Negative rights can be observed by, for instance, simply not interfering with someone else's speech or not harming a person unjustifiably; no action is needed. But positive rights are more difficult to respect, since they require an action or a series of actions that may take some effort, for instance, hiring teachers and building classrooms in order to provide a public system of education. Because negative rights take less effort to respect, it is generally worse to violate a negative right.[45]

So even if the jury is still out on whether the cost to business and industry is really worth the reduction in EHS risk from stricter laws, there seems to be *prima facie* reason to favor stricter laws on grounds that this protects our negative right to not be harmed, which must take precedence over any positive rights of the nanotechnology industry – and it is not even clear what rights are at stake in business and industry. Another *prima facie* reason for some individuals may be based on the aforementioned "people over profits" slogan or intuition.

We should note that the other harms objection, by itself, does not dispute the soundness or logic of the stricter-law argument. Rather, it attempts to show that P3 should not be accepted by shifting the focus to a contest between benefits and harms, suggesting that more damage than good will be caused by stricter laws, which we are taking here to be at best an open question or stalemate. Therefore, in our analysis, the stricter-law argument appears to survive this objection for the time being.

Further, we should also note here that the self-regulation objection and the other harms objection are incompatible, i.e., it is logically inconsistent to hold that stricter regulation would hinder a fledgling nanotechnology industry *and* that self-regulation is the answer. Any self-imposed regulations nevertheless represent more regulations than which currently exist. But if it is also believed that more regulation would stunt industry growth, then self-regulation too must impede industry progress – unless self-regulation is a hollow or token gesture to appease regulators and the concerned public.

Future harms objection[46]

If there is something reasonable, but not completely convincing, about the other harms objection, then we can perhaps strengthen it here by pushing its time horizon further out, giving the argument more consideration. In doing so, we can suggest that the preceding objection really did not consider enough harms: it looked only at immediate or short-term harms associated with stricter laws. But nanotechnology is something that is forecasted to give humanity profound benefits once it matures, and we have yet to consider those goals in pursuing nanotechnology. If we slow the industry down

---

[44] Ibid.
[45] Ibid.
[46] We thank Tihamer Toth-Fejel for research assistance for this section.

today, will that prevent or hinder us from realizing these benefits later – benefits that may plausibly outweigh EHS risks that exist either today or in the future?

In the following, we briefly present some of the risks of moving too slowly in nanotechnology that could be advanced by the future harms objection, which is not an objection we commonly see but is a view held or implied by at least some nanotechnology advocates. The risks here are in losing significant advantages that could be gained from nanotechnology, particularly competitive disadvantages with other nations that could develop nanotechnology more quickly (i.e., unencumbered by stricter laws). Taken together, these risks raise the stakes involved in the stricter-law debate and may present a more compelling challenge. Let us consider economic benefits, military implications, and reversing environmental and health risks.

Starting with economic benefits, nanotechnology is predicted to be anywhere from a $1 trillion to a $3+ trillion industry by 2015.[47] While other countries – including Brazil, China, South Korea, India, Israel, Russia, and many European nations – proceed at full speed ahead, the US and any other nation may lose significant economic benefits if it is not among the leaders in nanotechnology. Depending on the scale of economic benefits a nanotechnology industry or industries can provide, many lives in the US may be saved or made better, given that poverty is one of the greatest determinants of life expectancy. Further, the jobs lost from stricter US laws might not simply be future jobs not arising as a result of the stifling of new lines of business, but might be existing jobs lost overseas if other countries take the lead and develop those commercial innovations.

So the problem is that, while the US may pride itself for being democratic in recognizing and considering various interests among its population, from business owners to environmentalists, other governments may not be constrained by this guiding principle and can push nanotechnology research and products ahead, unencumbered by laws and regulations that may be stricter in the US. This may lead to a loss of economic benefits on a larger scale than previously considered, since in this scenario, jumpstarting the nanotechnology products industry then would take more than firing the research and manufacturing facilities back up; we would then need to play catch-up in competing with other nations who may have a significant, potentially insurmountable, head start.

Turning now to military issues, we can expect nanotechnology to have significant military applications, and indeed the military is a key driver of nanotechnology research for many nations, including the US. These predicted

[47] William Sims Bainbridge and Mihail C. Roco (eds), *Societal Implications of Nanoscience and Nanotechnology* (Washington, DC: US National Science Foundation, March 2001), p. 3. Lux Research site, "Nanomaterials State of the Market Q3 2008," available at www.luxresearchinc.com/ (accessed August 21, 2008); also available at www.industryweek.com/ ReadArticle.aspx?ArticleID=16884&SectionID=4.

innovations include new offensive capabilities (such as energy, robotic, and stealth weapons) and defensive measures (such as stronger, lighter armor and better detection and jamming capabilities). The military is also leading the charge to develop medical advances that can be used on the battlefield and better information systems for intelligence gathering as well as control and command centers. With nanotechnology, the production speed of military assets can be increased to give a sizable advantage in numbers alone, let alone efficiency. (See Chapter 9 for more extensive discussion, particularly §§9.2–9.3.)

Nanotechnology, then, has the potential to take a military force into the next generation and beyond. And to the extent that a balance of military powers around the world is needed to maintain some semblance of global security or peace, nanotechnology could disrupt this balance, if it is developed unevenly by current military powers. Never mind how a nation would feel to lose its position or influence in global affairs, a more worrisome question for many is: what would be the effect of a *non*-democratic government having the most advanced nanotechnology capabilities or developing them first? Does the failure to heavily invest in nanotechnology-based weapons expose that nation to a foreign attack that cannot be answered in kind, but rather only with less effective and conventional means? The potential loss of security and lives needs to be added to the list of harms, if stricter laws are also applied to or affect military developments.

Finally, consider that the current debate over stricter laws is often grounded in concerns that nanotechnology products today may be harmful to the environment and our health. But it is also important to note that nanotechnology is also expected to enable us to *reverse* many conditions that afflict our environment and health (see Chapter 8 for extensive discussion). For example, nanotechnology can be used to purify water in developing world countries that desperately need clean water (see §7.3).

Likewise, it can help ameliorate the large environmental impact of dirty industrial processes, including those having to do with energy generation. New and additional sources of energy, such as cost-effective solar energy, will reduce or eliminate pressure on current natural resources (see §7.4). Chevron recently announced work on nanotechnology that can convert tar found in sand into usable oil, which would serve a dual purpose of cleaning up the environment.[48] In the distant future, nanotechnology may be able to rebuild our depleting ozone layer or create nanobots that can "eat" oil spills and other contaminants.

Beyond environmental benefits, nanotechnology is being applied to agriculture to better feed the hungry and to medicine to save more lives. So if

---

[48] Jack Uldrich, "Nanotech of the North," *The Motley Fool*, March 26, 2006, available at www.fool.com/investing/high-growth/2006/03/06/nanotech-of-the-north.aspx (accessed August 21, 2008).

we are worried today about EHS risks in nanotechnology products, we should also keep in mind the EHS risks that they could mitigate or solve with a sufficient time horizon, including today's risks.

To analyze this objection: it is more robust than the preceding one for a number of reasons. First, it adds longer-term benefits (or harms from not pursuing nanotech) to the list of those that should be considered for a more complete picture of nanotechnology's social and economic impact. This, in turn, lends support to the precautionary principle objection: given these other considerations, risk aversion seems less to be a reasonable strategy, to the extent that potential benefits seem to far outweigh potential harms in aggressively pursuing nanotechnology.

But if this is such a compelling argument, why do we not see more people (explicitly) advance it? One reason seems to be that this sort of objection requires making mid- and far-term speculations about nanotechnology, which is always a risky business, especially if it also raises other ethical and societal concerns. For instance, nanotechnology's role in the military might resurrect questions from the Cold War about mutually assured destruction and first strikes. And many of the more interesting predictions about nanotechnology revolve around "molecular manufacturing" (see §§1.4 and 4.2). Roughly speaking, this is an advanced form of nanotechnology that involves building designer objects one molecule at a time, raising the possibility of creating virtually any object we want, from food to weapons; however, this is very much an area that many or most mainstream scientists are reluctant to speculate about or otherwise openly dismiss.[49] And if molecular manufacturing were more widely predicted to be plausible, it might open a Pandora's box of potentially disruptive and harmful effects on global trade and therefore politics. These are all complex considerations that would need to be addressed. Again, all this is speculative, so it is unclear what the possible far-term harms are and their probability, even if we are confident about far-term benefits.

The future harms objection also forces us to confront the unpleasant question of what the limits are to our right not to be unjustifiably harmed. Is it morally permissible to risk the health of, say, 1,000 or even 10,000 manufacturing workers and consumers today, if we can save 100,000 or 1,000,000 other lives later through the aggressive pursuit of nanotechnology? Also, speaking of future generations, if federal funding is a zero-sum game – i.e., funding nanotechnology now takes away from the budget in another area – what about lives today that could have been saved with the funding currently diverted to nanotechnology, which is more an investment in tomorrow? These are questions that have no simple or universally accepted answer, much less one that a legislator would want to tackle.

---

[49] Rudy Baum, "Nanotechnology: Drexler and Smalley Make the Case For and Against 'Molecular Assemblers'," *Chemical & Engineering News* 81.48 (2003): 37–42.

As such, this particular objection, as with the other harms objection, again depends on other factors, namely rights, that cannot easily be reconciled on an accounting ledger of benefits versus harms. So if we are to extend the time horizon in the other harms objection to make the future harms objection, then it seems only fair that we must consider long-term harms of *not* having stricter laws as well – again, leading us back to a probable stalemate in the debate.

Though the future harms objection appears to be stronger than its predecessor, the numbers involved are too difficult to quantify and forecast, as well as difficult to process in the framework of human rights. Further, though this objection is related to the other harms objection, the objectors seem to represent very different positions: to the extent that mainstream scientists and nanotechnology executives support the other harms objection, they may be reluctant to speculate about nanotechnology's promise beyond the near future, fragmenting support for the future harms objection.

Of course, none of this speaks to the objection's soundness or logic. *If* we can reasonably project overall benefits and harms across time, and *if* the benefits sufficiently outweigh the harms, and *if* the relevant human rights are not unjustifiably violated, then the future harms objection could be defensible. But these variables are perhaps too speculative to nail down with much confidence; therefore, the premise P3 that it was designed to attack seems to survive, or at least it has not been convincingly shown that it should be rejected.

Better-us-than-them objection

The final objection we will consider here is not one we see explicitly in nanotechnology-related literature, but it is one that comes up in other guises, such as whether to allow controversial biotechnologies like human reproductive cloning. The argument proceeds as follows. Many democratic nations, such as the US, occupy a fortunate position in the world where they can afford to be reflective about matters of ethics and philosophy. We have that luxury, but many other countries do not; they are embroiled in a more desperate fight for daily survival of their people, e.g., providing clean water, cheap energy, nutritious foods, and so on. And if the means became – and at some point will become – available to them, the chances are good that they will pursue and exploit nanotechnology (or other emerging technologies, such as autonomous military robotics) without such strict regulations, if any regulations at all, to impede research and development, perhaps for the base reasons of national glory or military superiority and even at the expense of their own citizens and environment.

Do we really want nanotechnology to be dominated by other nations of whom we are already suspicious (again, so the argument goes)? Even if we can take the higher moral ground and lay aside our national prejudices, it does not change the fact that the other country will probably not –

and not care. Imagine then how the world and beyond might look, if that foreign country were to be the one that controls nanotechnology, which could be the key to controlling literally everything. So if not only for this reason, we must keep our lead in developing nanotechnology; we do not want to live on a planet where control has been seized by less savory international leaders. For the same reason, even if we think our nano-future is bleak anyway, it could be worse if the wrong nations were to be the ones who shaped it the most.

Thus we must "own" nanotechnology and proceed full speed ahead, without stricter regulations to impede us; and when we do, we can take a rest and become reflective again, returning our attention to EHS matters. By deferring that moment of moral questioning from now to then, we would then be in a time and a place when we can do something about our angst and any harm previously caused. We can give nanotechnology to other countries, if we are so concerned about justly distributing tools that can help humanity. We can try then to build that utopia we had only read about. Even if we cannot do any of this, this possible world seems much nicer than the one where, say, a non-democratic country had its way with the world, to the extent that our utopia is more utopian and our values more valuable (to us at least).

That is why *our* nation – whichever it is – needs to dominate nanotechnology research, even if some problems are caused along the way, which appear to be smaller problems within a much larger picture. This is not, all things considered, ideal, but prudence dictates such a regulative approach: we would rather not be in a dangerous competition with other countries, but we are. And according to this objection, that is the difference between philosophy or academic ethics and real-world ethics.

To analyze this objection: as with the previous two objections, this one disputes premise P3 in our formalized argument. It does not deny that there may be EHS harms from nanotechnology (i.e., it does not dispute conclusion C1), but it asserts that EHS harms are *not* an unacceptable consequence if stricter laws would cause greater harm, thereby questioning conclusions C2 and C3. In fact, it argues that greater harm would be caused by *not* aggressively pursuing nanotechnology and allowing other nations to take the lead; therefore, EHS harms are the lesser of two evils and should be preferred and accepted over the alternative.

This is a very pragmatic – and forgivable – position to take on the role of ethics in society and especially in a democracy. But where risk aversion may not always be the best strategy, as the previous objection asserted, being pragmatic also might not be the best course of action or the right thing to do.

For instance, consider the ban on human cloning that exists in some countries. It would be pragmatic to argue that at some point, somewhere in the world, someone will clone a human being. If this is an inevitable event, then it would be better if we (our nation) were the ones who cloned a human first; we could at least ensure that safeguards were in place, that the clone could be treated humanely, that we would put any knowledge we gained to good use, etc.

But if this line of reasoning is unconvincing with respect to human cloning, then it seems to fail again when applied to nanotechnology. Even if there are greater benefits to be derived from nanotechnology than from human cloning, the analogy still holds: the admission of greater benefits merely modifies the utilitarian ledger rather than shifts the line of argument. Or, if it does change the argument and fails as an analogy, then the objection simply collapses back to a future harms objection. Furthermore, as we discussed in the previous two objections, this objection ignores the role of human rights, the rights of citizens today to not be harmed; or it at least is willing to sacrifice these rights for a future benefit, which is a controversial position to adopt.

Finally, there is no guarantee or even reasonable assurance that if we pursue nanotechnology without restriction, then our nation will have the lead; it is very much still an open field. So without some safeguards in place at the national level if not also globally (such as treaties to limit the threat of mutually assured destruction, in the case of an arms buildup), there is still potential for catastrophe to occur, especially if we move forward recklessly. And a more sensible or alternative solution to the situation motivated by the objection seems to be that we should advocate greater regulations and oversight – or at least cooperation as a first step – on a *global* scale, if stricter regulations only at the national scale would impede that particular nation.[50]

The belief that something is inevitable, whether nanotechnology or Armageddon or even individual deaths, does not seem to be a good enough reason to rush toward it, especially if we can buy some time by moving a bit slower – precious time needed to perhaps develop safeguards to mitigate any associated negative impacts.

6.4   An Interim Solution?

A full defense or analysis of the preceding objections, particularly the last three, is beyond the scope and goals of this chapter or even book, but there seems to be enough reason to believe that they are not entirely without merit. At the same time, there does not (yet) seem to be enough there to believe that the stricter-law argument should be rejected. So an interim or compromise solution may be needed now to cover both contingencies.

Moreover, even if stricter laws and regulations are ultimately justified, there are good reasons to think that they cannot be enacted anyway, or at

[50]   What that global regulation might look like, including under what authority, is also a discussion beyond the scope of this book, though necessary for a full treatment of the issue.

least that they face stiff resistance from lawmakers and regulatory agencies, particularly in the US. Clarence Davies, the author of the Woodrow Wilson International Center report that sparked the ongoing stricter-law debate, even admits that: "In the U.S. political system, it has never been easy to pass new laws regulating commercial products. In the current political climate [George W. Bush's administration], it is close to impossible."[51] Changing regulatory policy is likewise a formidable challenge: institutional inertia aside, the expanding nanotechnology sector (comprising many industries) has much at stake here and is well funded and well connected.

That is to say, the US legislative and regulatory systems are notorious for being complicated and mired in debate, so barring an urgent need – which many believe has not yet been established for nanomaterials – it is not unreasonable to believe that new laws or stronger regulations would be slow to appear in the near future, even if needed and even with a new administration more favorable to governmental interventions and regulations. We do not deny that regulations are currently evolving, but we take this pessimistic stance to err on the side of making a conservative prediction, thus enabling us to move forward with suggesting some interim solutions or incremental steps to addressing EHS risks.

There is a sense with many that the nanomaterials found in today's products have not been established yet as a clear and present danger, which may be part of the public, legislative, and regulatory hesitation to propose dramatic changes to current rules. Under this reasoning, rushing new laws or regulations through until more facts are revealed may be the same kind of mistake as rushing nanotechnology products into the marketplace without fully considering their impact on health, environment, or even society and ethics. Even if we can all agree that it would be better to have more information in order to make better-informed decisions, creating mandatory reporting requirements still faces daunting hurdles, though some are being considered – and indeed, some have already been enacted – at local, state, and federal levels.

But we can acknowledge this position *while at the same time* being prepared to adopt new regulations if and when more studies show that nanomaterials are indeed harmful and that new laws are warranted. That is to say, even if we are not ready to call for stricter policies now, we can and perhaps should have a contingency plan or "plan B" developed, discussed, and ready to be implemented should more compelling evidence be presented in favor of stricter laws. (The trick here, of course, is to specify the details of such a "plan B," which too is beyond the scope and goals of this chapter.) If we adopt a wait-and-see attitude toward nanomaterials, then it is incumbent upon us to aggressively conduct safety testing.

However, a critical point in the nanotechnology regulatory debate is that current testing methods may be inadequate against nanomaterials and

---

[51] Davies, *Managing the Effects*, p. 10.

products, so it is not clear that more testing will get us far if we do not improve those methods. Therefore, it is also incumbent upon us to aggressively develop new testing methods in order to conduct EHS testing effectively.

The proposed solution, then, is that rather than the nanotechnology industry being slowed down through more regulations, as some claim it would be, regulatory planning as well as EHS testing and research need to run faster and catch up – just as experts have called upon ethics to do.[52] Starting a serious dialogue today with policymakers would help compensate for the slow time-to-action for creating new laws, particularly if we are just idly waiting for more research to come out that would compel action. And continuing to support and fund research into nanomaterials safety is critical to evolve the safety standards that exist in current laws and regulations. Progress is being made in this area, as previously noted researchers and others have shown by developing new testing models to evaluate the safety and health risks of engineered nanomaterials.

Further, if we can improve testing methods, then we may not need new laws or stronger regulations, at least in the meantime. Where current laws and regulations require materials to pass certain safety and health standards, we may be able to simply evolve and raise those standards as scientific understanding and testing methods evolve, as opposed to erecting new regulations. This would only require that current policies recognize and utilize the latest advances in safety and health testing, which is a reasonable expectation.[53]

Without developing new testing methods, it does not seem that new or stricter laws can address risks posed by nanomaterials anyway, if current methods fail us. Therefore, incorporating these new standards is the salient point in the recommendation to create stricter laws and regulations. Or to put it another way, if current testing methods are inadequate to show that nanomaterials are harmful, and we know that they are harmful at least in some cases, then any new or stronger law that is still based on these current methods does not seem to add much value, just more barriers to business.

For instance, would harsher civil and criminal penalties or more detailed environmental impact reports cause company executives to act any differently, if no different methods were available to support or refute previous claims of product and materials safety? They would most likely run more

---

[52] Abdallah S. Daar, Anisa Mnyusiwalla, and Peter A. Singer, " 'Mind the Gap': Science and Ethics in Nanotechnology," *Nanotechnology* 14.3 (2003): R9–R13.

[53] The WWICS report criticizes some of the relevant regulations as being unclear or having loopholes or failing to apply to critical industries, such as cosmetics, so improved materials testing admittedly would not solve those deficiencies. However, those problems exist independently from nanotechnology: they are not specific to nanomaterials and have been issues for some time now. As such, they present a broader challenge in the field of regulatory reform and therefore are not so much addressed in our discussion here.

of the same tests to arrive at the same conclusion. This would seem to be an instance of GIGO or "garbage-in-garbage-out" where, without new testing methods, we are using inadequate processes that inevitably generate inadequate conclusions.

If it makes sense to push harder for better testing methods, that still leaves a problem of a stopgap measure in the meantime, since new funding takes a significant amount of time to disburse and research often proceeds slowly as well. One solution is to accept the proposal for the nanotechnology industry to regulate itself, as an alternative to doing nothing. To repeat a key point in our discussion of the self-regulation objection above, the objection itself does *not* dispute the soundness of the stricter-law argument. In fact, it agrees that more regulations are needed, but holds that the nanotechnology industry should be the one to create and implement them, since the industry knows nanotechnology the best and has a direct interest in sustaining the field. (The EPA program for voluntary industry reporting is also a reasonable first step.)

At best, self-regulation will eliminate EHS risks in nanotechnology to at least some degree; at worst, it seems that it would not create any additional EHS risks but at least represents a good-faith effort to mitigate those risks. If and when governmental regulations are needed, that process can be informed by the prior exercise of self-regulation.

Of course, the interim solution proposed above is a conceptual framework, and many real-world details still need to be worked out. For instance, are there any examples of "plan B" approaches to suggest that such a proposal can actually work? What are the specific steps we would need to take to strengthen pre-regulatory planning, methods for testing materials, and toxicology testing? Do we need (paradoxically) a legal basis for ensuring that this greater focus on EHS risks and testing actually occurs, for example, by stipulating that some percentage of all nanotechnology research funds will go toward these areas?

How do we know that more and faster study of the EHS aspects of nanotechnology can keep up with the full-throttle research and development (R&D) and commercialization of nanotechnology in not just the US but also abroad? And how much more funding is needed for ethics and risk to catch up with R&D? If other nations do not focus as much on EHS risks, would the US (or any other nation that adopts such an interim solution) need to compensate with even more funding, given that nanotechnology ultimately knows no national boundaries and impacts the entire world, especially given a global economic ecosystem? And if additional funding is warranted, where would that come from? Would it be diverted from other programs that are working on current cures for current ills, outside of nanotechnology's risks (which seem to be future risks, as opposed to actually harming people or the environment right now)?

These are all good questions, and we do not intend to present a complete solution here; but we merely hope to provide a starting point for discussion toward a feasible solution while the nanotechnology regulation debate rages on. Moreover, to the extent that nanotechnology is a highly interdisciplinary area, we would expect that collaboration among lawmakers, scientists, ethicists, economists, and so on would be needed to account for the complicated issues arising from nanotechnology – more than just what a few ethicists can achieve here.

6.5   Putting the Pieces Together

Though we have tried to avoid making difficult comparisons of nanotechnology's possible benefits with its possible harms, it seems that what is known now – and not just speculation, albeit educated – is that nanotechnology products *today* provide only incremental value additions or changes to existing products, i.e., they represent "better mousetraps" and not yet the revolutionary products predicted. On the other hand, the risks that nanomaterials pose *today* may be severe, possibly including the death of animals and people.

Therefore, we conclude that there is reason to think that current laws do not fully account for nanotechnology, if potentially hazardous nanotechnology products are reaching the marketplace. Nanotechnology, though technically not a "new" science, nevertheless introduces new materials that may defy current testing and safety standards not designed with nano-sized particles in mind. And research already indicates that nanomaterials are hazardous to the environment and human health, which is made all the more troubling considering that some nanomaterials come into direct contact with human beings, such as that in sunblock rubbed into one's skin.

But here's an important caveat: even if current laws are inadequate, would new or stronger laws be enough to fill that gap? In other words, the regulatory debate has been centered on the question of whether we need *more* regulation; but the more relevant question may be, *why* are current laws ill-equipped to deal with nanotechnology? The answer, or at least the complete answer, might be not that we are missing some law or process, but that the testing methods and standards built into existing laws have not caught up with the pace of nanotechnology. This is to say that the more-or-less-regulation choice may be a false dichotomy, when the question really should be about what *effective* regulation would look like.

While cleaning up and streamlining our maze of regulatory processes would certainly be helpful in general, unless we can quickly advance methods to more effectively test for environmental, health, and safety risks in nanomaterials, new or stricter laws may serve only to slow down the industry through procedural changes rather than to improve our evaluation of nanomaterials through substantive or qualitative changes in how we approach such materials.

So our suggestion is, rather than causing the nanotechnology industry and business to slow down now – which risks being a kneejerk reaction to create more laws in the face of a problem – we can stimulate other areas to quickly catch up. Regulatory pre-planning needs to catch up with the growing number of studies that confirm nanotechnology's EHS risks, in case new laws are ultimately needed. And testing methods and standards need to catch up to better confirm the safety of nanomaterials, which could occur within the framework of existing laws, and screen out the products that are hazardous to our environment, health, and safety. At the same time, we can take the nanotechnology industry up on its offer to regulate itself; there does not seem to be any harm in that, as long as we closely monitor efforts to ensure the public concern over EHS risks is not coopted or left to languish, and especially if the alternative is to do nothing.

There are several advantages of such a solution. First, while benefits today in nanotechnology (i.e., better sunblock, better sports equipment, better pants, etc.) might not justify its risks, this might not remain the case in the future. And overburdening the nanotechnology industry with regulation, though well intended, may ultimately cause more damage than good. So we need to find a reasonable balance that responsibly promotes innovation in nano-technology while at the same time safeguards EHS interests – in addition to other interests that nanotechnology may run up against as the field matures, such as privacy.

Second, a compromise may be needed anyway, once we recognize that there are significant challenges in creating new laws or regulations, and that there is presently little public awareness of these issues that might apply pressure to policymakers. Neither business interests nor EHS interests will be going away any time soon, nor do we really want either to do so. But we have seen the undesirable effects of placing too much focus on either business or EHS, so finding a balance between these legitimate interests is needed, particularly in a democracy that values a diversity of opinions.

Third, such a solution may serve to *accelerate* the industry responsibly, giving us new confidence that our nanoproducts are safe or identifying the ones that are not. Research into new testing methods would also give us new insights into nanomaterials, perhaps even new applications.

This proposal is meant to serve only as the first step to better or more confidently safeguarding EHS interests. Subsequent steps will likely require greater public or "upstream" engagement as well as product lifecycle ana-lysis. As a final note, we have focused primarily on nanotechnology's impact on the environment, health, and safety as well as on possible economic harms from restricting the development and applications of nanotechnology. But a similar regulatory debate also arises in other contexts, such as in human enhancement applications (see §12.6). As we will discuss in subsequent chapters, nanotechnology may raise concerns related to privacy and military technology proliferation, and these concerns again invite the question of regulation and prudent public policy.

7

Equity and Access

One concern with nanotechnology has to do with whether it will be fairly distributed. In other words, nanotechnology will enable many products, processes, and platforms, and these will be acquired and used by various people, institutions, and governments. These distributions can be subjugated to ethical analysis, particularly insofar as those lacking capital might not have *access* to nanotechnology and, therefore, the distributions will be *unequal*. Whether these inequalities are morally perilous – i.e., whether they are *inequitable* – is, of course, a substantive question for social and political philosophy. Our purposes here are not to defend some particular position, but rather to show what some of the possible positions are and how those positions bear on the distribution of nanotechnology.

In this chapter we have two particular goals. First, the issue of distributive justice is hardly privileged in regards to nanotechnology: for any technology or, more generally, any element of some social product, there are questions about fair distribution. That discussion can abstract away from nanotechnology in particular. In §7.1, we will offer general accounts of theories of distributive justice, and these will not have anything particular to do with nanotechnology. To be clear, this is not to suggest that nanotechnology is immune from conversations about distributive justice, just that there is nothing *about* nanotechnology that is endemic to those theoretical frameworks. Nevertheless, this discussion is useful for thinking about some of the ethical dimensions of nanotechnology. Also, it will bear, if indirectly, on the subsequent sections in the chapter; we will tie everything back together in §7.6.

In those intervening sections, we will focus on nanotechnology and some of the impacts that it could have in developing countries; our particular emphasis will be on water purification, solar energy, and medicine. These discussions are meant to provide a nanotechnology context to complement the more general theoretical discussion in §7.1. Again, we will not defend a particular position on what fair distribution of nanotechnology requires, though we will suggest the different ways in which the theoretical positions can be applied to the particular contexts we present. An emphasis will be placed on the notions of equity and access.

7.1 Distributive Justice

The topic of distributive justice is far too expansive to adequately cover in a single section, though we just mean to introduce the idea, as well as to delimit some of the central positions; the interested reader may follow the references for more depth. The basic idea is that societies produce various things which, when aggregated, constitute their social product. These things are myriad and, most obviously, include goods like pencils, cars, and clothing. But in addition to these goods, there are other components of the social product, such as services like health care, security, and inspection/licensure. There are probably cultural products as well, such as art, music, and sports teams. Take all of these things together and designate them as some social product. Now society must figure out how this social product should be distributed. Principles of distributive justice guide this distribution.

Before turning to some of those principles, let us highlight two issues. First, we have to think about how to delimit societies properly. For example, the US has a social product, but so does the world. So which society is the relevant one for conceiving of the distribution (and the associative obligations)? The right thing to say here probably has something to do with governments: the US has a government in a way that the entire world does not. This is not to say that there are not international agencies, international laws, and so on, only to suggest that, as an institutional unit, countries seem to bear some sort of privileged role, conceptually if not normatively.

There are surely complications to this claim: for example, the country of the United States comprises 50 individual states. Why should these individual states not be taken as the appropriate objects of distributive schemes (as opposed to the entire country)? To some extent, they are. Consider property taxes, which are collected locally and then used to fund local projects. In these cases, there are still issues of justice when, for example, we ask whether revenues from the wealthy part of the state should (also) be distributed to less wealthy parts of the state, or whether those revenues should stay more local. Nevertheless, we propose to focus on the "highest level" of operative unity. So, for example, the US stands to both Georgia and Colorado in a way that, effectively (and especially economically), nothing stands to the US and Canada. It is also worth acknowledging that, ultimately, nothing in the following discussion really has anything to do with countries *per se*. Rather, we choose this unit for convenience, though the discussion can apply to other units as well (e.g., to the above property tax example).

Second, even if we identify countries as the appropriate units for discussion over distribution, it hardly follows that countries cannot have obligations to *each other* as well as to their own citizenries. In fact, most of us think that countries do have such obligations – even if we disagree about the appropriate extent – and this gives rise to the idea of *global distributive justice*.[1] Global distributive justice will be precisely what we are interested in when we get to the forthcoming discussions of nanotechnology and the developing world, but let us leave that aside for now.

Rather, consider Aristotle, who was the first to write about distributive justice. Aristotle's virtue theory comprised various virtues, which were unified by his "doctrine of the mean": vices of excess and deficiency.[2] So, for example, the courageous person feels neither too much fear nor too little fear; otherwise he would be either cowardly or rash, respectively. But this immediately poses a problem for justice, which is one of the four cardinal virtues that Aristotle carries over from Plato.[3] Certainly someone can fall short of justice (and be unjust), but it hardly seems possible for someone to be overly just. Given this complication, as well as other concerns, Aristotle goes on to devote an entire chapter (Book V) of *Nichomachean Ethics* to justice, while most of the other virtues are crammed into a single chapter. In Book V, he presents a taxonomy of justice which draws a distinction between general justice and particular justice. General justice has to do with obeying laws and with the relationship of the virtues to each other – this latter concern is very Platonic – but is not itself among the virtues. Particular justice, though, is a virtue, and it has two subcategories: distributive justice and rectificatory justice. Rectificatory justice has to do with righting wrongs and (though we must table some obvious complications) the offender should make restitution so that, roughly, the wronged is indifferent between the wrong never having occurred and the amends made through the restitution.

For our purposes, though, what matters is the other element of particular justice: distributive justice. In distributions, Aristotle thinks that individuals should receive neither too much nor too little, but rather what is fair. He proposes that this fairness be understood in terms of merit, though

[1]   Some of the principal work in this regard has been done by Thomas Pogge. See, especially, his *World Poverty and Human Rights* (Malden, MA: Blackwell Publishing, 2002); see also his "Severe Poverty as a Violation of Negative Duties," *Ethics and International Affairs* 19.1 (2005): 55–83. For a discussion of some of the primary criticisms of Pogge's work, see Tim Hayward, "On the Nature of Our Debt to the Global Poor," *Journal of Social Philosophy* 39.1 (2008): 1–19.

[2]   Aristotle, *Nichomachean Ethics*, 2nd edn, trans. Terence Irwin (Indianapolis, IN: Hackett Publishing Company, 1999), II.9.

[3]   See Plato, *The Republic*, 2nd edn, trans. G.M.A. Grube, rev. C.D.C. Reeve (Indianapolis, IN: Hackett Publishing Company, 1992).

he allows that different people might have different conceptions of merit. Whatever conception of merit is used, though, he thinks that the assignments must be *proportionate*.[4] So imagine that there are two investments A and B, whose investors will receive shares C and D: A should stand to C as B does to D or else proportionality has been violated. To take a simple example, imagine that A is \$500 and B is \$1,000; further imagine that $C + D = \$3,000$. Whoever put up B should then get \$2,000 and the guarantor of A should get \$1,000, since \$500/\$1,000 = \$1,000/\$2,000. While investment makes for a simple example, it does not have to be the metric to which Aristotle appeals (e.g., there might be some other form of desert); rather, all that matters to his account is the notion of proportionality.

Moving forward almost 2,500 years – and with apologies to John Locke, to whom we will return, and to Karl Marx, to whom we will not[5] – the next account of distributive justice that we shall consider comes from John Rawls. In fact, his *Theory of Justice* is (probably not even) arguably the most important theory of distributive justice ever developed.[6] Rawls is effectively responding to two different groups: classical liberals and strict egalitarians. Classical liberals, such as Locke and Edmund Burke, wanted to argue for some set of personal freedoms from the government – whether for the acquisition of property, against undue taxation, and so on. Rawls, though, was worried that these sorts of freedoms would allow inequalities that were morally problematic, and his welfare liberalism has effectively laid claim to the moniker of 'liberalism' altogether nowadays, relegating classical liberalism to the more marginal libertarianism. For our purposes, let us just focus on two key elements of Rawls' theory.

In the *Theory*, he argues that distributions should not be along terms that are arbitrary from the moral point of view, particularly those deriving from natural or social advantages. So, for example, imagine that someone, A, is born with more intellectual aptitude than someone else, B (natural advantage), or else is sent to better schools (social advantage); regardless, further imagine that A then does differentially better than B throughout life as a consequence of one of these primitive advantages. To simplify, just

---

[4] *Nichomachean Ethics* V.3.

[5] Marx (in)famously defended a system by which production was dictated by abilities and distribution was dictated by need: "From each according to his ability, to each according to his needs!" "Critique of the Gotha Program," in *Marx–Engels Selected Works*, vol. II (Moscow: Foreign Languages Publishing House, 1949). Available at http://www.marxists.org/archive/marx/works/1875/gotha/ch01.htm (accessed July 25, 2008). Absent some seemingly implausible assumptions about human nature, this would encourage individuals to find ways to do less while needing more. Not surprisingly, socialism has always faced empirical challenges.

[6] John Rawls, *A Theory of Justice* (Cambridge, MA: Belknap Press, 1971). A revised edition came out in 1999 (Cambridge, MA: Belknap Press), which incorporates changes that Rawls made for translations, though many scholarly citations revert to the original. The revised version has a somewhat different pagination but includes a chart, which shows how to convert. When possible, we will refer to sections rather than to pages such that the references will apply equally well to either version. References to page numbers will be to the 1999 version.

consider salaries, and imagine that A earns 25 percent more than B. Is this inequality justified? Rawls thinks no, because A does not *deserve* to be better off than B; the only reason that A earns more is because of something, let us say, completely outside A's agency. And, as importantly, B does not deserve to be worse off than A, and it is *unfair* for A to be better off despite B not having tried any less hard.

The correction of natural and social advantages therefore suggests some sort of strict egalitarianism: the idea here is that A and B should have the same salary since neither of them deserves to make more than the other. But this is a conclusion that Rawls rejects and therefore leads us to a second key element of his theory. Rawls rejects egalitarianism on the grounds that everyone could be worse off under it than they could be under inegalitarian distributions. Imagine, for example, that we have $100 to share and that, if A had $75, she could invest in a scheme that would return $150. Further imagine that an equal distribution of $50 to A and $50 to B would wholly preclude the investment (and return). And, finally, imagine that there could be some distributive scheme such that some of A's gained assets were returned to B. So if A got $75 and B got $25 *ex ante*, then A could get the $150 and give $50 of the return, keeping $100 for herself. Or the government could tax A and effect the same result. Regardless, A ends up with $100 and B ends up with $75, so *both* are better off than *either* would have been under strict egalitarianism; further, note that the consequent distribution is unequal.

Rawls therefore wants to allow inequalities, but only those of a certain sort. In addition to thinking that there should be fair equality of opportunity regardless, he also thinks that social and economic inequalities must be arranged so that "they are to be of the greatest benefit to the least-advantaged members of society."[7] This latter constraint is his so-called "difference principle," which is one of the most enduring parts of his program. The Rawlsian framework is meant to push beyond strict egalitarianism, though the inequalities it allows are still meant to be progressive in that much extant inequality would fail Rawls' difference principle and, therefore, lack moral justification. There are many more features of Rawls' system that warrant discussion – including reflective equilibrium,[8] his principles of justice,[9] the veil of ignorance,[10] and so on – as well as important contributions made by his later work.[11] However, for our purposes, these take us too far afield, so we shall not consider them here.

[7]   Ibid., p. 303.
[8]   Ibid., Section 4.
[9]   Ibid., Section 11.
[10]   Ibid., Section 21.
[11]   See, especially, *Political Liberalism* (New York: Columbia University Press, 1993). *The Law of Peoples* (Cambridge, MA: Harvard University Press, 1999) and *Justice as Fairness: A Restatement* (Cambridge, MA: Belknap Press, 2001) primarily update and summarize, respectively, earlier work.

Rawls' colleague at Harvard, Robert Nozick, articulated a response to Rawls in *Anarchy, State, and Utopia*.[12] Despite an obvious respect for Rawls in the book, Nozick's position is antithetical to Rawls' and stands as its primary rival.[13] Nozick builds off what he calls the "Lockean Proviso," which was the core of Locke's labor theory of property. The central idea derives from Locke's *Second Treatise on Civil Government*:

> Though the earth, and all inferior creatures, be common to all men, yet every man has a property in his own person: this nobody has any right to but himself. The labor of his body, and the work of his hands, we may say, are properly his. Whatsoever then he removes out of the state that nature hath provided, and left it in, he hath mixed his labor with, and joined to it something that is his own, and thereby makes it his property. It being by him removed from the common state nature hath placed it in, it hath by this labor something annexed to it, that excludes the common right of other men: for this labor being the unquestionable property of the laborer, no man but he can have a right to what that is once joined to, at least where there is enough, and as good, left in common for others.[14]

Nozick, sympathetic to Locke, propounds an "entitlement theory" which contains principles allowing for acquisition and transfer of property, as well as rectifications of injustice.[15] Nozick, unlike Rawls, is not worried that inequalities could result from natural and/or social advantages or, to the extent that they occur, Nozick does not think that we can intervene against those inequalities because those that have differentially high resources are entitled to them (so long as they were justly acquired and thereafter justly transferred). To make this clearer, consider Nozick's famous Wilt Chamberlain example. Chamberlain was a well-known professional basketball player who played predominantly during the 1960s.[16] Imagine that Chamberlain was to receive 25 cents from each of 1,000,000 fans who wanted to see him play. Assuming that they each had come across their

[12]   Robert Nozick, *Anarchy, State, and Utopia* (New York: Basic Books, 1974).

[13]   This claim should perhaps be qualified. There has been far less industry in developing Nozick's position than there has been regarding Rawls'; even Nozick did little further work on his own position following its original publication. (Rawls' work spanned over 40 years, from his "Justice as Fairness" essay in 1958 through its "restatement" in 2001, whereas Nozick's work took place almost entirely between the publication of *A Theory of Justice* in 1971 and the publication of *Anarchy, State, and Utopia* in 1974.) An exceeding majority of contemporary philosophers would find themselves aligned much more closely with Rawls than with Nozick, though, insofar as there is a primary rival to Rawls, it has to be in Nozick. There are other views out there, but Nozick's is undoubtedly the second greatest work of political philosophy in the twentieth century; whether there is an insufferable gap between his and Rawls' is something about which we might be more skeptical than others.

[14]   John Locke, *Second Treatise on Civil Government* (New York: Prometheus Books, 1986), V.26.

[15]   Nozick, *Anarchy*, pp. 150–2.

[16]   Ibid., pp. 160–4.

quarters justly and that the transfers suffered from no defects (e.g., coercion), then what could we say against this arrangement? Even if Chamberlain is benefiting from natural and/or social advantages (e.g., height and/or training), Nozick thinks that there is nothing morally problematic with him accruing the wealth.

Taking some liberties with the details, this then leads to a contrast between Nozick and Rawls: the Chamberlain scenario would violate Rawls' difference principle. The reason is that Rawls says inequalities are only justified to the extent that they make the least-well-off person better off, yet this scenario is making the least-well-off person worse off – materially at least – such that Chamberlain can be made better off. In other words, take whichever fan (or fans) is the least well off and have this person lose resources such that Chamberlain be made better off; this would fail the difference principle. Rawls, with his emphasis on the least well off, is usually taken to be defending some sort of welfare program. Nozick, on the other hand, defends entitlements against interventions (e.g., as welfare programs would require) and therefore advances a libertarian program.

Having now briefly seen some of the key features of Aristotle's, Rawls', and Nozick's accounts, let us tie these together before we move on. This section highlights some of the principal accounts of distributive justice. Aristotle thinks that distributions should be proportionate to merits, though these merits could be any number of things. Rawls thinks that distributions can only be unequal insofar as the least well off are made better off. Lastly, Nozick disagrees: he thinks that inequalities can be justified even if they arise from Rawls' natural and social advantages and do not improve the situation of the least well off.

These discussions of distributive justice transcend any particular contexts, including nanotechnology. This section digresses from the central theme of the book, but it is important to understand some of these theoretical frameworks, for at least two reasons. First, understanding distributive justice can help us to think about broader contexts than just nanotechnology. Similarly, by thinking about these broader contexts, we can appreciate that, in some sense, nanotechnology does not have uniquely profound implications for distributive justice, at least on the theoretical level. Second, though, when we consider particular ways in which nanotechnology raises issues of equity and access, we have these frameworks at our disposal. Without further delay, let us return to nanotechnology.

7.2   Nanotechnology and the Developing World

As we said in the introduction to the chapter, nanotechnology will enable various products, processes, and platforms. Certain people, institutions, and governments will have access to these, while others will not. In the remainder of this chapter, we will consider some particular contexts wherein these issues of equity and access arise, focusing on international contexts, particularly those involving developing countries. We do recognize that there are clearly other contexts in which nanotechnology and distributive justice will have something to do with each other. Consider, for example, human enhancement, which we discuss in Chapter 12 (see also §9.2). Imagine that there is some cognitive enhancement enabled by nanotechnology. Of course, who has access to these technologies will be an issue and, in particular, who can afford them. But structurally, there is nothing different between that case and the case in which some of the best universities in the US cost $40,000+ per year and only certain families have access to them. There is probably not even much difference substantively insofar as the top universities offer, at least in some sense, "cognitive enhancement." So, from a distributive justice standpoint, we do not see much to say about these sorts of contexts, other than what would follow from the discussion in §7.1. This is not to say that there are not important ethical issues, or even that distributive justice is not an issue in those cases. Rather, the point is just that there is nothing endemic to nanotechnology that makes these cases interesting.

For the remainder of this chapter, we will contemplate some contexts in which nanotechnology does raise questions about equity and access and, potentially, that are not obviously suggested in other contexts. We are actually fairly skeptical about these latter contexts, though there is little room for skepticism on the former. As mentioned earlier, the focus will be on some of the promise that nanotechnology has for developing countries. One reason for this choice has to do with the dramatic needs that some of those countries have and the immense challenges that they are facing. When we talk about cognitive enhancement, for example, this is a technology that will only affect a small number of people, at least in the short term. Depending on estimates and definitions, though, developing countries are home to as many as 3 billion people. The scope of whatever technologies bear on those countries is therefore extremely broad. Another reason we focus on the benefits of nanotechnology for the developing world is that many of the needs of this sector pertain to basic human needs; nanotechnology is disanalogous in this respect with more highly circumscribed technologies. Certainly some people, for example, think that we have a right to cognitive enhancement, but few would argue that it corresponds to any sort of basic human need. Our primary concern in what follows will be with water purification and medicine, each of which correspond to basic human needs. We will also discuss solar energy which, while not directly corresponding to a basic human need, can have critical implications in other regards (e.g., pumping water or powering medical devices). For these reasons, scope and need, the context of developing countries and nanotechnology is especially useful to consider; in this discussion, we will draw from important work done by Todd Barker and his colleagues, as well as work by Joachim Schummer.[17]

Before we talk about nanotechnology and developing countries, it is worthwhile to make a few comments about what we take developing countries to be, since this is not an uncontentious issue. Many countries, including the richest ones, are still developing according to some indices, and there are other indices against which the poorest countries are either stagnant or regressing. The metrics for measuring levels of development range from simple per capita gross domestic product (GDP) to more complex indices, which incorporate sustainable conditions of living, including political stability and equality among the populace.[18] Following Schummer, we understand developing countries as those that have low or mid scores on the widely accepted human development index (HDI) used by the United Nations Development Programme. The HDI combines per capita GDP with life expectancy and educational standards. According to this index, the least developed countries are all in sub-Saharan Africa, followed by others in South Asia, the Middle East, East Asia, and Latin America.[19]

Aside from statistical and geographical features, the less developed countries share various other features. Historically, they tended to be colonies, and tend to still have various connections (e.g., economic, political, and/or military) to the colonizing power.[20] Many of these countries are rich in natural resources; undoubtedly this is one of the reasons that they were colonized in the first place. Much of their populations lack provision for their most basic needs, including food, water, sanitation, and health care; this last despite ravaging epidemics like AIDS and malaria.[21]

Since 2000, the United Nations Millennium Development Goals (MDGs) have set targets for trying to mitigate these concerns, most of which afflict those in the developing world. Recently, many reports – whether from governments, scientific communities, or other institutions – have indicated that nanotechnology could play a significant role in addressing these targets. Some of the potential that nanotechnology offers is through:

[17] Todd F. Barker et al., "Nanotechnology and the Poor Opportunities and Risks for Developing Countries," in Fritz Allhoff and Patrick Lin (eds), *Nanotechnology and Society: Current and Emerging Ethical Issues* (Dordrecht: Springer, 2008), pp. 243–63. See also Joachim Schummer, "Impact of Nanotechnology on Developing Countries," in Fritz Allhoff et al. (eds), *Nanoethics: The Ethical and Social Implications of Nanotechnology* (Hoboken, NJ: John Wiley & Sons, 2007), pp. 291–307.
[18] Schummer, "Impact," p. 291.
[19] United Nations Development Programme, *Human Development Report 2006: Beyond Scarcity Power, Water, and the Global Water Crisis* (New York: Palgrave Macmillan, 2006), pp. 263–73. Available at http://hdr.undp.org/en/media/hdr06-complete.pdf (accessed July 27, 2008).
[20] Schummer, "Impact," p. 291.
[21] UN Millennium Project, *Innovation: Applying Knowledge in Development* (2005). Available at http://www.unmillenniumproject.org/documents/Science-complete.pdf (accessed October 13, 2008).

more efficient, effective, and inexpensive water purification devices, energy sources, medical diagnostic tests and drug delivery systems, durable building materials, and other products. Additionally, nanotechnology may significantly increase developing countries' production capacities by enabling manufacturing processes that create less pollution and have modest capital, land, labor, energy, and material requirements.[22]

Developing countries are clearly recognizing this potential, as many of them are making substantial investments in nanotechnology research and development. Some developing countries, such as Brazil, China, India, and South Africa, already have national nanotechnology programs, and many more countries are expanding their nanotechnology research and development capacity. In these countries, as around the world, the investment in nanotechnology has been increasing substantially in recent years.[23] Let us now directly consider some of the ways in which nanotechnology could have an impact on developing countries.

7.3 Water Purification

As recently as 2002, over a billion people lacked access to safe drinking water, and over 2.5 billion people lacked access to adequate sanitation.[24] This lack of access, in turn, gives rise to approximately 5 million deaths per year, 85 percent of which are among children in poor Asian and African countries.[25] These are just staggering numbers, made even more dramatic by the fact that drinking water and sanitation are things that very few of us in developed countries ever need to worry about. Nanotechnology has been hailed as carrying much promise in regards to water purification and, if this promise can be realized, the positive effects could be substantial.

To consider the impact nanotechnology could have on water purification, let us consider somewhat more carefully some of the pollutants that make much of the water in the developing world unsafe. Broadly speaking, these pollutants come from two different sources. One problem is microbes, which come from humans due to insufficient sanitation. Another is heavy metals dissolved from minerals.[26] There are some specific microbial diseases,

---

[22] Barker et al., "Nanotechnology," p. 243.
[23] Fabio Salamanca-Buentello et al., "Nanotechnology and the Developing World," *PLoS Medicine* 2.5 (2005): e97, doi:10.1371/journal.pmed.0020097. Available at http://medicine. plosjournals.org/archive/1549-1676/2/5/pdf/10.1371_journal.pmed.0020097-L.pdf (accessed October 13, 2008).
[24] UN Millennium Project, *Innovation*.
[25] World Health Organization (WHO), "Water, Sanitation, and Hygiene Links to Health: Facts and Figures." Available at http://www.who.int/entity/water_sanitation_health/factsfigures2005.pdf (accessed July 28, 2008). See also UN Millennium Project, *Innovation*.
[26] Schummer, "Impact," p. 296.

such as schistosomiasis, trachoma, and intestinal helminthes, but the bigger worry is diarrhea, of which there are approximately 4 billion cases a year leading to over 2 million deaths.[27] Schummer, though, is skeptical that nanotechnology could be of much use in addressing these problems. According to the WHO, the death toll from diarrhea could be prevented by better sanitation, hygiene education, and improved water supply or water treatment. In light of this claim, though, Schummer thinks that it is hard to see how nanotechnology could be of much help, aside from competing with other water treatment options, such as chlorination. Rather, the principal problems here seem to be lack of basic infrastructure, facilities, and education.[28]

He goes on to claim that since the 1970s, many developing countries have focused on improving water supply by replacing surface water with underground water from wells; the focus was on this increased supply rather than on providing better sanitation and hygiene education. The problem, though, is that in many regions – especially Bangladesh, Nepal, India, Taiwan, Thailand, Argentina, Chile, China, and Mexico – the switch to the safer underground drinking water led to a higher concentration of arsenic and other heavy metals. Since arsenic, like other heavy metals, readily binds with iron hydroxide, various simple but effective filters have been developed, though these have yet to draw substantially from nanotechnology. Schummer grants that micro- and nanoporous filter development can lead to improved removal of both microbes and other pollutants – including heavy metals – though he denies that these will be affordable and manageable by those who need them most in the near future. Some of the most promising technologies, such as using expensive carbon nanotubes in filters, are hardly going to be widely disseminated through the developing world; they are more likely to be used by, for example, the US military to purify water for medical purposes on the battlefield.[29]

More optimistic, though, are Barker et al. who point to a wide range of water treatment devices that incorporate nanotechnology which are either in development or already on the market. There are several different categories of these products, which include: nanofiltration membranes; zeolites, clays, and nanoporous polymers; nanocatalysts; magnetic nanoparticles; and nanosensors.[30] Let us consider some of these, while paying special attention to whether they could be cheap enough to make an impact in the developing world.

---

[27] UN Millennium Project, *Interim Full Report of Task Force 7 on Water and Sanitation* (2004), p. 22. Available at http://www.unmillenniumproject.org/documents/tf7interim.pdf (accessed October 13, 2008).

[28] Schummer, "Impact," p. 296.

[29] Ibid. See also "Vermont's Seldon Labs Want to Keep Soldier's Water Pure," *Small Times* April 26, 2004. Available at http://www.smalltimes.com/articles/article_display.cfm?Section =ARCHI&C=Profi&ARTICLE_ID=269416&p=109 (accessed July 28, 2008).

[30] Barker et al., "Nanotechnology," p. 246.

Nanofiltration membrane technology is already widely used to remove salts from water, as well as for removal of micropollutants, water softening, and wastewater treatment. These membranes selectively reject substances, which means that they can be used to remove harmful pollutants while, at the same time, retaining important nutrients in water. The costs of membrane technology are currently high, though nanotechnology should be able to drive these costs down, thus increasing access to this desalinization and treatment technology. Early successes with these membranes included the filtering of polio viruses as well as *Escherichia coli* and *Staphylococcus aureus* bacteria. The membranes also allow faster water flow than traditional filters. As for whether this technology will make an impact in developing countries, a South African university has used them to treat salinated water in some rural communities.[31] This is obviously a promising start, though more widespread usage would have to be necessary before considering the technology a success.[32]

Zeolites, clays, and nanoporous polymers are also materials used for nanofilters. Some of these have been used for many years, but recent improvements in scientists' ability to manipulate the nanoscale allows for greater precision in designing the materials, especially the pore size of the membranes. Zeolites are:

> microporous crystalline solids with well-defined structures. Generally they contain silicon, aluminum, and oxygen in their framework and cations [positively charged ions], water, and/or other molecules within their pores. Many occur naturally as minerals and are extensively mined in many parts of the world. Others are synthetic and are made commercially for specific uses or produced by research scientists trying to understand more about their chemistry. Zeolites can be used to separate harmful organics from water and to remove heavy metal ions from water.[33]

Australia's Commonwealth Scientific and Industrial Research Organisation (CSIRO) has been able to develop a process that enables low-cost and local production of a synthetic clay called hydrotalcite that can be used to absorb arsenic, and maybe fluoride, from water. Hydrotalcite is already used in antacids and fertilizers and is synthesized by combining an ammonia solution with a mixed solution of magnesium or aluminum; the magnesium and aluminum solutions are easily produced from widely available materials. Production costs can be low, and existing production plants (e.g., for fertilizer) can be adapted to these purposes given similar production processes.

---

[31] Thembela Hillie, Mohan Munasinghe, Mbhuti Hlope, and Yvani Deraniyagala, "Nanotechnology, Water, and Development" (Meridian Institute, June, 2006), pp. 35–9. Available at http://www.merid.org/nano/waterpaper/ (accessed July 28, 2008).

[32] Barker et al., "Nanotechnology," p. 246.

[33] Ibid., p. 247.

This technology is then easy to apply: it can be added to water directly, or supplied in teabags that could be steeped in water prior to drinking.

Other possibilities exist, too, with polymers that can bind to contaminants in water. For example, researchers at the Los Alamos National Laboratory have developed nanoporous polymers that can reduce the concentration of common organic contaminants in water to safe levels of parts per trillion.[34] As for whether zeolites, clays, and polymers will be used for water treatment in developing countries, the two central issues will be price and access. Some of these technologies, such as the clays, already seem especially easy to create locally. Zeolites already occur naturally, though the synthetic ones will probably be more effective. The polymers have yet to be widely made. Technically, though, these products offer great options, so long as they can be disseminated.[35]

Nanocatalysts comprise enzymes, metals, and other materials whose catalytic capacities are enhanced by their nanoscale dimensions or from nanoscale structural modifications. These catalysts promote chemical reactions of other materials without becoming permanently involved in the reaction. Through control of a material's size and/or structure, catalysts can be developed that are more reactive, more selective, and longer lasting. Therefore, lower quantities of catalysts are necessary, which, hopefully, will reduce the overall cost.[36] Titanium dioxide (TiO_2), for example, and iron nanoparticles can be used to degrade organic pollutants and to remove salts and heavy metals from water. The UN Millennium Project is optimistic that these catalysts can be used to treat heavily polluted and heavily salinated water for drinking, sanitation, and irrigation.[37]

Some applications of nanocatalysts have already been successful. For example, the US Environmental Protection Agency (EPA) has field-tested nanoscale zero-valent iron (NZVI) to treat a site contaminated with trichloroethylene. NZVI functions as both an adsorbent and a reducing agent, thus causing the organic contaminants to break down into less toxic carbon compounds and the heavy metals to agglomerate and stick to the soil surface. The Inframat Corporation is developing a material based on highly porous nanofibrous structures that can be used to remove arsenic from drinking water; these work by combining a nanofibrous manganese dioxide (MnO_2) process with a granular ferric hydroxide ($Fe(OH)_3$) adsorptive process. Another company, Environmental Care, uses titanium dioxide (TiO_2) to trigger a chemical process, which converts contaminants into carbon

[34]   Mihail C. Roco, R. Stanley Williams, and Paul Alivisatos (eds), *Nanotechnology Research Directions: IWGN Workshop Report. Vision for Nanotechnology Research and Development in the Next Decade* (Baltimore, MD: Loyola College, 1999), Section 10.7.3. Available at http://www.wtec.org/loyola/nano/IWGN.Research.Directions/chapter10.pdf (accessed July 28, 2008).
[35]   Ibid., pp. 247–8.
[36]   Ibid., p. 248.
[37]   UN Millennium Project, *Interim Full Report*.

dioxide and water. There are a number of other promising applications of nanocatalysts as well, though their impact on developing countries will again come down to access and pricing.[38]

Magnetic nanoparticles are another possibility for water treatment: these can bind with chemicals without the use of secondary adsorbent materials. Furthermore, they can be customized with surface coatings such that they are suitable as nanocatalysts for chemical decomposition. Once the adsorption or catalysis has taken place, the nanoparticles can then be removed from the water magnetically. For example, Rice University is developing a technique to introduce magnetic nanoparticles to water such that they can bind with contaminants like arsenic. Once the nanoparticles have bound to the contaminants, they can be removed with magnetic filters. A company in India, Tata Chemicals, has developed filters embedded with magnetic nanoparticles; the costs of these could be as low as 5 percent of other extant technologies. Furthermore, local materials, like sand and rice husks, could be used in the development of these filters. And a university in Brazil has developed an application with magnetic nanoparticles that can remove oil from water. Insofar as some of this research is already taking place in developing countries like India and Brazil, magnetic nanoparticles may well play an important role in water treatment in the developing world.[39]

Finally, nanosensors can be used for the detection of contaminants. These sensors can develop single cells or even atoms, thus making them much more sensitive than other sensory apparatus. A European project, BioFinger, is developing a portable and inexpensive water analysis system; it is handheld and uses disposable microchips that cost approximately €8. Another example comes from the University at Buffalo, where researchers are developing a handheld sensor that can detect the presence of toxins; this makes use of nanosensor technology. Finally, the EPA is sponsoring research to develop a 1 cm^2 nanoreactor which is capable of the detection and remediation of lead, cadmium, arsenic, hexavalent chromium, and copper.[40]

So we have now seen various ways that nanotechnology can help with water purification. What are the implications for these technologies in the developing world? Schummer seemed pessimistic, particularly considering cost and availability.[41] It is worth noting, though, that his essay was written a year before Barker et al.'s, so perhaps this more recent work includes new possibilities. Regardless, it is hard for us to speculate on the

[38] Barker et al., "Nanotechnology," pp. 248–9.

[39] Ibid., p. 249.

[40] Ibid., p. 250.

[41] Schummer is also concerned whether lack of infrastructure or education would allow the developing world to take advantage of these technologies even if they were otherwise to be available. This is a separate discussion that we shall not pursue here, at least insofar as we remain skeptical on the issue of availability.

extent to which these nanotechnology-enabled options will be inexpensive and widely available. In §7.6 we will make some comments about some of the implications regarding global distributive justice if these options are not inexpensive, or at least not as inexpensive as the developing world would be able to afford. But whether the developing world, without assistance from the developed world, will have access to these implements is hard to say. Certainly we expect the prices to fall for many of these options, though so many of them are still in research and development stages that even preliminary estimates regarding cost are challenging. Insofar as some of this research and development is being done in the developing world – we provided examples above of work in India and Brazil – we have reason to be encouraged. But, overall, we think that agnosticism is the most appropriate stance to take for the impacts that nanotechnology will have for clean water in the developing world.

7.4   Solar Energy

In addition to lacking access to clean water, much of the developing world lacks access to electricity. One estimate puts the number of people living without electricity at around 2 billion, most of whom are in rural areas of developing countries.[42] Access to electricity is not essential for living, though it is a major step in development. For example, it can replace inefficient sources of energy, as might be used in lighting. And, furthermore, it enables and facilitates important processes and infrastructures.[43] For example, electricity can be used to pump water for human and agricultural use. It can be important in health clinics, from the direct powering of medical devices to the refrigeration of medicine.[44] Refrigeration more generally helps to preserve food, which could make rural communities less subject to the vicissitudes of weather or famine. Electricity enables communication, which could have important implications for health as well as commerce. So, while not essential, electricity nevertheless drastically increases the possibilities available to communities.

Rural communities in the developing world that lack access to electricity make use of traditional biomass energy, such as wood, crop residues, and dung. There are various drawbacks to these energy sources, though. First, they are comparatively inefficient: much of the produced energy is lost without being able to contribute to any energy needs. Second, there are negative health consequences for using these sorts of energy, including burns

---

[42]   World Energy Council, "The Challenge of Rural Energy Poverty in Developing Countries" (London, 1999). Available at http://217.206.197.194:8190/wec-geis/publications/reports/rural/exec_summary/exec_summary.asp (accessed July 30, 2008).

[43]   Schummer, "Impact," p. 297.

[44]   Barker et al., "Nanotechnology," p. 251.

and respiratory problems due to indoor pollution. And, third, depending on the source of the biomass, the utilization of these energy sources can degrade the environment as well as resource bases.[45]

Cheap solar-powered electricity would be a drastic improvement over biomass fuels: the former is more efficient, safer – in terms of both immediate risks and more longitudinal health effects – and less environmentally perilous. And, furthermore, many developing countries have great potential for solar power given their tropical or semi-tropical locations; these areas have substantial solar exposure. The reason that developing countries have not yet had access to solar power is that the photovoltaic cells, made from glass and silicon, are too expensive and delicate. Nanotechnology, though, might enable the production of cheap photovoltaic films that can be unrolled across the roofs of buildings; it may even be possible to paint solar-power films onto surfaces.[46,47]

Even Schummer, who was pessimistic about the significance that nanotechnology might have for water purification, is optimistic about nanotechnology's potential for providing power to the developing world, particularly to areas of high solar radiation. As he points out, one reason that photovoltaics are promising is that much of the developing world's need for power is through its rural residents; photovoltaics do not depend on centralized power plants and grids which, of course, these rural communities do not have. Furthermore, solar energy is sustainable, which is an improvement over at least some uses of biomass.

Schummer worries that there might be some cultural barriers to electricity, particularly in communities that have never before had it. There will, for example, be basic educational needs, such as would be necessary for dealing with "cables, switches, fuses, transformers, and rechargeable batteries, in addition to the electric devices for which the whole setting is built up."[48] In other words, communities would need to learn not just about the devices that they wish to power, but also about how basic electrical infrastructure works. Nanotechnology, while perhaps making the electricity more accessible, will not be able to help with these cultural and educational challenges.

That said, though, there does seem to be a lot of promise on the technological side. As mentioned above, nanotechnology should be able to drive down the costs of photovoltaics, thus making them more accessible. A US

[45] Ibid.

[46] Ibid.

[47] Photovoltaics are, most basically, solar cells packaged in modules such that sunlight energizes electrons, moving them into higher energy states and, therefore, creating electricity. They can be connected up in arrays such that multiple cells can contribute to electrical current. The electricity is given off in direct current, though it can be passed through an inverter to create alternating current.

[48] Schummer, "Impact," p. 297.

company, Nanosolar, has developed nanotechnology-based solar panels that print photovoltaic cells directly onto flexible plastic and foil. These panels could be as efficient as traditional silicon panels, yet produced at one-fifth of the cost. Global energy company BP, in collaboration with the California Institute of Technology, is developing solar cells made from an array of nanorods; this approach should allow for greater efficiency than traditional solar cells. Finally, solar energy can be harvested during the day to be used at night, but it has to be stored in the meantime. Nanotechnology can also contribute to the effectiveness of the associated rechargeable batteries and fuel cells. Altair Nanotechnologies, for example, is using a proprietary nano-titanate that can be used in lithium ion batteries. These batteries will charge very quickly, up to 80 percent of their maximum capacity in under a minute. And, furthermore, they can be recharged more often than conventional batteries: conventional lithium ion batteries can recharge approximately 750 times, whereas the nano-titanate batteries can charge as many as 9,000 times.[49]

Technologically, there seems little doubt that nanotechnology has much promise in being able to bring electricity to many of the 2 billion people that currently lack it. But, again, the issue will simply come down to pricing and to whether these developing countries can afford the technology. Imagine, for example, that Nanosolar can actually bring solar panels to market at one-fifth of their traditional cost. A 15 watt solar panel costs about $100; this produces enough electricity to be able to charge a battery. One-fifth of that is still $20. Many of the poorest developing countries have per capita GDPs of under $1,000. Could these countries afford $20 solar panels? And, of course, one 15 watt solar panel would not generate much electricity. For example, even a small American house probably uses at least 600 kilowatt hours per month, which is 20 kilowatt hours per day. Using some standard assumptions about sun exposure and efficiency, it would take 1,000 15 watt solar panels to generate that much electricity daily.[50] Even with the discounted panels, it would cost $20,000 to power the house. There are a lot of flaws with this estimate, not least notably that countries in the developing world would have far lower electrical needs; a significant percentage of our electricity goes to comparatively trivial uses (e.g., televisions, air conditioning, and so on). But even generating 1 kilowatt hour per day (5 percent of our consumption) would cost $1,000, which, again, could be the per capita annual income for someone living in a developing country. So, while there is room for optimism, some realism might be appropriate as well. If the costs can keep falling – e.g., down to 5 percent of conventional technologies rather than 20 percent – then solar energy would

---

[49]   Barker et al., "Nanotechnology," p. 251.
[50]   Solar Panel Estimator. Available at http://files.blog-city.com/files/M05/102402/b/solarcalc3. html (accessed July 30, 2008).

be more economically viable for the developing world. But there are uses that could be manageable even at existing costs. For example, a village might want a refrigerator to preserve medicine; this could be a critical and effective use of electricity.[51] Maybe nanotechnology could enable these sorts of applications at affordable prices. But, in terms of broader applications, the cost could still be prohibitive.

7.5 Medicine

In Chapter 11, we offer extended discussion of the possibilities that nanotechnology has for medicine. But, in this chapter, we will briefly comment on the implications that nanomedicine – i.e., the application of nanotechnology to medicinal purposes – has for the developing world. We will defer any substantive discussion of nanomedicine until that later chapter, but can approach the present task without that discussion. Just to foreshadow, we will identify diagnostics, drugs, and delivery as three areas in which nanotechnology will have an impact; the alliteration should provide a useful mnemonic. For now, let us consider directly some of the impacts nanomedicine could have for the developing world.

Undoubtedly, one of the greatest potentials lies in the treatment of HIV/AIDS. This disease is among the worst epidemics in human history. In 2005 alone, over 4 million people became infected with HIV and almost 3 million people died from AIDS-related disease. An estimated 40 million people are currently living with HIV. The least-developed countries, especially those in sub-Saharan Africa, are those most affected by the epidemic; over 30 percent of adults in Swaziland, for example, are infected.[52]

The UNAIDS report claims that "The steady growth of the AIDS epidemic stems not from deficiencies of available prevention strategies but rather from the world's failure to use the highly effective tools at its disposal to slow the spread of HIV."[53] Schummer elaborates: "Since vaccines will likely be unavailable for many years, the primary prevention tools against sexual transmission are condoms and safer sex education. Yet from Catholic Church policy to male preference of condom-free intercourse, many social factors have prevented effective implementation."[54]

---

[51] A refrigerator, though, is hardly frugal when it comes to electricity; most refrigerators consume 5–10 kilowatt hours per day.
[52] Schummer, "Impact," p. 298. See also UNAIDS, *Report on the Global AIDS Epidemic* (2008). Available at http://www.unaids.org/en/KnowledgeCentre/HIVData/GlobalReport/ 2008 (accessed July 30, 2008).
[53] UNAIDS, *Report on the Global AIDS Epidemic* (2006), p. 124. Quoted in Schummer, "Impact," p. 298. Also available at http://data.unaids.org/pub/EpiReport/2006/2006_ EpiUpdate_en.pdf (accessed October 13, 2008).
[54] Schummer, "Impact," p. 298.

Since abstinence and condoms have had (extremely) limited efficacy against the AIDS epidemic, new ideas are in high demand. Nanotechnology offers much promise in this regard, particularly as pertains to vaginal microbicides. In addition to the effectiveness of these microbicides, they also empower women to protect themselves from the disease; for example, against men who are unwilling to use condoms. An Australian company, Starpharma, has developed a dendrimer called SPL7013 that can be used as a microbicide;[55] dendrimers are branching molecules that have been researched since the 1970s, though this is now being labeled as nanotechnology.[56] This product prevents HIV infection by binding to receptors on the surface of the virus; because those receptors are then already bound, they are thereafter unable to bind to receptors on cells in the human body.

SPL7013 could be extremely effective in fighting HIV, particularly as its utilization does not suffer from some of the challenges that, for example, abstinence and condoms do. But could it be effective in the developing world? Schummer is again skeptical, citing the high production costs of dendrimer-based products. As of the time of writing, SPL7013 is still only in phase I clinical trials, meaning that it could be years before it is made available. So, as excited as the nanomedia has been about this product, its availability could still be years away, and the pricing could be high. There are other, cheaper, microbicides under development, though most of these do not incorporate nanotechnology; furthermore, some of them are already in phase III trials.[57] Also, this sort of product would have to compete against other microbicides, some of which are certainly going be cheaper: these include "soap (sodium lauryl sulfate, also called the 'invisible condom'), cellulose sulfate, and lemon juice (all currently in [phase] I to III [trials]), as well as . . . a bunch of other prevention methods specifically tailored to the needs and customs of poor countries."[58] Whether the pricing will be viable for the developing world, and whether SPL7013 is superior to its competitors, are the key questions, and ones that will have to wait its progress through clinical trials. Like Schummer, we are skeptical that this product could be priced very close to some of its alternatives

---

[55] Ibid. For an associated scientific study, see Yong-Hou Jiang et al., "SPL7013 Gel as Topical Microbicide for Prevention of Vaginal Transmission of SHIV$_{89.6P}$ in Macaques," *AIDS Research and Human Retroviruses* 21.3 (2005): 207–13.

[56] Bethany Halford, "Dendrimers Branch Out," *Chemical and Engineering News* 83.24 (2005): 30–6. Available at http://pubs.acs.org/cen/coverstory/83/8324dendrimers.html (accessed July 31, 2008).

[57] Michael M. Lederman, Robin E. Offord, and Oliver Hartley, "Microbicides and Other Topical Strategies to Prevent Vaginal Transmission of HIV," *Nature Reviews Immunology* 6 (2006): 371–82.

[58] Schummer, "Impact," p. 298. See also R.V. Short, "New Ways of Preventing HIV Infection: Thinking Simply, Simply Thinking," *Philosophical Transactions of the Royal Society B* 361.1469 (2006): 811–20.

and, therefore, we wonder whether it could have a substantial impact in the developing world.[59]

While much of the media attention has focused on SPL7013, there are some other applications of nanotechnology that might bear on medicine in developing countries. For example, nanotechnology could enable diagnostic devices that are smaller than conventional ones; they could also be very accurate and simple to use. A UK nanotechnology firm is developing a portable and inexpensive device that could be used to test for malaria – in addition to avian flu, *E. coli*, meningitis, and some cancers – which is most prevalent in developing countries. This device uses quartz elements (as in wristwatches) to test for the presence of specific disease markers in blood samples. It can be powered with batteries and, therefore, used in the field; this is an improvement over some conventional diagnostics that require submission to a laboratory and often require days or weeks to produce results.[60]

Finally, nanoporous materials can be used to improve drug delivery. By making the nanopores only slightly larger than the molecules of the drugs, the diffusion rate of the drug can be maintained at a constant rate, independent of how many molecules remain in the capsule. This, potentially, could make a big impact in the developing world since its rural inhabitants are likely to live far from hospitals. If the drug flow can be more precisely controlled, this could allow the same capsules to medicate for longer as well as to increase their efficiency (i.e., by not dispersing the drug at a faster rate than its maximum efficacy). Both of these advantages, of course, would also help keep prices down since fewer drugs would be required. Unlike some of the other worries that we have had about pricing, there is no obvious reason to think that nanomembranes could not be effectively and economically integrated into drug delivery mechanisms; therefore, this technology could make an impact in the developing world.

7.6   Nanotechnology, the Developing World, and Distributive Justice

In §§7.3–7.5 we have discussed various ways that nanotechnology could be applied to problems in the developing world, particularly as pertains to water purification, solar energy, and medicine. We think, though, that the results of those discussions are highly equivocal: *technologically* there

---

[59] Schummer also seems skeptical as to whether dendrimer technology should be properly characterized as nanotechnology, particularly given that much of this research has been taking place since long before the contemporary attention to nanotechnology. He also points out that some of these alternatives, like soap or cellulose, could be tagged as nanotechnology, strictly speaking.

[60] Barker et al., "Nanotechnology," p. 252.

is little doubt that nanotechnology can make a difference in developing countries, but *economically* it is far less clear whether these solutions will be viable. Just because the developing world needs water, energy, and medicine and just because nanotechnology stands to underwrite some solutions to those needs, it hardly follows that *it will*. Rather, those solutions still have to be cost-effective and, specifically, cost-effective on the sorts of scales that are affordable in developing countries. And, given the extremely low per capita GDPs of many of those countries, the affordability criterion is highly precarious.

In the last section of this chapter, we want to assume some skepticism about the affordability, within developing countries, of the prospects discussed in the preceding sections. Nothing critical hangs on this skepticism and, if it is unwarranted, then all the better. For example, if it turns out to be the case that developing countries can afford nanotechnologically enabled solar panels, that would be fantastic and, hopefully, those solar panels would make important contributions to their communities. But what if those communities cannot afford the technologies? What is the moral status of their inability to access them? And what is the moral status of the differential access that some countries have to these technologies as against the developing ones? Is this an instance of inequity? Or merely inequality? In other words, does this differential access violate some norm of justice or fairness? These are a lot of questions; the best that we can do – whether for lack of space or certainty – is to raise them.[61] So let us now return to §7.1, which, by now, probably seems a long-ago digression in need of integration with the rest of the chapter.

Assuming that the developing world cannot afford these technologies, does the developed world have any moral obligation to do anything about it? Let us approach this question by considering an extremely famous essay written by Peter Singer.[62] At the time of writing (1972), Singer observed that many of those living in East Bengal (roughly present-day Bangladesh) were dying from lack of food, shelter, and medical care. Singer argued that the affluent – by which, let us say, he means to include most of us in the developed world – have a strong moral duty to provide for those in need; he further argued that the sorts of sacrifices that would be morally required of us would be fairly extensive. In much of the essay he defends his position against objections, such as that the Bengalis are far away, that the government should do the work rather than us individuals, that charitable giving will collapse economies, and so on. Those details are not important for our purposes, but suffice it to say, his response to these

---

[61] For more extended discussion – though not pertaining to technology in particular – see Pogge, *World Poverty*, and Pogge, "Severe Poverty." See Hayward, "On the Nature," for discussion of some of the primary criticisms of Pogge's work.

[62] Peter Singer, "Famine, Affluence, and Morality," *Philosophy and Public Affairs* 1.1 (1972): 229–43.

anticipated criticisms is very convincing.[63] But, given our discussion, let us consider what might be taken to be Singer's core argument. It can be represented as follows:

P1   Suffering and death from lack of food, shelter, and medical care are bad.

P2   If it is in our power to prevent something bad from happening, without thereby sacrificing anything of comparable moral importance, we are morally obligated to do it.

P3   In the case of the Bengalis, we can indeed prevent suffering and death from lack of food, shelter, and medical care without sacrificing anything of comparable moral importance.

C   Therefore, we are morally obligated to provide the appropriate aid.

This is a very simple argument that is nevertheless quite powerful. We offer some quick points of clarification. First, the "sacrificing anything of comparable moral importance" bit in P2 and P3 just means that if helping others creates moral hazards comparable to those that the aid would remedy, then the obligation does not follow. (Singer presents it as a sufficient condition, but we think it is most plausibly read as a necessary condition as well.) So, if saving one person requires putting ourselves in peril, then that would constitute a sacrifice of comparable moral importance and, therefore, is not required. Second, the motivation for P3 is that much of the death facing the Bengalis is easily preventable. For example, much of that death was coming from conditions such as diarrhea; those deaths can (often) be prevented by the administration of salt tablets or other very inexpensive treatments. Quite seriously, a couple of dollars could save a life (if not multiple lives) and the thrust of Singer's essay is that, therefore, we are obligated to provide this aid.

Let us now return to the possibilities for nanotechnology that we discussed in this chapter. Electricity is not essential for human life, though it can provide important support in other regards (e.g., pumping water or powering medical devices). Even setting aside electricity, though, consider water purification and AIDS prevention. Without these, people will die. Millions of people. Granting that nanotechnology can contribute to solutions to these problems and, for the moment, supposing that only nanotechnology can offer these solutions, does the developed world have an obligation to provide these things to the developing world? Singer would clearly say yes.

Return to our discussion from §7.1 and consider our theories of distributive justice. Remember that Aristotle's account was for proportionality based upon merit, though merit hardly need be univocal. If, for example,

---

[63]   A well-known critique of Singer's position is given by Garrett Hardin, "Lifeboat Ethics: The Case against Helping the Poor," *Psychology Today* (September 1974): 38–40, 123–4, 126.

distribution were conceived of in a Marxian sense – by which need was the primary metric for merit – then Aristotle would have us giving to developing countries. But then there are standard worries, such as that if we gave away some of our own resources, then we would be less motivated to work hard and that creativity and development would lapse. The reason we give long patents for pharmaceuticals, for example, is that absent the time to recoup research and development money (as well as to make a profit) the pharmaceutical companies, allegedly, would not pursue new and risky drugs, some of which thereafter become staples in our arsenals against disease (and some of which fail). Aristotle could, then, hold that the merits go to those who foster an environment in which they might be able to provide for others; if these are the ones to whom the merits are due, then distribution (e.g., to developing countries) is not required.

We can get more traction if we go to Rawls and Nozick. Remember that Rawls would only allow inequalities so long as they were to the benefit of the least well off. So, to simplify, this would mean that the developed countries could amass differential advantages over the developing countries *so long as* some remediation were made to those developing countries. Just to keep Rawls simple – it gets somewhat more complicated when we consider global distributive justice – we might imagine that some sort of welfare system would be appropriate such that the resources of the developed countries were shared, in part, with the developing countries. How extensive should this sharing be? This is a complicated issue and, among other things, gets into the veil of ignorance construct that we did not discuss. But, suffice it to say, a Rawlsian framework could provide for substantial, if not necessarily onerous, redistribution to developing countries.

Nozick, on the other hand, could easily oppose such a redistribution. This redistribution would, in some sense, be applied as a tax: some portion of the developed countries' resources would be garnished to be distributed. Nozick famously likens taxes to slavery insofar as taxes are rarely, at least in some sense, consensual and, quite literally, those who pay them are working "for someone else" for some number of minutes out of each hour.[64] And, for Nozick, maybe the developed world does benefit from the sort of natural and social advantages that Rawls aims to correct: maybe it has more hospitable environments, greater traditions of education, and so on.[65] Nevertheless, it (and its citizens) might be entitled to its comparative

---

[64] Of course, some of the benefits of the tax revenues return to those who pay them in the form of government services, for example. But this is not always the case, as with the childless parents whose property tax nonetheless pays for schools, or the art hater whose taxes go, in part, to museums.

[65] See, for example, Jared Diamond, *Guns, Germs, and Steel: The Fate of Human Societies* (New York: W.W. Norton, 1997). Diamond is interested in why certain countries (e.g., in Europe) have gone on to colonize the rest of the world rather than vice versa; he tells this story by appeal to what could be readily accounted for by Rawls as natural and social advantages. The book is highly controversial, but certainly worth reading.

riches. On Nozick's account, then, the developed world might not have a moral obligation to assist the developing world, despite what hangs in the balance.

Where does this leave us? That depends on the preferred account of distributive justice, and we have hardly defended any particular one in this chapter. Rather, our goal has been to show what some candidate accounts are, and then to develop some of the implications that nanotechnology could have for developing countries. Again, if the technologies turn out to be affordable, then the developing world will be able to provide for itself. A principal worry ethically, though, is if this proves not to be the case and developing countries lack access to the technologies that they need the most. In this scenario, issues of equity do arise, and we have to think carefully about what justice and fairness require of us.

Unit III

Ethical and Social Implications

8

Environment

Of the nine National Nanotechnology Initiative (NNI) Grand Challenges, one explicitly names environmental improvement and another deals with energy (directly related to some of the most pressing environmental issues of modern times). However, nearly all technologies have some impact on the environment, and anticipating the impacts can alleviate or help solve problems that stand in the path of technical progress.

The executive summary to the report on the NNI Workshop on Nanotechnology and the Environment begins with the following statement:

> Nanotechnology has the potential to significantly affect environmental protection through understanding and control of emissions from a wide range of sources, development of new "green" technologies that minimize the production of undesirable byproducts, and remediation of existing waste sites and polluted water sources. Nanotechnology has the potential to remove the finest contaminants from water supplies and air as well as continuously measure and mitigate pollutants in the environment. However, nanotechnology may pose risks to the environment and human health, and these risks should be examined as the technology progresses.[1]

Because of the size and controllability of the properties, mechanisms, and aims of the techniques and tools that nanotechnology enables, nanotechnology poses some interesting, unique, and exciting opportunities for the environment (from manufacturing techniques that lessen the problem of pollutants to technologies that can utilize and remake nature in fantastic ways). However, it also poses some new problems that risk extreme damage to

---

[1] *Nanotechnology and the Environment: Report of a National Nanotechnology Initiative Workshop*. Available at http://www.nano.gov/NNI_Nanotechnology_and_the_Environment.pdf (accessed August 15, 2008).

the environment, from new materials/devices that could enter into plant and animal cells and disrupt their natural processes to the unintended consequences of harvesting minerals and elements from the Earth even more precisely.

8.1　Society, Technology, and the Environment

Humanity's relationship with the environment is complex, to be sure, and full of seemingly competing interests and contested histories of interaction. In its most simplistic form, the contradictions of the societal relationship can be seen as the conflict between having dominion over the environment[2] and being given charge of its stewardship.[3] At one extreme, the environment is regarded as a resource that can be used in any way that may benefit the human condition. At the other extreme, humanity is the steward of the environment, charged with its care and expected to do nothing that permanently alters or disturbs this natural state. In Judeo-Christian tradition, this duality is noted at the very creation of man, with man having dominion over the living things and the Earth but being told to "replenish the Earth" as well:

> And God said, "Let us make man in our image, after our likeness, and let them have dominion over the fish of the sea and over the fowl of the air and over the cattle and over all the Earth and over every creeping thing that creepeth upon the Earth." So God created man in His own image; in the image of God created He him; male and female created He them. And God blessed them, and God said unto them, "Be fruitful and multiply; and replenish the Earth and subdue it and have dominion over the fish of the sea and over the fowl of the air and over every living thing that moveth upon the Earth." And God said, "Behold, I have given you every herb bearing seed which is upon the face of all the Earth and every tree in which is the fruit of a tree yielding seed; to you it shall be for meat, and to every beast of the Earth and to every fowl of the air and to everything that creepeth upon the Earth, wherein there is life, I have given every green herb for meat."[4]

The back-and-forth of these two states has been part of the human discourse since the development of agriculture, cities, and the first technologies. The relationship that societies have with their environment has been a question

---

[2]　In Genesis 1:26 mankind is given dominion over the "fish of the sea and over the fowl of the air and over the cattle and over all the Earth and over every creeping thing that creepeth upon the Earth."

[3]　"[T]he End of Man's creation was that he should be [charged] to preserve the face of the Earth in beauty, usefulness, and fruitfulness." Sir Matthew Hale, *The Primitive Origination of Mankind* (London: William Godbid, 1677), p. 481.

[4]　Genesis 1:26–30, *The Bible*, The King James Version, ed. David Norton (New York: Penguin Group, 2006), p. 4.

much debated throughout human history. For example, in medieval Europe, timber was one of the most valuable resources, used as the fuel for heating buildings and ovens, but also important to almost every medieval industry:

> In the building industry wood was used to build timber-framed houses, water mills and windmills, bridges, and military installations such as fortresses and palisades. In the wine industry wood was used for making casks and vats. Ships were made of wood, as was all medieval machinery such as weavers' looms. Tanners needed the bark of the trees and so did the rope makers. The glass industry demolished the woods for fuel for its furnaces, and the iron industry needed charcoal for its forges.[5]

However, there was also concern at the time over the proper use of land. Fitz Nigel was a treasurer to the King of England who wrote in the 1170s that "If woods are so severely cut that a man standing on the half-buried stump of an oak or other tree can see five other trees cut down about him, that is regarded as waste. Such an offense even in a man's own woods is regarded as so serious that even these men who are free of taxation because they sit at the king's exchequer must pay a money penalty all the heavier for their position."[6]

Today, the collection of issues that are referred to in society as "environmental" issues is very diverse. Air pollution, natural habitat alterations, building materials (such as asbestos), and disposal of industrial waste are among the issues that are covered in discussing the environment. In general, the rubric of global warming or climate change refers to issues that seem to be related to a gradual alteration of the temperature, precipitation, or wind patterns. However, we also relate environmental issues to those problems associated with animal species. For example, endangered species, habitat management, and humane treatment of pets are all issues associated with an environmental stance. In general, though, when we refer to the environment, we are referring to the natural living world and the Earth.

Interestingly enough, it was a technology that seems to have enabled this global outlook on the environment. When the early space missions brought back photographs of the Earth as seen from space – a fragile, blue-green system amidst a harsh colorless black space – the importance of effective environmental management took hold in the greater public conscience. The images known as "Earth Rise" taken on the Apollo 8 mission and "The Blue Marble" taken on the Apollo 17 mission were prominent inspirations for the environmental movement. Al Gore featured each image in his documentary *An Inconvenient Truth*:

---

[5] Jean Gimpel, *The Medieval Machine* (New York: Henry Holt and Company LLC, 1976), pp. 75–6.
[6] Doris Mary Stenton, *English Society in the Early Middle Ages* (Baltimore, MD: Penguin Books, 1951), pp. 104–5.

[Earth Rise] is the first picture most of us ever saw of the Earth from space. It was taken on Christmas Eve, 1968, during the Apollo 8 mission . . . While the crew watched the Earth emerging from the dark void of space, the mission commander, Frank Borman, read from the book of Genesis: "In the beginning God created the Heavens and the Earth."

One of the astronauts aboard, a rookie named Bill Anders, snapped this picture and it became known as Earth Rise. The image exploded into the consciousness of humankind. In fact, within two years of this picture being taken, the modern environmental movement was born. The Clean Air Act, the Clean Water Act, the Natural Environmental Policy Act, and the first Earth Day all came within a few years of this picture being seen for the first time. The day after it was taken, on Christmas Day, 1968, Archibald MacLeish wrote: "To see the Earth as it truly is, small and blue and beautiful in that ethereal silence where it floats, is to see ourselves riders on the Earth together, brothers on that bright loveliness in the eternal cold – brothers who know now that they are truly brothers." . . .

[The Blue Marble] is the last picture of our planet taken by a human being from space. It was taken in December 1972 during the Apollo 17 mission – the last Apollo mission – from a point halfway between the Earth and the Moon.[7]

These images drove home the message that events and pollution that happen in one place on the Earth can have an impact on the environment not just locally, but across the entire globe. This connection was especially highlighted by the discovery of the hole in the ozone layer above Antarctica (where there are no permanent human settlements) caused by the use of CFCs in developed countries.

Society's dependence on and use of the environment are nothing new. Consider the technology of irrigation. The Mesopotamian civilization produced the world's first large cities. This society depended almost entirely on irrigation systems that allowed agriculture to develop to the point that individual farmers could produce more food than is required to feed a family. This sort of non-subsistence growth is necessary for the development of urban centers. Egyptian society relied on the natural irrigation of the Nile annual flood cycle to fertilize its farmland. The natural flooding cycle of the Nile and the fertility that the Nile provided the surrounding area both allowed for the stable development of Egyptian society. The fertile land enabled the Egyptians to develop agriculture above sustenance living and attracted animals for hunting, taming, or transportation. Further, the Nile itself was a convenient method for transporting goods and travel. This story of cities, cultures, and empires developing around water or being defined by other geographical and natural phenomena retells itself throughout history.

---

[7] Al Gore, *An Inconvenient Truth* (New York: Rodale Books, 2006), pp. 12–15.

The environment and changes within it can affect everything from crops and food availability to flight patterns of airplanes and shipping routes. Because the environment affects humans, protection from and alteration of it has always been an inspiration for creating and directing technology. This includes the development of shelter, irrigation, crop rotation, windmills, solar power, and many other technologies. Because humans affect the environment, the impact of technology on nature has always been a pressing issue. As technology develops, it affects and uses the environment in different ways.

This effect on the environment is not always negative or harmful. Labeling the impact as always harmful is not particularly helpful in understanding the role that technology has in an ecosystem. More important is considering the changes a technology can make on the environment, despite the imprecision of such an evaluation, so that we can begin to understand the undesired consequences and the intended benefits of using a technology.

The environment and environmental resources are also enablers of technologies, as raw materials for the technology itself, as materials required for the creation of the technology, or as the energy required for the use of the technology. Access to environmental resources for technological use has also been the cause of conflict throughout history. Around 424–422 BCE, the town of Amphipolis was the site of struggle between Athens and Sparta. It was important, according to Thucydides, because it provided access to gold and silver mines and "for the timber it afforded for shipbuilding."[8] In more modern times, oil of course provides much of the energy needs for the world. Though this energy has enabled scores of technological gains, it is not without its price and it has required diligence to look for more sources. In 1939, King Abdul Aziz of Saudi Arabia remarked "Do you know what they will find when they reach Mars? They will find Americans out there in the desert hunting for oil."[9]

Awareness of the negative impact of societies on the environment is nothing new either. In his incomplete dialogue "Critias," Plato describes the changes that the land around Athens had undergone over the course of 9,000 years due to both natural and societal changes:

> [I]n comparison of what then was, there are remaining only the bones of the wasted body, as they may be called, as in the case of small islands, all the richer and softer parts of the soil having fallen away, and the mere skeleton of the land being left. But in the primitive state of the country, its mountains were high hills covered with soil, and the plains, as they are termed by us, of Phelleus were full of rich Earth, and there was abundance of wood in the mountains. Of this last the traces still remain, for although some of the

[8] Thucydides, *Peloponnesian War*, trans. Richard Crawley (London: J.M. Dent & Sons, 1903), p. 216.
[9] Rachel Bronson, *Thicker than Oil: America's Uneasy Partnership with Saudi Arabia* (New York: Oxford University Press, 2006), p. 19.

mountains now only afford sustenance to bees, not so very long ago there were still to be seen roofs of timber cut from trees growing there, which were of a size sufficient to cover the largest houses; and there were many other high trees, cultivated by man and bearing abundance of food for cattle.[10]

The interplay and relationship between human societies and the environment is in a constant evolution. The environment is both highly valued, and used and altered by members of society with many different interests, talents, and values.

Technological use of the environment is most evidently seen in the exploitation of natural resources, particularly those resources that serve as energy sources. In modern times, that has been oil. In the past, other resources provided energy sources and using these resources required new technologies. Water mills are an early technology that utilized the flow of rivers to perform work. Early water mills were improved by Roman engineers by the end of the first century BCE.[11] By the Middle Ages, water-powered mills were being used to grind corn, forge iron, and perform other works.[12] Utilizing water in the environment as an energy source is common throughout history. More recently, hydroelectric power – converting water flow to electricity through the use of water turbines – is effected by damming rivers. This raises the level of the water behind the dam, giving it a high potential energy. The water is then led down through pipes: "In the course of its passage down the steep pipes, the falling water rotates turbines. The turbines in turn drive generators, which convert the turbines' mechanical energy into electricity. Transformers change the alternating current produced by the generators into a very high-voltage current that is suitable for long-distance transmission."[13] In fact, water-based power represents one of the three principal sources of energy in the modern world, along with nuclear power and the use of fossil fuels.

As suggested above, technology, while sometimes a solution to environmental problems, tends also to cause them; the replacement of one set of technologies with another almost always has unintended environmental consequences. It is important to recognize this and to avoid seeing one set of technologies as a panacea, especially when they involve complex interactions with the environment. In the next few sections, we will discuss both the environmental risk and the promise of nanotechnology. This discussion

---

[10]   Plato, *Critias*, The Internet Classics Archive. Available at http://classics.mit.edu/Plato/critias.html (accessed August 1, 2008).
[11]   "Waterwheel," *Encyclopædia Britannica*, Standard Edition (Digital) (Chicago: Encyclopædia Britannica, 2008).
[12]   In fact, in the thirteenth century, the people of Toulouse, France formed the Bazacle Milling Company in which shares of the mills on the Garonne River were traded at fluctuating values, depending on the economic conditions and the production of the mills – like a modern stock exchange. This entity survived until it was nationalized in 1946.
[13]   "Hydroelectric power," *Encyclopædia Britannica*.

is not meant to be exhaustive, but is rather intended to introduce some key ways in which nanotechnology will interact with the environment.

8.2 Environmental Risks of Nanotechnology

Every new technology brings with it new challenges and new impacts on the environment. Potential problems with nanotechnology are numerous. The small size of nanoscale materials and devices increases the potential for exposure to the materials, makes clean-up difficult, and exacerbates the problem of toxic materials crossing the cell membrane. In a 2006 review article, Nel et al. outlined some specific mechanisms through which nanoscale materials interact with cellular tissues. For example, because of the small size of nanoscale particles, they will have incomplete crystal planes on their surface. This could disrupt the electronic configuration of the tissue, creating surface groups that could react in odd ways with the environment. Also, nanomaterials are often functionalized with chemical groups that make them catalytically active or make them hydrophobic (repelled by water) or hydrophilic (attracted to water). These chemistry changes can cause the nanomaterial to alter the surface of the cell.[14] Not limited to causing problems by themselves, nanoparticles can act as carriers for contaminants such as metallic particles. These metallic particles can enhance chemical changes in the cell, acting as catalyst particles.

The comparatively large surface-to-volume ratio of nanomaterials increases their chemical reactivity and this could have numerous unanticipated consequences. Furthermore, we have virtually no knowledge base of the lifecycle of structures and devices developed with nanomaterials and a precautionary approach – see especially §5.3 – seems to require that we step lightly into mass use of materials that we know little about.

Because, historically, materials have had such a high profile impact on the environment or health, the study of the environmental/health impacts associated with nanotechnology and nanoscale materials is attracting a large degree of interest. The toxicity of nanoscale materials is also important to their disposal at the end of life. Toxicity of nanomaterials is being heavily researched to ensure that nanotechnology products will not cause undue harm to the environment:

> The unique features of manufactured nanomaterials and a lack of experience with these materials hinder the risk evaluation that is needed to inform decisions about pollution prevention, environmental clean-up and other control measures, including regulation. Beyond the usual concerns for most toxic materials, such as physical and chemical properties, uptake, distribution, absorption,

---

[14] Andre Nel et al., "Toxic Potential of Materials at the Nanolevel," *Science* 311.5761 (2006): 622–7.

and interactions with organs, the immune system and the environment, the adequacy of current toxicity tests for chemicals needs to be assessed to develop an effective approach for evaluating the toxicity of nanomaterials.[15]

Even though we are not entirely sure what characteristics of nanomaterials will generally be important for toxicity, we can identify characteristics that are expected to have a significant impact thereon. These include the particle size, structure/properties, coating, and particle behavior under certain situations.[16]

A lifecycle approach to nanotechnology covers understanding nano-materials from their production through use and disposal of the product. Lifecycle approaches to managing technology attempt to understand the impact that the technology will have and how to manage the challenges of sustainability and anticipating future environmental issues. Thinking about the end of a product at the time that it is created or conceived could lead to a process that is more efficient and less harmful to the environment – allowing for easy disposal and/or biodegradability of the material. The lifecycle approach can also be applied to risk assessment. The fundamental properties concerning the environmental fate of nanomaterials are not well understood, since there have not been many studies. Important issues are biodegradation, possible interactions between nanomaterials and other contaminants, and the applicability of current environmental models to nanomaterials. Understanding the unique waste management and handling issues of nanoscale materials will help us understand what new (if any) regulatory measures are necessary to manage the disposal and handling of nanomaterials.

This is more complicated than it might seem at first. Nanoparticles will most likely be an important part of most consumer products, and regulation over their disposal will prove difficult. Nanomaterials are already in use in products such as stain-resistant pants, sunscreen, tennis equipment, electronic devices, paints, cosmetics, and ink.[17] Nanomaterials do not act the same in environments as larger materials do. For example, airborne nanomaterials remain airborne longer and resist falling to the ground more than larger materials.[18] This is because small particles such as nanoparticles

[15] "Nanotechnology Toxicity," US Environmental Protection Agency: National Center for Environmental Research. Available at http://es.epa.gov/ncer/nano/research/nano_tox.html (accessed August 14, 2008).
[16] Joyce S. Tsuji, "Research Strategies for Safety Evaluation of Nanomaterials, Part IV: Risk Assessment of Nanoparticles," *Toxicological Sciences* 88.1 (2006): 12–17.
[17] National Nanotechnology Institute, "Nanotech Facts." Available at http://www.nano.gov/html/facts/The_scale_of_things.html (accessed August 15, 2008).
[18] R.J. Aitken, K.S. Creely, and C.L. Tran, *Nanoparticles: An Occupational Hygiene Review.* Report prepared by the Institute of Occupational Medicine for the Health and Safety Executive. 2004. Available at http://www.hse.gov.uk/research/rrhtm/rr274.htm (accessed September 23, 2008).

follow the laws of gaseous diffusion in air, meaning that the rate of diffusion increases as the particle diameter decreases. Interestingly, the smallest nanoparticles agglomerate with each other in the air, forming larger particles and then precipitating out of the air. Larger nanoparticles can remain suspended in air for longer times, from days to weeks given certain environmental conditions. Inhalation of these particles is therefore of particular interest. Carbon nanotubes in particular have been of interest because they have shown some similarities to asbestos. Nanotubes seem to have a similar shape to asbestos fiber and some studies have suggested that carbon nanotubes may remain in the lungs of laboratory animals, sticking to tissue walls after inhalation. Exposing the lining of the chest cavity to long multiwalled carbon nanotubes has resulted in inflammation and the formation of lesions.[19] However, while nanoparticles remain airborne longer than larger particles – and thus are more susceptible to inhalation – nanoparticles that have been deposited onto a surface tend to remain adhered to the surface, therefore posing less risk.

A similar dual story can be told with nanomaterials in soil. On the one hand, nanoparticles could be rendered immobile in the soil by being adsorbed onto soil particles. However, nanoparticles are small enough to fit between soil particles and might diffuse farther through soil. A system is needed in order to determine the various toxicity levels of different nanoparticles. Biodegradability of nanoparticles is another aspect that needs to be specifically considered. Many nanoparticles are made out of inorganic materials that are inherently non-biodegradable. Other studies have shown that carbon buckyballs can be metabolized.[20] Further, in some cases, biodegradability is vital to the material's purpose, such as nanomaterials that are used in drug transport.[21] As with bulk materials, different nanoscale materials will have different environmental responses. Nanoparticles might also have different environmental responses than their chemically equivalent bulk counterparts.

8.3 Nanotechnology Solutions to Environmental Problems

From an engineering perspective, there are two general approaches to solving environmental problems. One is using technology to fix problems that already exist, and the other is to use technology to prevent anthropogenic

[19] Craig A. Poland et al., "Carbon Nanotubes Introduced into the Abdominal Cavity of Mice Show Asbestos-Like Pathogenicity in a Pilot Study," *Nature Nanotechnology* 3 (2008): 423–8.
[20] T.R. Filley et al., "Investigations of Fungal Mediated (C60–C70) Fullerene Decomposition," *Preprints of Extended Abstracts Presented at ACS* 45 (2005): 446–50.
[21] T. Madan et al., "Biodegradable Nanoparticles as a Sustained Release System for the Antigens/Allergens of Aspergillus Fumigatus: Preparation and Characterization," *International Journal of Pharmaceutics* 159 (2005): 135–47.

injuries to the environment. In general, nanotechnology's benefits to the environment fall into both of these two categories. In the first, nanotechnology can be used for environmental remediation. This could be accomplished using nanomaterials of oxidants, reductants, and nutrients that show promise for transforming and encouraging the growth of microbes. Again, the greater chemical reactivity can help. For example, the *in situ* remediation of chlorinated solvents (like trichloroethylene) usually produces undesirable byproducts. Nanoscale bimetallic particles have been shown to eliminate all of these.[22] In the second approach – the prevention of anthropogenic injuries to the environment – nanotechnology can develop new materials that can replace toxic or rare materials. It has the potential to reduce the total volume of material used in products, and also to reduce the energy use of products. The NNI has identified three primary research and development applications for nanotechnology in the environment: measurement in the environment, sustainable materials and resources, and sustainable processes. It is also possible to imagine clean-up and other applications that nanotechnology can provide.

When helping solve environmental problems, technology can be applied in a variety of ways. The first step to solving a particular problem is to measure it. Thus, sensing technology is paramount for environmental technologies. Because of the high surface-area-to-volume ratio of nanoscale materials, they represent an excellent resource for developing new sensing systems. In general, what are needed are real-time biological and chemical sensors that can perform detection on a continuous basis and with high sensitivity and selectivity. 'Sensitivity' refers to a particular sensor's ability to respond to very small amounts of the element that is being detected. 'Selectivity' refers to a sensor's ability to distinguish between one species of chemical and another. Nanoscale materials promise to provide both high sensitivity and high selectivity. In fact, direct electrical sensing of biological and chemical species using one-dimensional nanostructures such as silicon nanowires and carbon nanotubes has been demonstrated several times.[23]

Many of the techniques for realizing one-dimensional nanostructures as sensing devices utilize the functionalization of the surface of the nanostructures

---

[22]　Daniel Elliott and Wei-Xian Zhang, "Field Assessment of Nanoscale Bimetallic Particles for Groundwater Treatment," *Environmental Science Technology* 35.24 (2002): 4922–6.

[23]　See the following: Jacob N. Wohlstadler et al., "Carbon Nanotube-based Biosensor," *Advanced Materials* 15 (2003): 1184–7; Jing Li et al., "Carbon Nanotube Sensors for Gas and Organic Vapor Detection," *Nano Letters* 3 (2003): 929–33; Qui Pengfei et al., "Toward Large Arrays of Multiplex Functionalized Carbon Nanotube Sensors for Highly Sensitive and Selective Molecular Detection," *Nano Letters* 3 (2003): 347–51; Moonsub Shim et al., "Functionalization of Carbon Nanotubes for Biocompatibility and Biomolecular Recognition," *Nano Letters* 2 (2002): 285–8; Yi Cui et al., "Nanowire Nanosensors for Highly Sensitive and Selective Detection of Biological and Chemical Species," *Science* 293 (2001): 1289–92.

with some other chemical species. This principle is often shown in the lab with biotin and streptavidin.[24] Biotin can be chemically attached to proteins, macromolecules, and nanoscale materials through a process caused biotinylation. Streptavidin preferentially binds to biotin. This binding will typically change the electrical properties of the surface of the nanoscale material (and, thus, the entire nanomaterial) in a distinctive manner. Monitoring this readout could enable researchers to detect a very small number of binding events. The same principle can apply with other chemical species. Note, of course, that chemical (environmental and biological) responses are often caused not by a single molecule (or species of molecule) but by combinations of molecules. A collection of nanowires could each have its surface functionalized to detect a different species and provide a differing electrical signal. The collective effects of these materials are being developed for different types of chemical sensing, producing, in effect, an "artificial nose." This more advanced sensing technology will help provide more accurate measurements of chemicals in the atmosphere. Sensing devices based on these nanomaterials are small and unobtrusive.

A single sensor (or set of sensors) can work in highly localized situations, such as to monitor the environment of an individual room. In addition, such sensors are small enough that they could unobtrusively be placed in such a way as to monitor the air quality or environmental pollutants of a larger swath of area, such as a city. Already, nanomaterials have shown the potential to play a critical role in detection, including the detection of pathogens with piezoelectric cantilevers,[25] the transport of electrical and optical signals through designed nanostructures,[26] and molecular detection using functionalized nanoparticles.[27]

Technological development typically happens in developed countries like the US where the economy can afford the upfront cost of research and development of new materials. Nanotechnology itself often requires an enormous amount of capital upfront for the basic development of materials and devices. For instance, just for laboratory characterization of nanomaterials, highly specialized (and expensive) microscopes are required – like atomic force microscopes, electron microscopes, and diffractive characterization techniques.

[24] Biotin is an acid that is widely available in nature, especially abundant in egg yolk, beef liver, and yeast. It helps in the formation of fats and carbohydrates. Stretavidin is a type of avidin. Avidin is a specific protein that is readily found in egg whites and combines with biotin. Because of their ready availability and the preferential binding that they show, they are often used in laboratory settings.

[25] Jeong W. Yi et al., "*In Situ* Cell Detection Using Piezoelectric Lead Zirconate Titanate Stainless-Steel Cantilevers," *Journal of Applied Physics* 93.1 (2003): 619–25.

[26] M. Quinten et al., "Electromagnetic Energy Transport via Linear Chains of Silver Nanoparticles," *Optics Letters* 23.17 (1998): 1331–3.

[27] K.K. Caswell et al., "Preferential End-to-End Assembly of Gold Nanorods by Biotin–Streptavidin Connectors," *Journal of the American Chemical Society* 125.46 (2003): 13914–15.

However, many environmental problems disproportionately affect developing countries.

Global warming, potential rising of the ocean waters, air pollution, lack of access to clean water, spread of disease because of poor environmental conditions, and access to and responsible use of natural resources are all issues that are particularly acute in the developing world. This is because of the population growth, the lack of technologies and resources to alleviate the strain caused by these problems, and the lack of good governance and mature economies to work responsibly and proactively to solve environmental problems.

Indeed, nanotechnologies that could have an impact in these areas are being developed, even if the immediate eye is not toward use in developing world applications. An example of nanomaterials being developed for environmental applications is self-assembled monolayers on mesoporous supports (SAMMS).[28] Detecting and removing metals and chemicals from water, the atmosphere, the ground, or other areas is important to helping clean up anthropogenic injuries to the environment. SAMMS use nanoporous silicas for scaffolding. The pores in the SAMMS provide extremely high surface areas for very low volumes. These pores can be functionalized with thiols and be designed to capture mercury, lead, arsenic, and radionuclides. Specific functional groups can be designed to target specific metals and species. It bears notice that:

> The SAMMS technology was originally developed for the US Department of Energy, which has identified the separation and removal of mercury from the environment as a key technology need at its Oak Ridge Site. Mercury releases are also a significant environmental concern for the coal and petroleum industries. Mercury contamination of water produced from petroleum production is a significant environmental concern. Globally, the regulations addressing mercury discharge in produced water are getting more and more stringent. In addition, mercury emissions from coal-fired power plants are a significant concern worldwide, with hundreds of tons of mercury being released each year into the air we breathe. In the mining industry, there are problems with mercury (and other heavy metals like [cadmium] and [lead]) contaminating impoundments, run-off and streams. SAMMS can help to control all of these heavy metal releases. SAMMS have proven effective at removing [mercury] from contaminated oils, and is the only sorbent technology to have proven this capability. SAMMS has even been shown to be effective for removing mercury from contaminated chemical warfare agents (e.g. mustards).[29]

[28] Wassana Yantasee, "Nanostructured Electrochemical Sensors Based on Functionalized Nanoporous Silica for Voltammetric Analysis of Lead, Mercury and Copper," *Journal of Nanoscience and Nanotechnology* 5.9 (2005): 1537–40.

[29] "Description of the Technology: SAMMS Assembly," US Department of Energy: Pacific Northwest National Laboratory. Available at http://samms.pnl.gov/tech_descrip.stm (accessed August 20, 2008).

Technologies for sensing and for the removal of dangerous materials all utilize the increased surface-to-volume ratio that nanomaterials provide. As discussed in Chapter 3, this is one of the most important features that distinguish nanomaterials from their bulk counterparts. Their surface effects cause them to have unique properties that allow them to be used in these new types of applications. Their size allows for sensitivity, and their ability to functionalize their surface allows them to have high selectivity. It is not hard, given this example, to imagine using nanotechnology for the benefit of the developing world. The challenge, then, is one of encouraging the market of developed economies to produce applications that will be used primarily in the developing, impoverished world, aimed at alleviating local environmental problems there, and that will likely not reap very large economic returns. Again, see Chapter 7 for more discussion in this regard.

We discussed earlier the extensive use of environmental resources to power technology. Energy production and use are remarkably inefficient today, and not just because we leave lights on when we exit rooms. Basically, we create and convert energy at present through heat, both in Carnot engines[30] and in thermal processes that manipulate matter and create phase changes (e.g., as with a steam engine). A good example of energy efficiency is in lighting systems. A typical candle converts the combustive heat of a flame into light. As anyone who has ever gotten close to a flame knows, it lets off heat. In fact, if measured by light emitted versus energy expended, the overall efficiency of a candle is less than 1/20 of 1 percent. This measurement takes into consideration the wavelength of light emitted, counting only the light that is emitted in the visible spectrum. Much of this light also gets lost due to absorption.

Incandescent light bulbs are the most commonly used way at present to turn electricity into light. Here, too, energy efficiency is incredibly low: a 100 watt bulb with a standard 120 volt power source has an efficiency of about 2.5 percent. More recently, compact fluorescent light bulbs have gained popularity as being more efficient. Fluorescent light bulbs rely not on heating, but on inelastic collisions of electrons within the gas inside the bulb. Electrons are sent into the gas and, as an electron collides with an atom in the gas, it transfers energy into the atom's electrons. An electron that gains energy can temporarily "jump" to a higher atomic energy level. When the electron returns to its previous energy level it gives up the extra energy in the form of light. In fluorescent bulbs, this light is first emitted in an ultraviolet form. It is then converted to visible light with the aid

---

[30] Carnot engines are theoretical concepts that represent the maximum efficiency at which heat transfer can operate. A heat engine basically works by moving energy from a warm area to a cold area, and as it does so it converts some of that energy to a different form of energy such as electricity, light, or mechanical work. This can also be reversed and electricity can be used to move energy from a cold region to a warm region, such as with a refrigerator.

of a fluorescent coating. The efficiency of compact fluorescent lights can approach about 9 percent.

Light emitting diodes (LEDs) are often touted as an even better lighting solution. LEDs consist of semiconductor materials in which one side has an excess of electrons and the other side has an excess of holes.[31] When a voltage is applied across the material, these electrons and holes are forced together in the middle. Here they combine, and this combination releases energy by emitting light. The efficiency of LEDs has so far been shown to achieve values as high as 22 percent, though this number is increasing. Theoretically, this efficiency could approach 50 percent.[32]

Do we have an ethical charge to be as efficient as we reasonably can be in how we use energy? Certainly, there is an economic argument to be more efficient with energy. Energy use costs money, and using less of it to achieve the same task – let us ignore upfront costs – means that less money is spent. Therefore, more capital is available for other development and uses. The economic argument has within it an ethical component. From an equity perspective, it should be noticed that when goods, processes, and/or materials are more expensive, it is the least well off that go without them first. As energy prices increase, it is the least well off that most immediately have to give up other activities in order to pay for them. To the extent that energy use is a necessity – and in the modern industrialized world it is generally considered to be so – then decreasing the hardship of the less well off is certainly an ethical good. In this way, increasing efficiency is an undeniable ethical good. Furthermore, energy sources originate from the environment, and using energy for technological purposes takes away resources from the environment and makes them unusable to future generations. In order to fulfill our role as stewards of the environment, we are required to use non-renewable resources as efficiently as possible, both in order for them to be available to future generations[33] and in order to preserve the planet in a responsible manner. It could also be argued that those of us most able to become more efficient in our energy use have a responsibility to do so. This is especially true when this group intersects with the group of societies that are the biggest users of finite resources. Typically, this calculation is done by looking at a society's ecological footprint,[34] which is an attempt to measure the human demand of a population on the ecosystem of the Earth. Such a calculation allows us to investigate particular societies and to see the relative impact that they have on the larger population as a

---

[31] Holes are technically the absence of electrons in a place that would normally have them. To a rough approximation, holes act as positive electrons.

[32] Nadarajah Narendran, "Improved Performance White LED," *Fifth International Conference on Solid State Lighting, Proceedings of SPIE* 5941 (2005): 45–50.

[33] Earth is a finite planet with finite resources and decisions that are made about the use of those resources today have an impact on the availability of the resources in the future.

[34] William E. Rees, "Ecological Footprints and Appropriated Carrying Capacity: What Urban Economics Leaves Out," *Environment and Urbanization* 2 (1992): 121–30.

whole. Determining which societies have the strongest ethical imperative to embrace more energy-efficient technologies can be looked at as a combination of determining which societies are the worst offenders and which are the most capable of changing their energy use. It is not surprising that developed countries are the highest offenders; however, because of their wealth, these countries are also among those most able to change. Combined with an ethical charge to assist the developing world (and with this assistance, to help alleviate the problems associated with poverty), it is an ethical mandate to use as few finite resources as possible, in order to allow these resources to be used by others.

Nanotechnology has the ability to make energy use more efficient. Because nanotechnology can develop materials that are stronger, conduct energy, electricity, and heat better, and have more efficient processes, less energy is needed to achieve the same task with nanomaterials-based products. Nanotechnology also can enable the use of more efficient energy technology in a cost-effective manner. Because of the generally capitalist nature of modern economies, energy-efficient technologies also need to be cost-effective in order for them to be adopted. This requires that the efficiency be worth the upfront cost of the changing technologies. For example, solar cells represent a way in which a renewable energy source like the sun can be converted into energy. The concept is not new: plants routinely convert sunlight into energy through photosynthesis, and modern solar cells have been around since the mid 1950s when researchers at Bell Laboratories discovered that silicon doped with different species were electrically sensitive to light. Nanomaterials and nanostructured technologies have enabled solar cells to increase their efficiencies to the point that they are now in the research stage of being economically feasible (for certain conditions).[35] Table 8.1 – which is based on consumption from 2004 –

Table 8.1 Nanotechnology applications and energy savings

Nanotechnology application	*Estimated % reduction in annual US energy consumption*
Strong, lightweight materials in transportation	6.2
Solid state lighting (such as LEDs)	3.5
Smart sensors	2.1
Smart roofs	1.2
Energy-efficient separation membranes	0.8
Molecular control of industrial catalysis	0.2
Improved electrical conductance of transmission lines	0.2
Total	14.2

[35] Richard D. Schaller et al., "Seven Excitons at a Cost of One: Redefining the Limits for Conversion Efficiency of Photons into Charge Carriers," *Nano Letters* 6.3 (2006): 424–9.

illustrates estimated reductions in total US annual energy consumption with certain nanotechnology applications.[36]

8.4   Overall Assessments: Risk and Precaution

When it comes to the environmental impact of nanotechnology, there are potential benefits and potential injuries. In §5.1, we laid out four types of decision making with regards to knowledge of outcomes and the probabilities of the outcomes. To recap, the situations were: decision making with full knowledge of outcomes and probabilities; decision making with full knowledge of outcomes and some, though not all, knowledge of probabilities; decision making with full knowledge of outcomes and no knowledge of probabilities; and decision making with incomplete knowledge of outcomes (as well as their associated probabilities). With environmental impact of nanotechnology, we are almost certainly not making decisions in the first situation. The environmental system in which we live is complex and non-linear, and we know neither all the outcomes that can occur from introducing a new technology into the system, nor all the probabilities of the outcomes that we do know. At best, we are in the fourth situation: making decisions with incomplete knowledge of outcomes and their probabilities.

Conceptualizing the impact of nanotechnology on the environment in this manner helps to inform the issue in several ways. Most will admit that we can never know the entire impact that a novel technology will have on the environment. There will always be unforeseen consequences. This is especially true when discussing the environment, as many of the consequences occur in a time span that is too long to be reasonably measured in a laboratory environment before use of the new technology. Knowing that there are unforeseen consequences when acting in complex systems does not preclude us from acting on those systems in ways that have a reasonable possibility of assisting society. In fact, technology use has long been an iterative process where failure and injury in one generation of technology have informed the next generation. One needs to look no further than the technology of building structures, to see this technological evolution in practice. The history of designing bridges and buildings is a history of failures, which inform new designs and new engineering principles.

Furthermore, a balance between what assists one generation but may potentially harm a future generation must be drawn. Suppose we know of a currently unpreventable harmful consequence of using a technology but we also know that this harmful consequence will not realize itself for many generations. Suppose we also know that using this technology will greatly

---

[36] Marilyn A. Brown, "Nano-Bio-Info Pathways to Extreme Efficiency." Presented at the AAAS Annual Meeting, Washington, DC, 2005.

advance the current state of society in some way. It does not seem unreasonable to consider leaving the solving of this problem to future generations with future technologies.

Diligence is required to understand and measure the impact that technologies have on the environment and, as discussed in §5.3, to adopt reasonable precaution against new technologies. That is, if we are aware of certain damages to the environment by the use of nanotechnological products, then we should act as best we can to prevent them, even if we lack a direct scientific correlation between cause and effect. Preventing these damages may consist of disallowing the use of the offending technology, but it may also consist of setting up a layer of technology on top of the offending layer to correct it.

Our human history of using technology and materials in the environment gives us an initial wealth of knowledge concerning the attributes of technology to look for when evaluating a nanotechnology's potential impact on the environment. We know to be concerned about toxicity, and we know some of the properties of technologies that impact this (e.g., particle size, structure and properties, coating, and particle behavior). Furthermore, we are concerned about emissions into the atmosphere and monitoring for certain emissive species that are harmful to the environment (e.g., greenhouse gases). It is also prudent to encourage research into the impact of novel technologies in novel ways, such as potential containment issues of harmful nanomaterials, the dispersal of nanoparticles into the bloodstream of organisms, and the responsible disposal of nanomaterials.

Where does this leave us? Our goal has not been to provide an exhaustive list of the potential environmental benefits and harms that nanotechnology will bring. Anyone pretending to offer such a list demonstrates an extreme ignorance of the history of technology and the environment. Rather, our goal has been to show some ways in which technology has interacted with the environment, to show that the general discussion of technology and the environment is one with a long history, to give a general idea of the new questions that nanotechnology development adds to this discussion, and to offer some possible frameworks for assessing and proceeding with developing new technologies.

9

Military

9.1 The Military and Technology

In looking at how paradigmatic shifts in technology such as the development of nanotechnology have the ability to change society in various ways, it is imperative to look at how it changes the military. Advancing technologies create new weapons and new defenses. New technologies also create new targets for attack – such as water wheels in the Middle Ages and power stations in the modern era – and new methods of defending those targets. New technologies can render old systems obsolete and be the impetus for the development of new systems. The development of satellite surveillance forced many nations to develop underground bunkers and hiding places, and even underground nuclear testing areas. New military technologies can even change the entire nature of the world system and the rules by which states and other actors on the world stage act. There is no doubt that the development of nuclear weapons has fundamentally changed the nature of international politics and warfare, for better and for worse.

The introduction of new technologies to military and defense applications, and how they alter the battlefield, international relations, and society, has a long and rich tradition. A medieval example of this is the longbow. Developed in Wales, it introduced a new form of artillery fire and contributed significantly to the removal of the armored knight from the battlefield.[1] Furthermore, the introduction of the longbow made the cavalry less important. This, in turn, assisted in the decline of feudalism, as professional, non-noble archers became far more important in battles.

It is hardly debated that warfare has done much to shape the modern world. This is due in no small part to the technologies that have been

[1] "English longbow," *Encyclopædia Britannica* (2008). Available at http://www.britannica.com/EBchecked/topic/188247/English-longbow (accessed August 19, 2008).

developed to create more efficient, more effective, more valuable weapons and defense systems. In his book *War Made New*, Max Boot divides the past five centuries into four revolutions in military technology.[2] The first was the Gunpowder Age: gunpowder was brought from China and used in European artillery and by the cavalry, which led to the rise of the Western world. Next, the Industrial Revolution brought on rapid mechanization of fighting techniques with the Gatling gun and the steam engine. First used in the US Civil War, the Gatling gun was employed most effectively in wars by colonial powers against non-developed warriors, such as the Zulu in Africa. Third, the Second Industrial Revolution resulted in combat aviation and aerial bombing. The book finishes with the Information Revolution, which has brought about unprecedented battlefield communication, precision guided missiles, and smart bombs.

Each of these revolutions has also effected significant changes in society and in international affairs. The use of gunpowder, along with the longbow and other artillery technology, helped bring about the downfall of the feudal system in Europe. It allowed for European military victories in and colonization of much of Africa and the Americas, along with imperial control of large swaths of the rest of the world. The Industrial Revolution – with its rapid mechanization, the assembly line, and increases in productivity – led to increased urbanization, child labor, the rise of organized labor, and railways, which enabled speedy travel across large areas. Boot's Second Industrial Revolution brought with it commercial air travel. The Information Revolution has generated an unprecedented connectedness in the world through Internet and electronic communications, satellites, and computing technologies. Most of the technologies were explicitly developed for military use. The Internet, so central in everyday interactions, was first developed through the United States Department of Defense Advanced Research Projects Agency (DARPA, originally ARPA). The Department of Defense had several needs for advanced communication between computers, including how to maintain command and control over Air Force missiles and planes after a nuclear attack. The Department needed a decentralized network so that if any locations in the US were attacked, it would not lose control. Another need involved the vast scope of ARPA-sponsored research. A communication network that allowed researchers in various locations to use ARPA computers was necessary in order to share results quickly.

> For each of these three terminals, I had three different sets of user commands. So if I was talking online with someone at S.D.C. and I wanted to talk to someone I knew at Berkeley or M.I.T. about this, I had to get up from the S.D.C. terminal, go over and log into the other terminal and get in touch

---

[2] Max Boot, *War Made New: Technology, Warfare, and the Course of History: 1500 to Today* (New York: Gotham Books, 2006).

with them. I said, oh, man, it's obvious what to do: If you have these three terminals, there ought to be one terminal that goes anywhere you want to go. That idea is the ARPANET.[3]

In 1968, the bid for ARPANET was contracted out and the development of the Internet was under way.

In World War II, nuclear technology made its first appearance in the form of atomic weapons. Since then, nuclear weapons and nuclear power have had a powerful impact on society and politics, both in international relations and in society in general. In international relations, the specter of all-out nuclear warfare led to tense relations between the major nations of the world. However, it also is probably what kept the Cold War cold, instead of the situation resulting in a third world war. The policy involved was known as "mutually assured destruction." Because both sides (the Soviet Union and the US) had the counterstrike capability to destroy each other, neither was willing to launch a first attack. Indeed, military technology has had a significant role in shaping the modern world. With military nano-technology, this relationship is likely to continue.

In Chapter 3, we discussed the importance of material in nanotechnologies. For the development of military-related nanotechnology, the associative materials will be paramount. Materials have always been the enabling tech-nology of any era. As a society, we recognize the importance of this by naming eras in history after the prevalent material of the time: the Stone Age, the Bronze Age, and so on. Even more recent periods, such as the Industrial Age and the Information Age, have been enabled and are chiefly identifiable by the materials of the time – steel and silicon/semiconductors, respectively. However, because in nanotechnology the structure and device are wedded to the material used in such a strong way, nanomaterials develop-ment becomes central to all nanotechnology development.

In military applications, the parallels are also true. The Stone Age had stone-based tools, but also stone-based weapons. The Bronze Age enabled durable swords that could not have been constructed with brittle stone. More recently, steel enabled skyscrapers, but also tanks and other heavily armored vehicles. Silicon technology has facilitated information warfare and the introduction of integrated systems into weapons, including GPS and laser guided missiles. In this chapter, we will see the significant role that materials play in allowing for the development of new weapons and defenses, new military technologies.

However, military technology can also be thought of using various conceptual distinctions. First, we can think about the use of the nano-technology, which could be either offensive or defensive: it could be used,

---

[3] John Markoff, "An Internet Pioneer Ponders the Next Revolution," *The New York Times on the Web* December 20, 1999. Available at http://partners.nytimes.com/library/tech/99/12/biztech/articles/122099outlook-bobb.html (accessed September 1, 2008).

for example, in either weapons systems or defensive systems. Second, we could think about the scope of the technology, whether it directly services individual soldiers or something else, such as transportation, structures, and so on. Third, we can think about whether the applications of nano-technology can be confined to military uses or whether they will have some broader societal implementation, even if first catalyzed by military uses. In examining nanotechnology and the military, it is necessary to consider these dimensions, the interplay between them, and what impacts they might have. To begin, it is useful to discuss what specific types of technology are possible and/or are already being developed within each field.

9.2   A Nano-Enabled Military

In 2002, the US Army established an interdepartmental research center at the Massachusetts Institute of Technology called the Institute for Soldier Nanotechnologies (ISN). ISN was charged with developing ways to sub-stantially improve the survival and performance of US soldiers, by using nanotechnology. This technology would have a point of use in individual soldiers, by which we mean that the technology acts on, affects, or is implanted by an individual. Much of the activation of the technology occurs automatically, without any conscious input from the soldier at all. This can be accomplished with environmental, biochemical, and other sensors that trigger a response from the technology. So what are some aspects of a nano-enabled military? By looking at some of the technologies that are being discussed and those already in development, we can achieve a good idea of what a nano-enabled military looks like.

By simply making devices smaller and lighter, the soldier can be made more mobile and have a smaller logistical footprint. The average present-day soldier carries in excess of 100 pounds of equipment while on assignment. Much of this weight is due to the electronic equipment (including com-munication equipment) and the power supplies (usually batteries) used to power them.[4] Through the use of smaller, lighter equipment, this weight could be reduced dramatically without a sacrifice in functionality; the soldier could therefore move more quickly and/or further (in the same amount of time). Much of this weight reduction can be accomplished by reducing the scale of the power generators that the soldiers have to carry. For example, recently nanoscale power generation has been demonstrated by utilizing an array of piezoelectric nanowires. By converting mechanical, vibrational, or hydraulic energy into electricity, these "nanogenerators" can be used to power the electrical systems carried by the soldiers.

---

[4]   This electronic equipment, of course, is one of the principal reasons that present-day soldiers carry more than their historical counterparts. The Roman soldier, for example, is commonly claimed to have carried 45 pounds, which is less than half the contemporary soldier's load.

Other nanosystems being developed for soldiers utilize some of the novel properties that nanomaterials have when interacting with their environment. An example of this is mechanically active materials and devices. These nanomaterials are capable of dynamically changing their stiffness and mechanical actuation. Embedded in a soldier's battle suit, mechanical actuators can allow a transformation from a flexible material to a stiff armor nearly instantaneously. This can distribute an energetic impact, such as a nearby explosion. Further, such materials can be transformed into a cast that stabilizes broken bones. Contracting materials in the battle suit can apply direct pressure to a wound, serve as a tourniquet, or even perform cardiopulmonary resuscitation (CPR) on the soldier. All of these examples can, in theory, be activated by chemical and/or mechanical sensors in the suit without direct conscious input from the soldier, since electronic polymers can be used to create sensitive detectors of explosives, nerve gas, and even specific biological agents.

Many of these technologies are demonstrably transferable and useful to the non-military population of society; that is, they are dual-use technologies. Most law enforcement and emergency response teams would be able to use much of this same technology to protect police officers and first responders. Reducing the load weight of firefighters would allow them to move more quickly and agilely in rescue operations. Seemingly, especially from the description above, soldier nanotechnologies are defensive in nature and aimed at saving and protecting the soldier's life. It would be naive to imagine that nanotechnology is not being explored for use in weapons, however.

Nanoweapons have the potential to change the nature of warfare in very fundamental ways. Just as a modern, technologically enhanced military bears very little resemblance to undeveloped militaries, so too a military with nanoweapons will be very different from a conventional military. Nanoscale materials have the potential to weaponize intense laser technologies. Self-guided anti-personnel bullets are also plausible, as the ability to miniaturize the technology that guides missiles becomes likely with nanotechnology. Targeted strikes on buildings can become even more precise. The possibility of unleashing a "swarm" of nanoscale robots programmed only to disrupt the electrical and chemical systems in a building is a far more militarily desirable solution than bombing the entire building.

A specific example of nanoscale materials showing their impact in weaponry is with nanoaluminum. Bulk-scale aluminum contains aluminum atoms that cover roughly one-tenth of 1 percent of the surface area. Nanostructured aluminum contains aluminum atoms that cover roughly 50 percent. More atoms on the surface create more sites for chemical reactions to occur. This is used in conjunction with metal oxides such as iron oxide to create superthermites, which increase the chemical reaction time by three orders of magnitude. Therefore, greater amounts of energy can be released, creating more powerful conventional explosives and faster moving missiles and torpedoes (so fast, in fact, that they can bypass evasive actions).

In another application, nanotechnology may assist in making chemical and biological agents more difficult to detect. Some of the difficulties of chemical and biological warfare involve the effective delivery of the toxic agent. This could be overcome by utilizing what will be one of nanotechnology's initial big impacts on medical technology. As will be discussed in §11.3, current drug treatments are not very selective in their choice of targets: they go after the entire body instead of only targeting a specific organ system. Nanoparticles can be developed that help the body absorb the drug at selective sites. This delivery vehicle of a functionalized nanoparticle could be augmented with nanoscale machines capable of entering the specific targeted cell and thereon releasing the drug agent. Of course, this works in delivering harmful agents as well to the body. Because of the targeting, much less of the chemical or biological agent is necessary for it to be effective. A nano-enhanced chemical such as cyanide could be synthesized in smaller, less detectable amounts in small labs. The current bans on chemical and biological weapons prohibit existing weapons, not future ones. Development of nanotechnology-enabled weapons can undermine the numerous treaties and international regimes that regulate chemical and biological weapons by their development and difficulty of detection.[5]

While we discuss human enhancement more generally in Chapter 12 – as well as some of the associative conceptual issues – military applications thereof bear notice, and we want to canvass some of them here. Battlefield conditions often require soldiers to perform tasks that are outside normal human activity and, as such, take tolls on soldiers that human bodies are not typically designed for. A good example of this is sleep deprivation. As reported in a March 2008 JASON[6] report on human enhancement, "The most immediate human performance factor in military effectiveness is degradation of performance under stressful conditions, particularly sleep deprivation."[7] Sleep deprivation has a negative impact on alertness, physical ability, and the ability to reason. It is a factor that must be taken into account when planning battlefield activities. In battlefield situations, soldiers are usually sleep deprived for long periods. Consider:

[5] Andrew Oppenheimer, "Nanotechnology Paves Way for New Weapons," *Jane's* August 1, 2005. Available at http://www.janes.com/defence/news/jcbw/jcbw050801_1_n.shtml (accessed September 9, 2008).
[6] The JASON group is a truly remarkable group of scientists that perform wide-spanning research into potential technological developments for the Department of Defense. Though not as important now as they once were (most branches and agencies in the military have their own internal experts on emerging scientific issues and technologies), they are an example of scientists applying their knowledge and expertise in a civic forum. Their history is masterfully told in Ann Finkbeiner's *The Jasons: The Secret History of Science's Postwar Elite* (New York: Penguin Group, 2006).
[7] E. Williams et al., "Human Performance," *JASON: The MITRE Corporation* (March 2008). Available at http://www.fas.org/irp/agency/dod/jason/human.pdf (accessed September 4, 2008).

[Harris] Lieberman and coworkers studied soldiers in US Army elite units during a combat simulation field exercise. Wrist activity monitors showed that the soldiers slept about 3 hours per night over a 53 hour period. Twenty four hours after their initial deployment they displayed significant decrements in their cognitive function, including vigilance, memory, reaction time and reasoning. The observed decrement in ability was several-fold worse than individuals whose blood alcohol levels are above the legal limit. Although the combat exercise resulted in multiple stresses in addition to sleep deprivation (e.g. dehydration), the predominant effect leading to the performance decrement was sleep deprivation.[8]

In fact, encouraging sleep deprivation among enemies has long been a tactic that armies will use against their enemy to promote exhaustion. Though we have little understanding of why sleep is necessary, we know that loss of sleep leads to loss of mental effectiveness, readiness, and awareness of surroundings. It is a given that sleep deprivation will accompany any sustained military operation. Therefore, anything that can be done to improve how soldiers respond to lack of sleep will largely influence their ability to perform.

Human enhancement with nanotechnology can have a major impact here. Even simple detection of fatigue, lapses in attention, and changes in neurological behavior would allow battlefield commanders to have better knowledge about the forces under their command. Nanoscale sensors can be used to detect this fatigue in individual soldiers. This can be done first by monitoring the brainwave patterns, but it can also be done more simply by closely monitoring, in real time, muscle response, eye movement, chemical levels in the body, and other triggers and suggestions of fatigue. Further, DARPA is investing in what it refers to as "metabolically dominant" war-fighters that will "be able to keep their cognitive abilities intact, while not sleeping for weeks. They will be able to endure constant, extreme exertion and take it in stride."[9] Nanotechnology provides access to the chemical methods to enhance soldiers with minimal side effects. Though these types of enhancement will undoubtedly make soldiers more dominant war-fighters, it is unclear what impact long-term (even permanent) enhancement of this type would have on their ability to be human and readapt to society once the fighting is over. As more soldiers survive battle and return home, issues of mental health care and readjustment to peaceful society become even more important. Further, adjusting back to a non-enhanced life – for example, going from being "metabolically dominant" to needing

[8] Ibid.; internal reference is John W. Castellani et al., "Cognition during Sustained Operations: Comparison of a Laboratory Simulation to Field Studies," *Aviation, Space and Environmental Medicine* 77 (2006): 929–35.

[9] Michael Goldblatt, "Office Overview." Presented at DARPATech 2002 Symposium. Available at http://www.darpa.mil/darpatech2002/presentations/dso_pdf/speeches/GOLDBLAT.pdf (accessed September 7, 2008).

sleep – could be traumatic for the soldier. The challenges of reintegration of war-fighters into society have always been great, and nanotechnology will hardly abrogate these challenges.

Further, direct human enhancement of the war-fighter represents a distinct change from the battle enhancements that humans have implemented throughout history. It represents a new way of preparing for battle. Thus far in history, and today, we have enhanced soldiers by giving them a better battlefield (or better knowledge of the battlefield). Satellites provide full vision of the situation, telecommunications provide information about other soldiers on the ground, tanks allow military movement into new areas, body armor provides new protection, and so on. In short, we have built a better environment for soldiers. Nanotechnology-enabled human enhancement of soldiers changes how the preparation for the battlefield is done. Instead of building a better environment for the soldiers, it builds better soldiers for the environment.

9.3   A Nano-Enabled Defense System

Defense and protection from undue harm are, ostensibly, primary functions of any government. Nanotechnology use in defense and protection spans a wide range of applications and applies to entire overall defense and protective systems, not just individuals; it affects societies as a whole. One aspect of defense is border protection. Border protection includes not only standing guard at borders and regulating which individuals may enter into a particular country, but also controlling ports and scanning cargo to make sure that no harmful chemical or biological species are exposed to the population. It involves quarantining possible contagious diseases, which means effectively testing for them. And it involves performing these tasks with minimal interruption to the commercial and personal interests of the citizenry.

Defense also includes protection from threats against the populace. This is not limited to invasions by foreign armies. Threats can be as simple as anthrax powder delivered through the mail or a radiologically enhanced "dirty bomb" exploded from inside a suitcase. These threats are not always foretold by a warning and they require diligence and constant monitoring. In an open and free society, it is also required that this monitoring occurs with minimal disturbance to the populace, minimal invasion into individual privacies, and minimal interference with liberties.

Nanotechnology can provide some of this capability with environmental, chemical, and biological sensors. Sensing technology is one of the more useful and immediate applications of nanotechnology that is being developed. The smaller size of nanomaterials allows for a faster response time and greater sensitivity to species that they are tuned for. This is due to, as discussed before, the increased surface-to-volume ratio of nanomaterials.

Furthermore, the small unobtrusive size of nanoscale sensors also allows for the placement of sensors in unique locations without significant disruption to the people. A sensing network could easily be placed, for example, on traffic signals or lampposts throughout a city. The sensors would be capable of communicating in real time a chemical fingerprint of locations throughout the city. With proper monitoring, this information could be used to detect real-time chemical and biological threats. Further, it could be used to provide first responders to chemical and biological attacks with the information that they need to properly outfit themselves. With the threat of biological, chemical, or "dirty" bomb attacks, this type of technology can provide for quicker detection of harmful species and, therefore, quicker annihilation. A network of sensors could be placed relatively simply around major cities in order to monitor for harmful releases. However, any sensing and monitoring system carries with it wide privacy and civil liberties implications. Near ubiquitous surveillance systems exist in some societies and have caused some concern. These issues are explored in much greater detail in Chapter 10.

Public infrastructure protection is another major sphere of homeland defense. For example, water treatment centers are another potential "weak link" in the homeland defense system. As many have noted, they represent a significant target for dispersing harmful chemical species into the public space. Nanoscale sensors and filters that allow only "desired" chemical and biological species through provide an easy solution to this problem. This can be accomplished in a number of ways. First, magnetic nanoparticles could be functionalized to adhere to certain harmful chemical species. The magnetic nanoparticles could be placed early in the filtration system and then removed later with a magnet, thus removing the chemical threat. Another way of achieving more effecting filtering would use nanoporous materials. These are materials that would be engineered as sieves that would only allow water molecules through them.

Furthermore, the same "swarms" of nanoscale robots that act as a weapon can act to destroy "undesirable" biological, chemical, or even nanotechnological weapons. Because of their small size, nanoscale devices could be more easily made airborne and be dispersed in an area with potential chemical and biological threats. These devices could then neutralize the threat. This may sound like science fiction, but swarms of nanotechnology robots are already under development for applications as far off as space exploration with NASA's Autonomous NanoTechnology Swarm (ANTS) project.

Nanotechnology can also offer greater defense in the field of electronics. The danger of an electromagnetic pulse (EMP) is one that is very difficult to shield against completely. Most shielding that provides some protection for electronics requires significant loss in performance of what is protected. However, there is reason to believe that optical computing, DNA computing, and other nanotechnology-based computing options are more naturally resistant to EMPs. All-optical computing has the added benefit of being less sensitive to electronic eavesdropping. Because the optical signal is confined to a fiber more tightly than an electronic signal is truly confined to a wire, it is more difficult for an external piece of equipment to listen in.

9.4 Ethical Concerns

Warfare and military development have always been fraught with ethical concerns. As such, the technology of warfare has carried with it much of the weight of these ethical concerns. The most obvious and familiar example of this is the introduction of atomic weapons to the craft of war. The ethical debate over nuclear weapons and their development continues even now, more than 60 years after their advent. This debate does not limit itself to nuclear weapons, but instead extends to all nuclear technology. In the US, the concerns over nuclear technology have led to a very strong inclination against building nuclear energy facilities, despite their ability to provide electricity free of oil or coal. From a historical look at the impact of military technologies on society and the debates that have surrounded these technologies, several major ethical concerns can be examined. All of these are broad enough to be divided into smaller concerns, but it is the intention here to throw light on the broader questions that are being addressed.

It might be useful to start with the most drastic consideration in nanotechnology. That is, what the development of nano-assembler technology means for international security. One of the most extensive considerations of the effects of nano-assembler technology on international security has been given by Mark Gubrud in a talk at the Foresight Institute:

> The greatest danger coincides with the emergence of these powerful technologies: A quickening succession of "revolutions" may spark a new arms race involving a number of potential competitors. Older systems, including nuclear weapons, would become vulnerable to novel forms of attack or neutralization. Rapidly evolving, untested, secret, and even "virtual" arsenals would undermine confidence in the ability to retaliate or resist aggression. Warning and decision times would shrink. Covert infiltration of intelligence and sabotage devices would blur the distinction between confrontation and war. Overt deployment of ultramodern weapons, perhaps on a massive scale, would alarm technological laggards. Actual and perceived power balances would shift dramatically and abruptly. Accompanied by economic upheaval, general uncertainty and disputes over the future of major resources and of humanity itself, such a runaway crisis would likely erupt into large-scale rearmament and warfare well before another technological plateau was reached.
>
> International regimes combining arms control, verification and transparency, collective security and limited military capabilities, can be proposed in order to maintain stability. However, these would require unprecedented levels of cooperation and restraint, and would be prone to collapse if nations persist in challenging each other with threats of force.

If we believe that assemblers are feasible, perhaps the most important implication is this: Ultimately, we will need an integrated international security system. For the present, failure to consider alternatives to unilateral "peace through strength" puts us on a course toward the next world war.[10]

In his book *Engines of Creation*, Eric Drexler gave a similar analysis of the effect on the international world of the introduction of molecular assembler technology.[11] This type of analysis rests on the idea that the introduction of nanotechnology (more specifically, molecular assembly nanotechnology) will be revolutionary and not evolutionary in nature, as it can be argued that the other technology "revolutions" were. If, per above, we can get beyond the initial skepticism of a "transparen[t], collective security" that would "require unprecedented levels of cooperation and restraint" coupled with the call for "an integrated international security system," we can start to look at the meat of these types of analyses. An analysis of this sort is purely speculative. It rests on claiming that there is no history that we can look to in order to guess what will happen, and then goes on to do just that: guess what will happen. This seems a little unfair, and these types of analyses (i.e., those that claim there is no precedent, no prior history from which we can draw a lesson) are often represented by gloom and doom scenarios. At the very least, they tend to foretell the radical and swift overthrow of whatever current system is in place. These analyses ignore the fact that historically, systems tend to change gradually and incrementally, rather than all at once.

These types of analyses are useful, however, in that they make more moderate predictions seem more mainstream and more investigative. They also provide a starting point from which it is possible to work back to a more reasonable middle. Looking back over the history of military technological introduction, it is clear that it has actually been a process that gradually changes the international system, and is itself changed by the system in gradual ways. The Cold War stand-off was no more a cause of the doctrine of mutually assured destruction and nuclear proliferation than these doctrines were a cause of the Cold War. In fact, though many claim that the American–Soviet stand-off was unprecedented in its nature, there is history that seems to hint that such a stand-off was predictable. When two great nations, of distinctly different ideologies, join forces to defeat a common enemy, the result is the kind of stand-off in which the rest of the known world is forced to choose a side. Of course, this is what happened

[10] Mark Gubrud, "Nanotechnology and International Security." Presented at Fifth Foresight Conference on Molecular Nanotechnology at Palo Alto, CA, 1997. Available at http://www.foresight.org/Conferences/MNT05/Papers/Gubrud/ (accessed September 4, 2008).
[11] Eric Drexler, *Engines of Creation: The Coming Era of Nanotechnology* (New York: Broadway Books, 1987), ch. 11. Available at http://www.e-drexler.com/d/06/00/EOC/EOC_Table_of_Contents.html (accessed September 26, 2008).

with Athens and Sparta after the two city-states joined together to defeat an invading Persian empire. Immediately after the Greek victory over the Persians, Athens was, by far, the major power in the Greek world. However, this led to a showdown with the Spartans, who in virtue of their ideology were suspicious of Athenian intentions. This situation is strikingly similar to that which existed immediately after World War II. There were many aspects of the postwar era that make it distinctly different from the Greek world after the Persian War, from the existence of the UN to nuclear weapons. However, the same showdown occurred.

Similarly, it is possible to look at what effect the introduction of new technologies has had on warfare and the military and draw some lessons for what may come with the introduction of various nanotechnologies. One lesson is that a sudden, complete overturning of the current world system of states (with several non-state actors) is unlikely. The other end of the spectrum is equally unlikely. This camp says that nanotechnology will bring about an end to fights about resources, food, and other things that states go to war over and, thus, bring about peace. However, with this, it would be wise to remember the words of former international relations professor Hedley Bull, who said that although states at peace are thought of as the alternative to states at war, the typical alternative is "more ubiquitous violence."[12] Bull's warning was written in 1977, and history before and since tends to confirm this statement. Ubiquitous violence, flare-ups, small wars, police actions, and whatever other name is conferred on these actions taken by states have been the order of the day. The non-state actors, too, tend to increase their violent actions during times of supposed peace between states. Terrorism, rebellion, and genocidal acts grow more frequent.

It is useful then to look at what impact new technologies tend to have on these acts of violence, whether by a state or not. Many of the technologies are protective in their nature, as outlined above; they cause fewer civilians and combatants to die. Other technologies make precision guidance of weapons more precise and cause a weapon to have a higher fatality and specificity rate. What these add up to is quite interesting. New technologies, of which nanotechnology leads the way, make the craft of war and violence easier and less costly to participate in from an offensive point of view. On the attacking side, this is evident. Fewer soldiers die because of advances in protective and medical technologies. Fewer civilians are injured because of population-wide protection. Fewer citizens are involved in the military because more powerful and more precise weapons require fewer ground troops. Because of the higher precision of the weapons, the costs could possibly be lower on the defending side as well. So-called "collateral damage," from which the more developed, liberal states tend to shy away, can be

---

[12] Hedley Bull, *The Anarchical Society* (New York: Columbia University Press, 1977), p. 179.

minimized. Weapons can be made that strike only one building and do it with accuracy and precision. Further, nanotechnology allows for the simple shutting down of vital systems to populations such as water, electricity, and energy, as opposed to the utter destruction of the facilities in which they are housed. So, war-like acts are easier to inflict on enemies of a nanotechnologically enabled state. Acts that become easier usually become more ubiquitous in their nature.

However, it has also happened that democratic nations, in which the vast majority of nanotechnology research and development is being done, have a low tolerance for casualties in military actions. With nanotechnological developments making it easier to protect, defend, and otherwise shield soldiers and populations from taking casualties, this tolerance will probably become even lower. As death by violent acts becomes rarer, tolerance of it becomes rarer as well. This lessens the likelihood that long-drawn-out, high-casualty military actions will be tolerable to the population of the technologically advanced nations, and they will probably decrease. By nature, prognostication is imperfect, but combining these last two probabilities, it seems likely that "small wars" in which technologically advanced nations perform "police actions" on less developed regimes will become more and more frequent, in no small part because of nanotechnology.

This, then, raises challenges to the 'just war' tradition that entering and fighting a war is morally justified only under certain conditions, such as in self-defense and when combatants can be discriminated from non-combatants (which rules out weapons of mass destruction, nano-enabled or otherwise).[13] Specifically, and as noted by other ethicists but in the context of other technologies, innovations such as nanotechnological weapons and defenses may make it *easier* and therefore more likely that we will enter wars and conflicts, given that our technological superiority will reduce friendly casualties, i.e., make war more risk free.[14] We will not comment much on those objections except to note that they are common and can be – and have been – applied to many new technologies for their period. If the objections were right, then they would rule out not only the introduction of new weapons but also incremental improvements in, say, personnel armor and even better battlefield medicine. We could not improve on offensive or defensive tools, because they would make it easier, politically if not also economically, to engage in armed conflict, and this is presumably undesirable. On the other hand, if armed conflict is an inevitable fact of the human condition, then it is difficult to blame defense organizations for developing new strategies, tactics, and tools that minimize civilian and combatant lives on either side of a fight.

---

[13] See, for example, Michael Walzer, *Just and Unjust Wars: A Moral Argument with Historical Illustrations*, 4th edn (New York: Basic Books, 2006).
[14] See, for example, Rob Sparrow, "Killer Robots," *Journal of Applied Philosophy* 24.1 (2007): 62–77.

There are still multiple ideologies among states on the world stage and militarily; these regimes tend to regard each other with suspicion. Even if nanotechnology systems for military use are developed for purely defensive reasons, this can have a destabilizing impact on relations between great powers. Defensive systems render offensive systems moot. In recent history, there are examples of the development of purely defensive systems causing much concern to an opposing force. An example of this can be seen in the desire of Russia (previously known as the Soviet Union) to stop the development of a missile shield by the United States and other Western powers. Defensive weapons upset the status quo between states (i.e., they upset the power balance) and, as such, they become objectionable to the powers that are rendered impotent by them. In much the same way that new offensive weapons that render defensive technologies obsolete can upset a security balance, so too can new defensive technologies that render offensive weapons obsolete. The great castles of medieval Europe were strictly defensive in their nature and they were nearly immune to armies of knights. The prospect of besieging a castle was bleak enough to deter most opponents from trying. However, with the advent of gunpowder the castle became obsolete. Changing the balance of power of weaponry can have wide-reaching implications in international relations, even when it is toward the defensive end. To be sure, more powerful defensive weaponry can protect a citizenry from attacks, but an enemy that finds itself extremely resistant to attacks will feel at greater liberty to act in a belligerent manner. If an action becomes easier to do and brings fewer negative consequences, then it will become more common.

The international stage is not the only arena in which a nanotechnological revolution in the military will have an impact on society. It seems likely that the amount of health and mental care necessary to provide to members of the armed services will increase. As nanotechnology allows for stark increases in the ability to save a life, injuries that once were life-threatening or led to certain death become treatable. Illnesses and chemical attacks become less threatening. Furthermore, a much higher percentage of soldiers will live through military actions and will, as such, be in some need of psychiatric care. In addition, it seems likely that as medical nanotechnology is able to fix more problems, further new problems will be noticed and require solution. Tailoring each treatment to individual patients based on their DNA and their environment again increases the actual care (though perhaps not the time) that each patient needs. Another issue that needs to be considered is that of a population that has an increasing number of members who volunteer to serve in the military. When the chances of death are lessened, the idea of military service becomes more attractive and a greater percentage of the population might be trained in the military.

In concluding this chapter, let us acknowledge that we have not discussed many ways in which nanotechnology might be applied to military purposes; we have chosen to focus on some key areas and to develop them rather than to be comprehensive. To wit, several frameworks for considering the impact of nanotechnology have been given. The first of these frameworks is in the nature of the technology being used: does nanotechnology portend incremental change, gaining its novel applications from its small size, imbued with even more novel properties? Or does nanotechnology portend abrupt change, revolutionizing not only the technology, but the speed of change as well? The second framework is in the nature of the perceived use of the nanotechnology: is it offensive or defensive? The third framework proposed is structured by the point of use of the nanotechnology: is it used by an individual person, a small group, or the entire society? Finally, the fourth framework dealt specifically with military nanotechnology: how transferable is the technology to non-military applications? As a note on the fourth framework, it should be acknowledged that it is not necessarily true that technology that is easily transferred to non-military applications will have a greater impact on society than other technologies. There are many other frameworks with which to consider nanotechnology; this is not meant to be exhaustive.[15] Many of them imply their own metric in examining the ethical use of the nanotechnology.

On the whole, nanotechnology will likely be transformative in its nature and will bring about a more highly enabled military. The proper use of nanotechnology in the military can only be discovered through an interactive discourse with an informed base, and we certainly encourage such discourse. The ethical use of nanotechnology in warfare needs to be integrated into and to be consistent with other doctrines governing the ethics of war. These can range from legal doctrines such as the Geneva Conventions, or ethical doctrines such as those arising from the just war tradition. Certainly these conventions and traditions have been and will be challenged, especially given the contemporary advent of terrorism. Nevertheless, whatever doctrines we ultimately endorse, we should be mindful of the implications they have for nanotechnology as applied to military purposes.

---

[15] For example, we have focused on wars and large-scale military actions and have not discussed non-state actors; defense, sensing, and monitoring applications speak to protection against non-state actors. However, nanotechnology will probably also have a large impact on the way in which non-state actors operate. Technology changes in one adversary create tactical changes in another. Furthermore, nanotechnology might make weapons cheaper to produce and more widely available to ill-meaning adversaries. This, in turn, requires more effective deterrent and defense capabilities.

10

Privacy

Privacy has emerged as one of the central areas in which nanotechnology is expected to have an impact. While privacy has long been a topic studied by, especially, ethicists and jurists, it has come under even more intense emphasis in a post-9/11 world. Since the advent of the war on terror, many restrictions on civil liberties have been countenanced as required for our safety; in the US, the Patriot Act is the most visible (and controversial) of these measures. As we will see below, it is a mistake to think that concerns over privacy are new, but it is undoubtedly true that they are even more acute given fears about terrorism. In this climate, nanotechnology's implications for privacy have generated much interest.

This chapter will discuss the relationship between nanotechnology and privacy. The catalyst for this discussion will be radio frequency identification (RFID); RFID, while not requiring nanotechnology, is nevertheless a technology that nanotechnology will redefine and the core application of nanotechnology that has privacy implications. We will talk about what RFID is and some of its background in §10.3. In §§10.4–10.6, we will consider three particular applications of RFID with privacy implications: item-level tagging, human implants, and RFID-chipped identification. The first two sections of this chapter, though, explore what privacy is and whether there is a moral (or legal) right to it. With numerous exceptions, much of the literature that pertains to technology and privacy – including not just nanotechnology, but other forms of technology, especially computers – has been fairly careless about specifying exactly what it takes privacy to be or why privacy should matter, either morally or legally. Neither of these two projects is transparent, so some prior discussion is important. We will therefore begin with a discussion of privacy's historical and legal background (§10.1), and then move on to a more contemporary discussion of its philosophical foundations (§10.2). §10.7 draws connections across these two parts of the chapter.

10.1   Historical and Legal Background

At least since Plato, privacy has been a theme of philosophical reflection. In the *Laws*, Plato wrote:

> [T]he old saying [is] that "friends have all things really in common." As to this condition, whether it anywhere exists now, or ever will exist, in which there is community of wives, children, and all chattels, and all that is called "private" is everywhere and by every means rooted out of our life, and so far as possible it is contrived that even things naturally "private" have become in a way "communized," eyes, for instance, and ears and hands seem to see, hear, and act in common, and that all men are, so far as possible, unanimous in the praise and blame they bestow, rejoicing and grieving at the same things, and that they honor with all their heart those laws which render the State as unified as possible, no one will ever lay down another definition that is truer or better than these conditions in point of super-excellence. In such a State . . . they dwell pleasantly, living such a life as this.[1]

The sort of community that Plato idealizes has little cause for privacy, and we might all agree that his utopianism – if we even see it that way – is unlikely to be realized. Aristotle, on the other hand, clearly defended distinctions between a public sphere (πόλις, *polis*) and a private sphere (οἶκος, *oikos*), which reverberates throughout his *Politics*.[2,3] For example, in Book 2, he is interested in distinctions between public and private property[4] – especially including land[5] – and he clearly is not impressed with Plato's notion of sharing all in common.[6] Aside from private property, he also discusses private life[7] and the notion of a "private citizen."[8] Book 4 then introduces private affairs,[9] private interests,[10] and private contracts;[11] much of these are further discussed in subsequent chapters. But, aside from using the word 'private' – even as contrasted with 'public' – we do not receive a tremendous

---

[1]   Plato, *Laws*, *Plato in Twelve Volumes*, trans. R.G. Bury (Cambridge, MA: Harvard University Press, 1968), 739c–d.
[2]   Aristotle, *Politics*, trans. T.A. Sinclair, rev. Trevor J. Saunders (London: Penguin Books, 1992). References are to section numbers.
[3]   For more discussion, see Mary P. Nichols, *Citizens and Statesmen: A Study of Aristotle's Politics* (Savage, MD: Rowman & Littlefield, 1991). See also Judith A. Swanson, *The Public and Private in Aristotle's Political Philosophy* (Ithaca: Cornell University Press, 1992).
[4]   *Politics*, e.g., 1261b.
[5]   Ibid., e.g., 1267b.
[6]   Ibid., e.g., 1264b.
[7]   Ibid., e.g., 1273a.
[8]   Ibid., e.g., 1273b.
[9]   Ibid., e.g., 1293a.
[10]   Ibid., e.g., 1295a.
[11]   Ibid., e.g., 1300b.

amount of conceptual clarification as to what it is supposed to mean, or even much explicit discussion as to why it is a moral value.[12]

The notion of 'private' – whose relationship to 'privacy' still had yet to be clarified – returned with John Locke. In §7.1, we saw Locke's labor theory of (private) property, which held that:

> Though the earth, and all inferior creatures, be common to all men, yet every man has a property in his own person: this nobody has any right to but himself. The labor of his body, and the work of his hands, we may say, are properly his. Whatsoever then he removes out of the state that nature hath provided, and left it in, he hath mixed his labor with, and joined to it something that is his own, and thereby makes it his property. It being by him removed from the common state nature hath placed it in, it hath by this labor something annexed to it, that excludes the common right of other men: for this labor being the unquestionable property of the laborer, no man but he can have a right to what that is once joined to, at least where there is enough, and as good, left in common for others.[13]

Still, this hardly offered a systematic treatment of privacy; indeed, the word never even appears in the passage from Locke. Rather, it was exactly 200 years after Locke's 1690 publication of the *Second Treatise* that any sustained and explicit treatment was given to privacy.[14]

In 1890, two lawyers, Samuel Warren and Louis Brandeis, wrote an essay in the *Harvard Law Review* entitled "The Right to Privacy."[15] This began an important hundred years of legal thought on privacy, culminating with – or devolving into, depending on your outlook – *Bowers v. Hardwick* (1986), on which more below. Interestingly for our purposes, Warren and Brandeis were primarily concerned with how technology was eroding privacy. In 1890, though, the relevant technologies were photography and newspapers, both recently popularized. These technologies allowed for the dissemination of facets of life that were previously private, whether through images or words. For example, imagine that someone was committing adultery, and that photography and newspapers did not exist. If this person were unlucky enough to get caught, rumors might get spread by word of mouth, but, absent hard

---

[12] Aristotle does, for example, say that private property makes its bearers happy (ibid., e.g., 1263a), but this is hardly a systematic discussion as to the moral foundations of privacy.

[13] John Locke, *Second Treatise on Civil Government* (New York: Prometheus Books, 1986), V, §27.

[14] Note that John Stuart Mill, in his essay "On Liberty," defended an account about the appropriate reach of governmental authority; Mill thought that individuals should have freedoms from thoroughgoing governmental intervention. Surely this connects to our interest in historical thinking on privacy, though we shall omit Mill here for space constraints. See John Stuart Mill, "On Liberty," in *On Liberty and Other Writings* (Cambridge: Cambridge University Press, 1989), pp. 1–116.

[15] Samuel Warren and Louis Brandeis, "The Right to Privacy," *Harvard Law Review* 4 (1890): 193–220. Brandeis went on to be a Supreme Court justice.

evidence, they might stay rumors. And, furthermore, word of mouth can only travel so fast or so wide; this is not to say that it cannot travel quickly and widely, just that it is comparatively limited. Now take photographic evidence of the adulterous act – or at least something printable and sufficiently indicting – and circulate it, accompanied by appropriate testimony, to the readership of a newspaper. And thus these technologies portended a dramatic shift in the sorts of information one might be able to keep private about oneself.

Warren and Brandeis thought that there was a right to keep information about oneself private, by which they simply meant that it would not be disseminated to the public. There are complications to this thought, such as when we consider public officials; maybe we think that these officials forfeit some of their privacy. But, aside from complicating cases, there seems something straightforward about this. Imagine that you are eating breakfast inside your house and that some paparazzo climbs through the flowerbed of your house (which you own) and takes pictures of you eating breakfast, which she thereafter publishes. Something seems to have gone wrong here, and this is what Warren and Brandeis aimed to delimit. In particular, their account has come to be the progenitor of *informational privacy* accounts, which hold that we have a right to prevent information about ourselves being made public (against our will and, perhaps, with some exceptions). They thought that this was a moral right, as well as one that could be grounded in then-existing common law, and in some cases statutory law.

Following Warren and Brandeis' essay – even if not immediately – there was an increased recognition and expansion of the right to privacy. In a famous 1960 essay, William Prosser offered four separate actions against which the right to privacy offered protection; these were thereafter incorporated into the third edition of his universally acclaimed *Handbook of the Law of Torts*. The list included:

1   intrusion into one's solitude or private affairs;
2   public disclosure of private facts;
3   portrayal in a false light in the public eye; and
4   appropriation of one's identity or likeness.[16]

Warren and Brandeis were primarily concerned with 2, but Prosser's list of protections is broader as its scope includes more than mere informational privacy. For example, the protection against intrusion into one's private affairs has nothing, necessarily, to do with information; further note that

---

[16]   William L. Prosser, *Handbook of the Law of Torts*, 3rd edn (St Paul: West, 1964), pp. 833–42. See also William L. Prosser, "Privacy," *California Law Review* 48.3 (1960): 383–423. We thank Judith DeCew and Hans Allhoff for their assistance with the Prosser references.

even this protection is broader than a more limited and obvious one against mere physical intrusion.

Before turning to some of the philosophical grounds for privacy, let us continue with some of this legal thinking about privacy and, in particular, look quickly at some famous Supreme Court cases. The first is *Griswold v. Connecticut* (1965).[17] In this case, Estelle Griswold was the Executive Director of the Planned Parenthood League of Connecticut, and she was appealing a conviction under a Connecticut statute that forbade the dissemination of birth control; this statute was passed in 1879, but rarely enforced thereafter. Her conviction was upheld at the Appellate Division of the Circuit Court, as well as by the Connecticut Supreme Court. The US Supreme Court, though, overturned those previous decisions 7–2. The majority opinion held that "The Connecticut statute forbidding use of contraceptives violates the right of marital privacy which is within the penumbra of specific guarantees of the Bill of Rights."[18]

What makes this case so important is not just the ruling – i.e., that the Connecticut statute was unconstitutional, or the assertion of marital privacy – but rather the court's reasoning, which has gone on to form a substantive basis for a constitutional right to privacy. Critical here are the notions of 'penumbra' and 'emanation.' We saw the mention of 'penumbra' above, but also consider: "specific guarantees in the Bill of Rights have penumbras, formed by emanations from those guarantees that help give them life and substance."[19]

This language, almost metaphorical, literally means that the boundaries of some constitutional protections are shadowy and that those shadows are generated by some of the core values that undergird the associative protections. Privacy, while not specifically mentioned in any constitutional amendments, has nevertheless been implicitly protected – according to Justice William O. Douglas, who wrote the majority opinion – by explicit language that the founders used in other regards. The penumbras allegedly emanate from at least the following Amendments:

1 *First Amendment.* Congress shall make no law respecting an establishment of religion, or prohibiting the free exercise thereof; or abridging the freedom of speech, or of the press; or the right of the people peaceably to assemble, and to petition the Government for a redress of grievances.

2 *Third Amendment.* No Soldier shall, in time of peace, be quartered in any house, without the consent of the Owner, nor in time of war, but in a manner to be prescribed by law.

---

[17] *Griswold v. Connecticut* 381 US 469 (1965). Available at http://supreme.justia.com/us/381/479/case.html (accessed August 5, 2008).
[18] Ibid.
[19] Ibid.

3 *Fourth Amendment.* The right of the people to be secure in their persons, houses, papers, and effects, against unreasonable searches and seizures, shall not be violated, and no Warrants shall issue, but upon probable cause, supported by Oath or affirmation, and particularly describing the place to be searched, and the persons or things to be seized.
4 *Fifth Amendment.* No person shall be held to answer for any capital, or otherwise infamous crime, unless on a presentment or indictment of a Grand Jury, except in cases arising in the land or naval forces, or in the Militia, when in actual service in time of War or public danger; nor shall any person be subject for the same offense to be twice put in jeopardy of life or limb; nor shall be compelled in any criminal case to be a witness against himself, nor be deprived of life, liberty, or property, without due process of law; nor shall private property be taken for public use, without just compensation.
5 *Ninth Amendment.* The enumeration in the Constitution, of certain rights, shall not be construed to deny or disparage others retained by the people.[20,21]

Perhaps the Fourth Amendment has the strongest implications for privacy – even if those implications are hardly obvious – insofar as it protects people against searches (of our bodies and possessions) and seizures (of our possessions). This protection suggests a distinction akin to Aristotle's between the public and the private, such that the private cannot be accessed without good reason. Freedom to assemble or to practice religion (First Amendment), protection against the use of one's house for the sequestering of soldiers (Third Amendment), immunity against revealing incriminating information about oneself (Fifth Amendment; cf. Warren and Brandeis' informational privacy), and the retention of basic rights not expressly included in the Constitution (Ninth Amendment) are all provisions that, at least according to Douglas, suggest the existence and protection of a private sphere.

The ruling in *Griswold* set a powerful precedent in favor of a constitutional right to privacy, despite its controversy. The court that presided over that and other important and subsequent decisions has been accused of judicial activism; i.e., of creating laws that are not well founded in case or statutory law. Robert Bork, for example, has been one of the most

---

[20] Bill of Rights, US Constitution. Available at www.usconstitution.net/const.html (accessed August 5, 2008).
[21] The Fourteenth Amendment, which is not part of the Bill of Rights, has also been invoked to ground a constitutional right to privacy. Particularly important are its Due Process and Equal Protection clauses, which appear in this partial excerpt of the amendment: "No State shall make or enforce any law which shall abridge the privileges or immunities of citizens of the United States; nor shall any State deprive any person of life, liberty, or property, without due process of law; nor deny to any person within its jurisdiction the equal protection of the laws." Amendment 14, US Constitution. Available at www.usconstitution.net/const.html#Am14 (accessed August 5, 2008).

prominent critics of the Griswold ruling.[22] Bork saw that decision as being driven by social values, as opposed to good legal interpretation. As mentioned above, 'privacy' does not appear in any of the amendments that the Douglas decision cites; Bork therefore thinks that there is no provision for a right to privacy and that Douglas' invocation of penumbras and emanations to ground a right to privacy was just poor jurisprudence.

Nevertheless, the *Griswold* decision went on to have profound legal implications. In 1973, the court ruled in favor of Norma McCorvey (i.e., "Jane Roe") in overturning a Texas law against abortion;[23] *Roe v. Wade* – Henry Wade being the district attorney representing Texas – has gone on to be one of the most important and controversial legal rulings in US history. Douglas was still on the court, but the majority opinion was written by Justice Harry Blackmun. This decision was based strongly on the *Griswold* precedent, as well as interpretations of the Ninth Amendment; much of the principal legal argumentation in this case pertained to privacy. Since *Griswold* and *Roe* – or at least until the post-9/11 climate – there have been continual legal expansions in rights to privacy, with the notable exception of *Bowers v. Hardwick*.[24] This case was very similar to Griswold except that, rather than involving distribution of contraception to a married couple, Hardwick was arrested under a Georgia anti-sodomy statute for being engaged in oral sex with another man, despite this taking place in the privacy of his own home. Hardwick sued Bowers, the attorney general of Georgia, and the case went to the Supreme Court. However, the Supreme Court upheld the Georgia statute, to the consternation of many civil libertarians. This court, of course, was not the Douglas court, and Justice Byron White, who wrote the majority opinion, denied that a right to sodomy was deeply rooted in the US Constitution. Despite the limits that *Hardwick* would seemingly have signaled for privacy, this decision was effectively reversed in 2003 when they ruled that a Texas statute – otherwise quite similar to Georgia's – was found unconstitutional under the equal protection and due process clauses of the Fourteenth Amendment.[25]

Given some of this history and these cases, there has been an expansion of privacy rights in the US, going back to at least 1890. Since 9/11, though, these protections have been challenged given the war on terror.[26]

---

[22] Robert Bork, *The Tempting of America: The Political Seduction of the Law* (New York: Simon and Schuster).

[23] *Roe v. Wade* 410 US 113 (1973). Available at http://caselaw.lp.findlaw.com/scripts/getcase.pl?court=US&vol=410&invol=113 (accessed August 8, 2008).

[24] *Bowers v. Hardwick* 478 US 186 (1986). Available at http://caselaw.lp.findlaw.com/scripts/getcase.pl?court=US&vol=410&invol=113 (accessed August 8, 2008).

[25] *Lawrence et al. v. Texas* 539 US 558 (2003). Available at http://supreme.justia.com/us/539/558/case.html (accessed October 16, 2008).

[26] Rights, of course, have often been suspended during times of war. Habeas corpus was suspended during the Civil War, Japanese citizens were interned during World War II, and so on.

In particular, the US Patriot Act was passed in 2001 to authorize greater surveillance abilities in an effort to combat terrorism.[27] Most of these provisions were set to expire on December 31, 2005, though almost all were made permanent; only §206 and §215 were temporarily renewed (through 2009),[28] and some other provisions were slightly modified.[29] The debate over the Patriot Act's impositions on privacy pertains directly to the notion of security, which is one that we will consider directly in §10.6.

10.2   Philosophical Foundations

Having seen some of the historical and legal background on privacy, let us now turn to some of its philosophical foundations. In particular, we want to think more critically about the meaning and moral value of privacy, as well as its distinctiveness. Once we have given some consideration to these issues, we will then return to nanotechnology, well equipped for the ensuing discussion.[30]

In the previous section, we mentioned Warren and Brandeis' paper, "The Right to Privacy," which has given rise to the notion of informational

[27]   Uniting and Strengthening America by Providing Appropriate Tools Required to Intercept and Obstruct Terrorism Act (USA PATRIOT Act) of 2001, Pub. L. No. 107–56, 115 Stat. 272. Available at http://frwebgate.access.gpo.gov/cgi-bin/getdoc.cgi?dbname=107_cong_bills&docid=f:h3162enr.txt.pdf (accessed October 16, 2008).

[28]   For a list of these provisions, see Charles Doyle, "Patriot Act: Sunset Provisions that Expire on December 31, 2005" (Washington, DC: Congressional Research Service), available at http://www.fas.org/irp/crs/RL32186.pdf (accessed August 8, 2008). They include: §201 (wiretapping in terrorism cases); §202 (wiretapping in computer fraud and abuse felony cases); §203(b) (sharing wiretap information); §203(d) (sharing foreign intelligence information); §204 (Foreign Intelligence Surveillance Act (FISA) pen register/trap and trace exceptions); §206 (roving FISA wiretaps); §207 (duration of FISA surveillance of non-US persons who are agents of a foreign power); §209 (seizure of voicemail messages pursuant to warrants); §212 (emergency disclosure of electronic surveillance); §214 (FISA pen register/trap and trace authority); §215 (FISA access to tangible items); §217 (interception of computer trespasser communications); §218 (purpose for FISA orders); §220 (nationwide service of search warrants for electronic evidence); §223 (civil liability and discipline for privacy violations); and §225 (provider immunity for FISA wiretap assistance).

[29]   The Patriot Act was renewed and amended through two subsequent pieces of legislation. The first was the USA PATRIOT Improvement and Reauthorization Act of 2005, 109–333, H.R.3199. Available at http://thomas.loc.gov/cgi-bin/cpquery/R?cp109:FLD010:@1(hr333) (accessed August 8, 2008). The second was the USA PATRIOT Act Additional Reauthorizing Amendments Act of 2006, 109–70, S.2271. Available at http://thomas.loc.gov/cgi-bin/query/z?c109:s2271: (accessed August 8, 2008). For analysis, see Charles Doyle, "USA PATRIOT Act Reauthorization in Brief" (Washington, DC: Congressional Research Service, no date). Available at http://fpc.state.gov/documents/organization/51133.pdf (accessed August 8, 2008).

[30]   For a more extended introduction, see Judith DeCew, "Privacy," *Stanford Encyclopedia of Philosophy* (Summer 2008). Available at http://plato.stanford.edu/archives/sum2008/entries/privacy/ (accessed August 8, 2008).

privacy. The central idea behind this doctrine is that we have a right to keep personal information about ourselves from public dissemination. What information, though? If, for example, someone walks down a public street, that information – that she walked down the public street – would hardly seem protected. Imagine that some person B sees A walking down the public street and then tells some third person C. In this case, it does not seem as though B has violated A's right to privacy. Imagine instead, though, that B looks into A's window and sees A φ'ing, which B thereafter reports to C. Is this a violation of A's privacy? What if A had left the blinds up and B just happened to walk by and observe the act? Has A's right to informational privacy been somehow compromised by A not undertaking due diligence to protect the information? About the only time it seems remotely obvious that B has violated A's right is, for example, something like when A is inside her own house, closes all the blinds – thus ensuring protection from the casual passer-by – and B somehow sneaks in to observe A φ'ing, then passes that information along to C. It certainly seems like *that* is a violation of privacy, but the other cases are much more difficult; many of us lack clear intuitions in those cases.

This gives rise to the first challenge to informational privacy: *what* information is protected, and how can those protections be forfeited?[31] The answer cannot be that all information about us is protected from subsequent transmission. What about some other proposal such that all information that we ourselves choose not to make public is protected? So, for example, the person walking down the street is therefore making public some information (e.g., that she is walking down the street); and perhaps the person in the house who does not secure her blinds is also making some information public, if less obviously. When B sneaks into A's house, though, B is accessing information that A has not already made public, and perhaps therein lies the relevant differentia. But the notion of 'making public' is not all that transparent, either, at least because the concepts of 'making' and 'public' are not transparent. For example, imagine that A tells her brother something: has she made it public? Maybe it depends on the context under which the disclosure has been made. Surely explicit assurances of confidentiality are morally binding, but that is the easy case. What if no such assurance is asked for or given? Then has B violated A's privacy when disclosing? Structurally, it should not have anything to do with the nature of the information that is disclosed: surely some violations are worse than others, but whether a violation is a violation at all should not have anything to do with the nature of the information (other than whether its sharer wants it revealed and/or expects such confidentiality).

---

[31] Some of the following discussion receives more sophisticated treatment in Thomas Nagel, "Concealment and Exposure," *Philosophy & Public Affairs* 27.1 (1998): 3–30.

Aside from the notion of publicity, we can also wonder about how the information is gained; i.e., what it means to *make* something public. A might tell B something, though A might also hint at it, or even deny it. Or maybe B asks A, and A changes the subject, or blushes, or whatever. Merely because B can divine some information, it hardly seems that A has "made it" public. One temptation here is that it depends on whether A *wants* the information made public. But this cannot be right, either. Imagine that A is walking down the street with some embarrassing tear in her clothing. She might hardly want this to be made public, but it has been made public regardless. Of course, she *chose* to walk down the street, but that does not seem like the morally relevant feature, either. Imagine that she was coerced to walk down the street (e.g., by a gunman). In this case, and despite her lack of volition, she has nevertheless made information public. Therefore, volition alone does not seem like the central moral construct.

Accounts of informational privacy also have to specify *who* has this right, and it quite probably is not everybody. Consider, for example, an elected official who has a cocaine habit. Does he have a right to protect this information? Or, insofar as he is in public service, does the public have a right against his keeping the information private? Medical records, for example, are something that we often think should remain private; the protection of medical information is even provided for in the Hippocratic Oath (cf. sigh/somai). But consider a would-be President who has a history of (undisclosed) heart ailments. Certainly this information would be of interest to voters, who might reasonably be disposed against voting for him for just this reason. Does the candidate have a right to withhold it?

These are all hard questions and, to our minds, have never been clearly answered in providing a complete account of the doctrine of informational privacy. There certainly are some easy cases where we can identify violations of privacy, though the cases can quickly become more complicated. Any clear proposal as to the value and limits of informational privacy needs to be able to handle those cases and, while such a proposal could be delivered, we have yet to see an unqualified success in the literature. Maybe, though, this is an unfair burden as there will always be hard cases; certainly there are some well-formulated accounts out there.[32] Nevertheless, specifying what information is protected, how it is made unprotected, and who is protected are all questions that a fully developed account of informational privacy will have to answer.

To us, informational privacy represents a fairly generic account of what privacy *is*: privacy is the ability to keep information about ourselves

---

[32] In addition to Warren and Brandeis' seminal paper, see Charles Fried, *An Anatomy of Values* (Cambridge, MA: Harvard University Press, 1970). See also W.A. Parent, "Privacy, Morality and the Law," *Philosophy & Public Affairs* 12.4 (1983): 269–88. A critique of Parent is offered in Judith DeCew, *In Pursuit of Privacy: Law, Ethics, and the Rise of Technology* (Ithaca, NY: Cornell University Press, 1997).

private.[33] As indicated above, we have various concerns about how an account like this is specified, but we nevertheless think that it is generally clear what such an account aspires to do. Some other commentators seem to think that informational privacy is an account of "the meaning and value" of privacy, of which there are others,[34] but we do not think this is quite right. Rather, nothing in the informational account of privacy has told us why privacy is a moral value: that account is completely descriptive. If there is a normative dimension to privacy – i.e., some moral value associated with it – then some further story needs to be told. And, to us, that is what these "other accounts" are doing: providing normative grounding for informational privacy, or perhaps some more sophisticated version thereof. So, just to be clear, we think that there are (at least) two central questions as pertain to privacy. First, we have to figure out what it is. And, second, we have to figure out whether it has any moral value. The above discussion regarding informational privacy was a rough attempt to make some progress on the first question, and we now turn to the second.

Judith DeCew discusses two potential moral grounds for privacy: human dignity, and the related concepts of intimacy and social relationships.[35] Starting with human dignity, DeCew points to an essay written by Edward Bloustein as a response to Prosser.[36] Recall that Prosser identified four different legal protections that privacy was offering, but the question then becomes how to unify these different protections. We could just say 'privacy' is what unifies them, though this is little more than semantic stipulation. Or at least it does not provide any further elucidation as to what myriad privacy protections have in common. Bloustein thought that these protections were all against intrusions that were demeaning to individuals and affronts to their dignity. So, for example, the revealing of private personal information could be demeaning, as could misrepresentation in the public eye.

We think that there are two significant problems with this position. First, the terms that it uses, 'demeaning' and 'dignity,' are hardly any more elucidating than the concept, 'privacy,' that they are meant to clarify. If we are trying to explain to someone what a *lubl* is and we say that it is a *wonet*, we have hardly done any good. At least since Immanuel Kant, the notion of human dignity has represented a common theme in moral philosophy and yet, despite that tradition, the concept is still opaque. Second,

---

[33] This is probably too narrow as, for example, privacy might also include "control over access to oneself, both physical and mental, and . . . control over one's ability to make important decisions about family and lifestyle": DeCew, "Privacy." See also DeCew, *In Pursuit of Privacy*. Nevertheless, for our purposes – and especially given the applications of nanotechnology that we will consider in subsequent sections – the notion of informational privacy will suffice.

[34] See, for example, DeCew, "Privacy."

[35] Ibid.

[36] Edward J. Bloustein, "Privacy as an Aspect of Human Dignity: A Response to Dean Prosser," *New York University Law Review* 39 (1964): 962–1007.

whatever human dignity is supposed to amount to, violations of privacy cannot exhaust its purview. Imagine, for example, that a prisoner of war is kept in extremely sordid conditions; presumably, this is also an attack on his dignity. Therefore, we have not learned anything distinctive about privacy by appealing to dignity, and this was part of the project on which Bloustein meant to embark. We will return to this later, but will add now that the reason this is confused, as above, has to do with a failure to separate the descriptive and normative dimensions of privacy.

A second moral grounding for privacy has to do with the relationship that it shares with intimacy and social relationships.[37] The idea here is that intimacy and social relationships would be impossible without privacy and, insofar as these things have moral value, then so does privacy as a necessary means to their realization.[38] There are different sorts of relationships but, for simplicity, consider casual and intimate ones. One big difference between these two relationships has to do with the amount of information that the participants in the relationship have about each other. In other words, the privacy or disclosure of information is central to how we understand relationships. Similarly, trust is an integral part of relationships, and trust is hard to understand without privacy: when we trust someone, we trust them, among other things, not to disclose the information with which we have entrusted them. This moral value can be bolstered by saying that personal relationships are an essential part of meaningful lives, at least for most of us. Insofar as privacy plays an important role in defining and delimiting those relationships, it therefore has moral value.

We think that this sounds plausible, and agree that privacy is important to our social lives. But notice, again, that privacy is not the only thing that matters for social lives; other things, like shared activities, also matter. By appealing to relationships, then, we have not learned anything unique about privacy, any more so than we did by appealing to human dignity. But now the flaw in the dialectic should be clear: moral values do not have to be distinctive. Lying, cheating, and stealing might all be wrong for the same reason: they cannot be universalized (cf. Kant). But this hardly means that we have not identified why any of these actions are wrong. Bloustein's account failed to separate normative and descriptive elements, which is one reason it ends up being problematic. The intimacy and relationship accounts, though, just tell us why privacy is wrong; they do not tell us what privacy is. In fact, they are completely compatible with, if not suggestive of, the accounts of informational privacy discussed above. If we separate the question of

---

[37] DeCew, "Privacy," treats these as separate moral grounds, though they seem similar enough to us that we shall treat them jointly.

[38] See Fried, *An Anatomy*; James Rachels, "Why Privacy Is Important," *Philosophy & Public Affairs* 4 (1975): 323–33; Robert S. Gerstein, "Intimacy and Privacy," *Ethics* 89 (1978): 76–81; and Jean L. Cohen, *Regulating Intimacy: A New Legal Paradigm* (Princeton, NJ: Princeton University Press, 2002).

what privacy is from why privacy matters, then we are in a better position to answer both questions.

Before moving into discussion of nanotechnology in particular, let us offer two final thoughts. As to what privacy is, we think it has at least something to do with control over information. As to why it matters, its relevance for intimacy and social relationships provides at least part of the answer. But this strikes us as a fairly convoluted way of saying something very simple, which is that disclosure of unwanted information can make people unhappy. The reason that revealing something embarrassing about someone is wrong seems to be just that: the person will be embarrassed. Maybe this ties into some broader matrix of intimacy and social relationships; maybe, for example, this person would lose friends if the information is revealed. But maybe not, and maybe the disclosure does not affect any relationships. Rather, imagine that it only affects the person about whom it is revealed. The disclosure can still be morally problematic, and it is the effects on the relevant person that matters.

But what if the disclosure does not cause any untoward consequences on anyone? So imagine that A spies on B and then reveals some information about B to C. Further imagine that, as it turns out, B just does not care. Or maybe, despite A's intentions, B is ultimately happy that C has found out. Has his privacy still been violated? And, if so, is this violation morally condemnable? In the above paragraph, we made the case that consequences matter, but certainly many people think that there is more to privacy than consequences, namely those that think that we have a right to privacy. Surely credible defenses of privacy could be mounted whether from either consequentialist or deontological accounts; we reviewed, in §10.1, some of the legal foundations of a right to privacy, though moral analogues to those arguments are easily on offer.

Finally, it is worth noticing that privacy has its critics, from at least two directions. First, we could object that privacy is just not that important or that, given various exigencies, privacy protections should be reduced or eliminated.[39] We will return to this discussion briefly in §10.6, though we certainly agree that privacy protections cannot possibly be inalienable and inviolable. Whether it is probable cause justifying a search warrant or terrorism justifying a wiretap, there have to be limits to privacy, even if some cases (e.g., national security) are more controversial. The second criticism, though, is more conceptual. Judith Jarvis Thomson, for example, defends a sort of reductionism about privacy in which she argues that any talk about privacy can be reduced to talk of other things and that, therefore, there is no reason to talk about privacy as a distinctive moral concern. If A

---

[39] See, for example, Amitai Etzioni, *The Limits of Privacy* (New York: Basic Books, 1999). A shorter treatment can be found in his "The Limits of Privacy," in Andrew I. Cohen and Christopher Heath Wellman (eds), *Contemporary Debates in Applied Ethics* (Malden, MA: Blackwell Publishing, 2005), pp. 253–62.

trespasses on B's property to spy on B, we could just say that the violation is of a property right, rather than getting mired in all this confused talk about privacy, which we can henceforth abrogate from our vocabularies; Thomson thinks that we have enough uncontroversial rights at our disposal that we can accommodate all of the relevant privacy discourse while nevertheless jettisoning the concept. This line is initially compelling, but we think that it cannot work. Imagine two cases, one in which A trespasses on B's property and another in which A does the same thing, spies on B, and thereafter discloses embarrassing information. It seems obvious to say that something *worse* happened in the second case and that, in addition to the violation of property rights, B suffered a privacy violation as well. Thomson retains the ability to say that *something* was wrong with A's spying, but it seems to us that she will not be able to make important differentiations as to what the wrongs are.[40] Our conclusion from this section is therefore that privacy is a coherent, if complicated, notion, and that its protection does have at least *prima facie* moral value.

10.3   Radio Frequency Identity Chips

Now that we have some historical, legal, and philosophical context for privacy, let us turn to nanotechnology. It is worth noting that many of the challenges to privacy will relate to new technologies in general, though our focus will be on nanotechnology in particular.[41] Jeroen van den Hoven has written extensively on how nanotechnology bears on privacy, and we propose to follow his work in this discussion; we will also consider an important essay by Vance Lockton and Richard Rosenberg.[42] Van den Hoven writes that there are various facets of technology where the implications and long-term prospects are unclear and that, in these cases, it is hard to do the ethics well since we are inherently more speculative. This is certainly

[40]   For other criticisms, see Thomas Scanlon, "Thomson on Privacy," *Philosophy & Public Affairs* 4 (1975): 315–22. See also Julie C. Inness, *Privacy, Intimacy, and Isolation* (Oxford: Oxford University Press, 1992).

[41]   For more general discussion, see Amitai Etzioni, "Are New Technologies the Enemy of Privacy?," *Knowledge, Technology, and Privacy* 20 (2007): 115–19. See also David D. Friedman, "The Case for Privacy," in Andrew I. Cohen and Christopher Heath Wellman (eds), *Contemporary Debates in Applied Ethics* (Malden, MA: Blackwell Publishing, 2005), pp. 264–75.

[42]   Jeroen van den Hoven, "Nanotechnology and Privacy: Instructive Case of RFID," in Fritz Allhoff et al. (eds), *Nanoethics: The Ethical and Social Implications of Nanotechnology* (Hoboken, NJ: John Wiley & Sons, 2007), pp. 253–66. See also his "The Tangled Web of Tiny Things: Privacy Implications of Nano-electronics," in Fritz Allhoff and Patrick Lin (eds), *Nanotechnology and Society: Current and Emerging Ethical Issues* (Dordrecht: Springer, 2008), pp. 147–62. Finally, see Jeroen van den Hoven and Pieter E. Vermaas, "Nano-Technology and Privacy: On Continuous Surveillance outside the Panopticon," *Journal of Medicine and Philosophy* 32 (2007): 283–97.

true, and there are some chapters in this book where that is precisely the situation in which we find ourselves; two perfect examples are the discussions concerning the military in Chapter 9 and human enhancement in Chapter 12. Privacy, though, is one area in which we already have a reasonably clear picture as to what sorts of implications nanotechnology will have, or even does have. When dealing with nanotechnology and ethics, it is almost a rare luxury to be able to speak in the present tense. The technological developments, as well as the dissemination of the technology, will be ongoing, but some of the issues are already obvious.

The core technology driving privacy concerns is RFID. In this section, we will discuss what RFID is, as well as some of the background behind its development. In subsequent sections, we will discuss three specific ways in which RFID has implications for privacy: item-level tagging, human implants, and RFID-chipped identification. Finally, in §10.7, we will discuss some steps that could be taken to protect privacy given these applications, as well as assess the potential those steps have for success.

RFID is, most basically, a technology based on the need for remote recognition of objects. This technology has been primarily used for inventory, toll collection, and access control; we will provide more details on some of these applications below. The basic idea behind RFID effectively originated in World War II in which UK radar operatives realized that they could detect incoming planes, but could not detect whether those planes belonged to friends or foes: the radar only showed the location of a plane, but no other details. Allied planes were then equipped with transponders, which identified the planes, and hence RFID was born.[43]

For our purposes, we will focus on RFID "tags," which incorporate nanotechnology, especially nanoelectronics. An RFID tag communicates a unique identification number to an electronic reader by radio waves, thus abrogating the need for either physical contact with the tag (e.g., such as would be necessary to read a bar code), or even a direct line of sight to the tag. These tags have a small microchip, as well as a radio antenna, which can transmit data from the chip to the reader. This reader also contains an antenna, as well as a demodulator, which can transform the analogue radio signal into digital data that can then be used by a computer.[44] Tags come in two basic varieties: passive and active.[45] Passive tags have no internal power source, but rather use the electrical current induced by the incoming radio signal. Active tags have their own power supply, in the form of a small battery. Since active tags have batteries, they tend to be bigger than passive tags, and are correspondingly more expensive to manufacture

---

[43] Vance Lockton and Richard S. Rosenberg, "RFID: The Next Serious Threat to Privacy," *Ethics and Information Technology* 7 (2005), p. 221.
[44] Ibid., p. 222.
[45] There are also semi-passive tags; these have a power supply, but that power supply only goes to the chip, and does not power the broadcasting of a signal.

since the battery cost must be added. Passive tags will work indefinitely and active tags, limited by battery life, are viable for approximately 10 years, depending on their type. All of these are quite small; passive tags as small as 0.05 mm^2 and thin enough to be embedded in a piece of paper have already been manufactured, though these require an external antenna.[46] At the time of writing, passive tags are priced at about 10 cents each, though it is widely expected that the costs could be as low as 5 cents each in the near future.

Passive tags can be read at distances of up to 10 m though, for security applications, this distance could be limited to as little as 10 cm. Active tags have a longer broadcast range and can currently broadcast as far as 1,500 m, though that distance is growing. Each RFID tag stores an identification number, and this is the number that is communicated to the reader. The number is usually either 96 or 128 bits long, and most tags contain few or no security measures to protect the data. More advanced – and more expensive – tags can have from 512 bits to 72 kilobits of memory, as well as various data protection and encryption schemes.[47]

RFID tags are already widely used, and approximately 40 million Americans have them in car keys, building access cards, toll payment devices, and so on. Car keys, for example, are increasingly fitted with tags that are then read by a transponder in the steering column; if the transponder does not recognize the key, the car will not start. This makes hotwiring a car impossible and therefore increases security. Speed passes, used to pay highway tolls, communicate an identification number to a reader which then subtracts the toll from a prepaid balance electronically. Tens of millions of pets have been tagged so as to facilitate their identification in animal shelters, and at least 20 million livestock have been tagged to track the outbreak of diseases.[48] RFID tags are hardly on their way; they are already here.

The most important catalyst for RFID technology, though, has been Wal-Mart, the world's largest public corporation. In June 2003, Wal-Mart announced that, by January 2005, each of its top 100 suppliers had to add RFID tags to all crates and pallets arriving at Wal-Mart distribution centers. Its 200 next-largest suppliers had until January 2006 to comply. The motivation for these directives was to improve Wal-Mart's inventory and tracking capabilities, particularly as lines of sight would no longer be needed to establish the location or contents of boxes. A spike in demand for RFID tags ensued, with the top 100 suppliers needing at least 1 billion tags per year to comply.[49] The dates for compliance were not all met, but only

---

[46] "World's Tiniest RFID Tag Unveiled," *BBC News* (February 23, 2007).

[47] Lockton and Rosenberg, "RFID," p. 222.

[48] Ibid.

[49] David H. Williams, "The Strategic Implications of Wal-Mart's RFID Mandate," *Directions Magazine* (July 29, 2004). Available at www.directionsmag.com/article.php?article_id= 629&trv=1 (accessed August 13, 2008).

due to RFID manufacturers being ill-equipped to deal with the increased demand. In a very short time, the impact of Wal-Mart's mandate on the RFID industry has been tremendous. Lockton and Rosenberg point to the following two consequences:

> First, a standard for RFID technology can be created; rather than develop many different, incompatible tags and readers, it is more sensible for the companies to use the retailer's preferred style of tags. [Second], due to the massive demand for them, manufacturers will be able to drastically reduce the price of each tag, from a range of 25–50 cents to 5–10 cents each. This will allow for the technology to be adopted by smaller companies, and used in more applications.[50]

It is certainly true that Wal-Mart has catalyzed the RFID industry, bringing with it a sense of standardization, increasing its output, and driving down its prices. Having now seen some of the basics for this technology, let us turn to consider its impact on privacy. As we said above, we will consider three arenas in which privacy may be affected: item-level tagging, human implants, and RFID-chipped identification.

10.4   Item-Level Tagging

The purpose of the Wal-Mart initiative was to tag crates and pallets that were delivered to its warehouses. In other words, the crates and the pallets were to be tagged, but not the individual contents they contained. As the prices for tags fall and as the infrastructure to use them improves, we can expect not just crates and pallets to be tagged, but even individual items. Industry experts seem to think that this application is at least five years away, but it could certainly be the case that, in the near future, nearly all manufactured products have RFID tags. The point of having tagged items would be for inventory control, and would improve on three limitations of traditional bar codes. First, bar codes need a line of sight to be read, and RFID tags do not. To inventory a rack of clothing, for example, it would not be necessary to physically access each bar code for a scan; rather, merely getting a reader within the proper range would quickly enable the inventory of the entire rack. Second, the readers could presumably be configured in relay such that continuous, updated inventory control were possible: readers could be configured to each rack of clothing and left there, and then the results could be continually sent to a computer database. Traditional scanning can update inventories, but those updated inventories would be wrong as soon as, for example, something was stolen. RFID-controlled scanning could ensure real-time, accurate inventories. Third, bar

---

[50]   Lockton and Rosenberg, "RFID," p. 222.

codes only identify the type of product, rather than any details about it. An RFID tag could contain an electronic product code (EPC), which can hold 96 bits; this is long enough to uniquely identify every product ever made. The EPC contains a unique serial number, which could incorporate size, weight, expiration date, manufacturing location, shipping details, or whatever other information manufacturers or vendors wanted.[51] For example, rather than having to go through looking for expired milk manually, a grocery store might configure a computer such that it is automatically alerted as to the number and location of relevant containers.

Some privacy concerns have already been realized by item-level tagging. Procter & Gamble tagged lipstick such that a consumer picking it up activated a hidden camera; this camera then broadcast images to Procter & Gamble employees for marketing research. Gilette tagged razors such that hidden cameras photographed consumers who retrieved the razors, and these photographs were then sent to the company. This sort of surveillance is obviously morally perilous, and the practices have been stopped following outcry from consumer advocacy groups.

The bigger concern, though, is with consumer tracking. Stores using RFID tagging to help with inventory control will also be able to use the tags for security: readers can be positioned at the store's exits and detect any tags coming and going from the store. After an item is purchased, its EPC might be sent to the security system such that the system allows that item to pass, while still remaining activated against all unpurchased items. But now there are going to be tags on those items, even after they are removed from the store. Some of the tags will probably be on the packaging, but some will not, for at least a couple of reasons. First, for security purposes, it would make more sense to tag the items than the packaging, lest a shoplifter simply dispose of the packaging and thereafter be able to bypass security. Second, some items do not come with packaging, such as many clothes. Benetton, for example, proposed to weave RFID tags into all of its clothes, though this initiative was abandoned after consumer protest. Imagine, though, that some tags were put into clothes, particularly clothes that were worn on a regular or daily basis, like shoes. Or into other items that people would likely have with them often, like wallets or watches. Or car keys, many of which already have RFID tags, as mentioned above.

When an individual carrying one of these items enters a store, it could be through the same security reader that patrons use to exit the store. The RFID tags that are brought into the store would be read by the same security system, though it would obviously not be activated given that none of those EPCs would be in the store's active inventory – at least insofar as this is not a shoplifter returning stolen items to the store! Nevertheless, those EPCs could be recorded by the reader. So suppose that this person has a

---

[51] Ibid., p. 223. More of the technical details for EPCs are available at http://www.epcglobalinc.org/standards/tds/tds_1_4-standard-20080611.pdf (accessed August 15, 2008).

wallet, and that wallet has an RFID tag with an EPC. When the person enters the store, that EPC could be a *de facto* identifier of that person: even if no personal identification is known, that person's EPC would be a unique and regular identifier of his shopping habits. So a drugstore could tell when some person was coming to the store, and even what that person was buying in the store; this latter could be possible by associating the deinventoried EPCs with the reidentified wallet as the person exited. Among other marketing advantages for a store, they could put together more longitudinal purchasing records for individuals because the wallet EPC could be used to connect up all the purchases that were made on separate occasions. They could also gain information about whether individuals come to the store at the same time or different times, on the same day or different days, and whether they use the same branches or different branches. If stores networked and shared this information, a fairly detailed plan of someone's commercial activity could be developed. Given that marketers are always keen to have new and more information, they could surely find some implementation for these sorts of details. Or, even outside marketing, private investigators might try to access the information, or even legitimate law enforcement might make a claim to it.

Given our current presumptions, the information would be effectively anonymous, but there would be ways around that, too. For example, imagine that a store's database reveals that some consumer – as identified by his wallet – turns up every Tuesday at 6:00 p.m. to buy spaghetti; this could be following some child's scheduled soccer game and in keeping with a family tradition. While the identity of the consumer is not (yet) known, it would no longer be hard to figure it out. And once it is discerned, a wealth of information about that person might thereby be revealed. Note also that payments might be made electronically, as with credit cards. This information could be linked to the purchased items, which could then be linked to the EPC in the wallet. And therefore, the next time that wallet enters the store, some store employee could be notified of the holder's name so that he can be greeted personally. What an improvement for personalization and customer service! And oh, by the way, should I get you the same spaghetti sauce that you bought last Tuesday? Or do you want the one you got three weeks ago? That toothbrush that you bought last October should really be replaced, and we are sorry about that expired milk we sold you two weeks ago. See you next Tuesday, Mr Johnson.

Sound far-fetched? Technologically, it is almost trivial, as hinted at above. Scan someone who leaves the store for RFID tags, which the store is going to do anyway because of its security system. Link the EPCs from the sold products to the other EPCs that are detected. Regarding those other EPCs, wait for them to recur in scans. The wallet and the car keys are going to turn up a lot more often than some pair of jeans. Take the most commonly occurring EPCs, and use those as identifiers for the customer; thereafter, link them to purchases, as well as backwards to past purchases.

Now you have a full customer profile. Integrate credit card information into the profile and you have even more, even enough to do targeted direct mailings.

Note that, in addition to enabling consumer tracking, these RFID tags can pose security risks: for example, anyone with a reader and a database of product codes could figure out what someone has in a bag. Money could even be implanted with RFID tags in order to prevent counterfeiting, but an unscrupulous person with the appropriate technology could then read those tags. On a crowded bus, for example, someone with a reader might be able to figure out which passenger has the most money in his wallet, therefore determining targets for robbery. Given the proliferation of RFID technology, these readers will hardly be difficult to access or to afford.

10.5   Human Implants

A second way in which nanotechnology could affect privacy is through human implants of RFID tags; most commentators find this application is even more worrisome than item-level tagging. Human implants have been possible since 2001, when Applied Digital Solutions (ADS) introduced VeriChip. VeriChip is encased in a glass container approximately the size of a grain of rice, and is injected into the fatty tissue below the triceps. The chip contains a 16-digit identification number that can then be recognized by a reader; the technology is effectively the same as with item-level tagging. Implantation, including the cost of the chip, is approximately $150. There have been three uses of VeriChip which have received considerable media attention, and which we will briefly review.

First, the Mexican Attorney General had his staff implanted with chips such that they could gain access to secure areas; if a scanner does not match someone's number to one in a database, they are not able to pass. This story has been especially bizarre insofar as it has been so commonly misreported. The Associated Press, for example, reported that 160 staff had been implanted, and this number has been repeated just about everywhere it is discussed.[52] Nevertheless, the actual number of staff implanted seems to have actually been under 20.[53] Furthermore, the purpose of the implants seems to be confused, with some people reporting that it is so that staff can be tracked after kidnapping.[54] But given the range of VeriChip's broadcast

---

[52]   Will Weissert, "Microchips Implanted in Mexican Officials: Attorney General, Prosecutors Carry Security Pass under Their Skin," *Associated Press* (July 14, 2004). Available at http://www.msnbc.msn.com/id/5439055/ (accessed August 14, 2008). Cf. van den Hoven, "Nanotechology: RFID," p. 256 and Lockton and Rosenberg, "RFID," p. 224.

[53]   Thomas C. Greene, "Anti-RFID Outfit Deflates Mexican VeriChip Hype: Only 18 Volunteers to Date," *The Register* (November 30, 2004). Available at http://www.theregister.co.uk/2004/11/30/mexican_verichip_hype/ (accessed August 14, 2008).

[54]   See, for example, van den Hoven, "Nanotechology: RFID," p. 256.

is approximately 10 cm, this means that you could not locate the kidnapped victim using the implant. All of this, of course, is one of the hazards of nanohype: the real story seems to be that around 20 people were implanted so that they could access secure areas.

There are also clubs that implant VIP patrons with chips, with Baja Beach Club in Barcelona being the one that has gained the most media attention.[55] Patrons of this club are often in bathing apparel and, therefore, have no obvious way to carry money with them. If they are implanted with VeriChip, then their 16-digit identification number can link to their debit account such that they can literally pay for drinks by offering their triceps; the chips also offer access to VIP areas. The implantations take place on site, and after patrons have signed a waiver. The chips can be surgically removed when they are no longer desired.

Finally, the US Food and Drug Administration has approved VeriChip for implantation into humans for quick access to medical records.[56] While this infrastructure is not yet well developed, the idea is that a person could literally be scanned when entering a medical facility and all of her medical records, which would be linked to her VeriChip identification number, would then be accessible. If she were traveling, for example, her physical records might all be somewhere else. Or maybe she would be unconscious or otherwise debilitated in such a way that she would be unable to provide appropriate identification, were it not already on her person. The chip, though, linked to a database, would solve these concerns and therefore engender more effective medical care.

Turning now to privacy, the first thing to note is that, at the time of writing, the 16-digit identification codes on VeriChip are not encrypted: anyone with a scanner and who can get within range (again, approximately 4 inches) can access an identification number. ADS has been very careful to claim that their databases are secure, which, assuming that is true, means that it would be impossible to get any personal information from that database even if the identification code could be obtained. But this hardly seems to solve all of the problems.

First, VeriChip can be cloned, and directions on how to clone the chips are all over the Internet. What this means is that, if someone can access the broadcast from the chip – by using a reader – then the "unique" number can be reprogrammed into another chip. So if the Mexican Attorney General is sitting next to someone on a crowded train, his triceps could be

---

[55] See, for example, "Barcelona Clubs Get Chipped," *BBC News* (September 29, 2004). Available at http://news.bbc.co.uk/2/hi/technology/3697940.stm (accessed August 14, 2008). Another Baja Beach Club in Rotterdam offers a similar option, as does Bar Soba in Glasgow. By publication of this book, the number will almost certainly be even greater.

[56] "FDA Approves Computer Chips for Humans: Devices Could Help Doctors Store Medical Information," *Associated Press* (October 13, 2004). Available at http://www.msnbc.msn.com/id/6237364/ (accessed August 14, 2008).

scanned, the chip cloned, and access gained to whatever secure areas he has access. To be sure, cloning the chip takes some equipment and technical ability, but it is not complicated.

Second, imagine that someone wanted to gain access to a secure area, but did not have the ability to clone the chip. Rather than his keys, for example, now the Attorney General *himself* would be the means to gain the desired access. Most of us would surely rather be robbed for our keys than for our arms, and VeriChips users are at some inherent risk given that the chips cannot be removed non-surgically.

Third, VeriChip has, as mentioned above, been suggested as a means for protection against kidnapping, though this just is not viable: VeriChip is a passive chip which only responds to the presence of a scanner and, furthermore, it has a very limited read range. It is therefore almost impossible that the chip could be of any use in tracing kidnapping victims, unless an extensive network of readers were set up and, furthermore, in such a way that they would necessarily be close to our triceps. ADS has developed another chip, which is of more use for personal security: this one includes a global positioning system that broadcasts its location through a cell phone carried by the user. However, it is hard to see how this would be of any use at all. If kidnappers knew that these chips existed, all they would have to do is disable the cell phones that would be making the transmission; surely this is one of the first things that they would check. Even worse, though, they might just go about searching for implanted chips to remove. Since the chips are always implanted in the same place, it would hardly be difficult to check.[57] It seems to us that the would-be kidnap victim could not be better off with this chip and, in fact, could be worse off.

Some of the uses are probably pretty innocuous, as with the implanting of chips for club patrons. It is unlikely that it would be worthwhile to clone a chip for a drink tab, or that there would be any point in abducting someone to gain access to VIP areas. Simple security measures, such as having the associative debit accounts kept at relatively low levels, would solve whatever worries we might have about this application. There do not seem to be any health risks, as the glass casing for the chip is non-reactive with the human body. The only concern might be the limited lifespan of popular clubs, but the chips can always be removed easily through extremely minor surgery.

The linkage of medical records to chips, though, does raise privacy worries, but they seem surmountable, at least to us. The primary point to make here is that the unique identification code does *nothing* unless one has access to the appropriate medical databases. This is different from, for example, the security case where cloning the chip could facilitate access to a secure area. Medical databases, while not invulnerable, have substantial

---

[57] Lockton and Rosenberg, "RFID," p. 224.

protections on them. The problem is that the chips are not encrypted, but this does not matter so long as, even with the code, the ill-intentioned person cannot access the information. We wonder, though, what the big advantage to using these chips for medical purposes actually is. The argument is supposed to be that it provides a quick way to access data, and that this will lead to improved medical outcomes. But how much faster, really, is using a chip than a driver's license? The argument cannot be simply that not everyone has identification on them, as only a few thousand people have been implanted with VeriChip. And how much information is actually going to be available if someone is scanned? Our medical histories are, for most of us, scattered among disparate providers and are hardly unified. Those who talk about chips being the panacea for medical information would first have to figure out a way to get all of this information integrated and in an accessible way. However, then the chips become unnecessary since, once the databases exist, you hardly need a chip to access them. So now we would seemingly be back to the argument that not everyone carries identification with them, which hardly seems powerful enough to justify everyone getting chipped. Rather, it seems to us that the focus should be on the infrastructure for collecting and making available medical information, rather than on whether it is accessed with a chip or with a driver's license.

10.6 RFID-Chipped Identification

Third, let us consider RFID-chipped identification as another threat to privacy. Item-level tagging is still a few years away, and the widespread human implanting of chips is even more distant. RFID-chipped passports, though, are already in use and have been issued by the US government since August 2005; other countries around the world are also either adopting or considering adoption of this technology. RFID-chipped driver's licenses have, at the time of writing, already been adopted by several US states, while other states are considering them. Since 9/11, the US government has become increasingly concerned with secure identification, particularly as against the threat of forgery. While many known or potential terrorists might be denied entrance to the country on their passports, they might be able to gain entrance with alternative passports. For this reason, the US government wanted to create more secure passports, and the embedding of an RFID chip was taken to be the answer. This chip contains all of the information printed on the passport, as well as a digital photograph and at least one biometric marker (e.g., a fingerprint or facial characteristic).[58]

The principal worry with the passport has to do with the identity of its bearer being intercepted: that passport would carry name, address, date of

---

[58] Lockton and Rosenberg (2005), p. 225.

birth, and perhaps other information that would be valuable to, among others, identity thieves. Since the passport reading technology will have to be available in so many countries, it is difficult to believe that the readers will be that hard to come by. The reading range is still pretty short (10–20 cm), but in a crowded setting it would be easy to get close enough to scan someone. Particularly when Americans are traveling abroad, they are likely to have their passports with them at most times as the passport is the only widely accepted international form of identification. Picking out an American in, say, Tokyo would just not be that difficult and it might reasonably be assumed that they are carrying their passports and, therefore, whatever information is necessary to steal their identities.

Also, scanning for identities could make it easier to select targets for pick-pocketing or kidnapping. Different areas of the country are more affluent than others, and knowing someone's address would convey some information about their likely wealth. With tools like Google Maps, it is easily possible to get physical images of the houses in which people live, thus giving even more information about their wealth. A criminal could identify someone, follow him to a hotel, go back and look up the potential victim's house online, and thereby choose wealthier victims to target.

RFID-chipped passports also make it easier to track someone, though the short read ranges are pretty limiting. For example, a bar, frequented by foreigners, could have a reader near its door such that it could have a regular list of the identities of its clientele. This information could then be used to tell when someone might be likely to turn up at the bar, and, again, background research could be done on these people in advance such that they might be targeted. Or maybe someone wants to go someplace where they would rather not be recognized, say a house of ill-repute. The pro-prietors might be able to access his passport on entrance, and thereafter be able to extort him lest they reveal that he had been there. Maybe the passport even includes emergency contact information (e.g., his wife) so that this would not be very hard to accomplish. If the technology were incorporated into a domestic driver's license, then all of the above worries would extend beyond foreign travel.

When the US first proposed to use RFID-chipped passports, there were going to be no protections on the data, so any of the above scenarios might have been possible. Various privacy advocacy groups, though, objected to the lack of security, and measures were thereafter implemented. First, the chip requires reader authentication which, in principle, means that unauthor-ized users are not able to access the information on the chip. Second, the information is encrypted, and it can only be decrypted by scanning the "machine readable zone" (MRZ), which is the information printed inside the passport and under the identifying information. Therefore, line of sight is required to read the passport and remote access is impossible. And, finally, a Faraday cage is incorporated within the passport's cover in a metal mesh such that the passport cannot be read unless it is opened; this again goes toward protecting line of sight requirements and preventing remote access.[59] Given these protections, the above worries no longer apply, but we point them out to highlight some of the challenges that RFID-chipped passports did pose to privacy – indeed as mobilized advocacy groups against unprotected prototypes – as well as the challenges that will still have to be dealt with in other countries.

Unlike RFID-chipped passports, RFID-chipped driver's licenses are not protected, though this is by design. In some of the US states where RFID-chipped driver's licenses have been adopted, such as Arizona, New York, and Washington, there are substantial issues with border control. These issues include not just keeping out ill-intentioned people, but also allowing well-intentioned people to be able to cross borders as quickly as possible, particularly as pertains to effective commerce. The solution reached in these states was to have identification numbers on the chips in the driver's licenses, but to allow these to be read remotely, at ranges as great as 20–30 feet (cf. the physical contact that must be made with the MRZ on the passport). If you imagine cars going through a customs plaza, this means that the second car in line, for example, could be pre-scanned before reaching the customs agent. When the car reached that agent, the information would already be on the screen and could have already been checked against various databases (e.g., for a criminal record). This would save a tremendous amount of time as compared to the physical transfer of identification.[60]

The privacy worry with these plans has to do with the fact that the RFID chips in the licenses allow for remote and unsecure access. What the chips have, though, is not personal information, but rather an identification number, that is then linked to personal information within some secure database. Assuming that database really is kept secure, are there any legitimate privacy concerns? It seems as though these licenses would still allow for tracking: so long as the identification numbers could be remotely and unsecurely accessed, then tracking would be possible. Someone with a reader could access a number and then track when that person (or at least that license) turned up at other places, assuming that the surveillance was in place at those other places as well. To modify the above example, imagine that a private investigator is trying to figure out whether someone is frequenting a particular brothel. If the target lives in a state using RFID-chipped licenses with unprotected identification numbers, the investigator would just have to scan the target to get the number, even if the number could not

---

[59] We thank Vance Lockton for discussion of these security features. See also "The US Electronic Passport: Frequently Asked Questions #12." Available at http://travel.state.gov/passport/eppt/eppt_2788.html#Twelve (accessed August 15, 2008).

[60] Again, we thank Vance Lockton for useful discussion. For some details of the plan in Washington State, see Kristen Millares Bolt, "New Driver's License Ok'd for Border: Gregoire Signs Test Program to Allow Non-Passport Travel," *Seattle Post-Intelligencer* (March 23, 2007). Available at http://seattlepi.nwsource.com/local/308864_border24.html (accessed August 15, 2008).

otherwise be used to access any particular information. The investigator could then implant a reader at the brothel, maybe near the entrance; since the read range for these chips is 20–30 feet, the location of the reader might not even be all that important. The investigator could have the scanning history transmitted to him offsite, thus effecting continuous surveillance without actually having to be present.

Despite much of the controversy about the new passports and privacy, it really seems as though appropriate security measures have been put into place; it further seems as though there has been a distinct advantage to these passports insofar as they are more difficult to forge. The licenses, particularly because they can be accessed remotely, might not be perfect. But, again, all that anyone could intercept is an identification number, not any personal information. That number could conceivably be used for tracking, but it is hard to imagine cases like the above becoming very common. And even if there were some surveillance cases enabled by the licenses, they are serving an important function as pertains to the flow of commerce in some US border states; for example, nearly $40 million worth of goods crosses the border per day at Blaine, Washington.[61] Furthermore, they make it harder to forge fake identification and therefore help to secure the country. Being able to maintain commerce and to secure our borders is important and, to our minds, probably more important than a few isolated cases of surveillance. Also, the issue of the licenses is somewhat complicated since, by June 2009, the Western Hemisphere Travel Initiative (WHTI) will require Americans to show a WHTI-approved identification to renter the US from both Canada and Mexico.[62] Passports meet WHTI guidelines, though some of the RFID-chipped licenses do not. Presumably there will be enough motivation for legislators in US states using those licenses to ensure compliance, though this is something that still has to be worked out in some states; by the time this book is published, we expect that it will be.

10.7 Is RFID a Threat to Privacy?

In the three preceding sections, we have considered specific ways in which RFID might be thought to pose privacy risks; we have focused on RFID as the application of nanotechnology most germane to privacy. In this last section, we want to discuss some responses to these risks and, in the process, make connections with the ideas presented in §§10.1 and 10.2.

When discussing the philosophical foundations of privacy in §10.2, we focused on the notion of informational privacy. There are more comprehensive accounts of privacy – for example those that include control over

---

[61] Bolt, "New Driver's License."
[62] Western Hemisphere Travel Initiative. Available at http://travel.state.gov/travel/cbpmc/cbpmc_2223.html (accessed August 15, 2008).

one's life and one's decisions – though we think some of these features are more properly understood as elements of autonomy than of privacy.[63] Nevertheless, our emphasis on informational privacy was a conscious decision meant to foreshadow the applications of RFID that we would discuss in subsequent sections. If any of those applications is a threat to privacy, then this is plausibly because it threatens to make information available. Item-level tagging, for example, could reveal what items someone has bought, or could be used to track that person, thus making available information about their location and habits. The principal worry with human implants is also informational: the identifying number inside the implanted chips could be used to track someone or, if linked to medical databases, could put that person's information at risk. And, with RFID-chipped identification, even if the passports are secure, some of the licenses could reveal identification numbers that could also be used for tracking.

How serious are these threats? Our position is that they are not that substantial, or at least that they can be fairly readily mitigated. Item-level tagging is, to us, the most worrisome, even if most commentators focus on the other two. Let us save that one for last and start with human implants. The first thing to note about human implants is that the VeriChip has a read range of approximately 10 cm. To scan someone, you would have to know that he is implanted, and then be close enough to scan him. And then what? For the medical databasing functions, these numbers merely identify an individual, but do not offer any other information; the information is in a secure database. So long as the database is kept secure, there is no threat. And, even if there were a worry with the databases, the worry would already exist for existing medical records and would not raise any new concerns with RFID.

As we mentioned above, at the time of writing, VeriChip can be cloned; surely this is problematic. If someone were using RFID to access secure areas, then cloning the chip would also allow access to those areas. But this just does not seem hard to fix. First, access could require some sort of biometric verification in addition to chip authentication. A retinal or fingerprint scan, for example, would be adequate to render cloned chips worthless. These scans, though, raise the question of what the point of even having the chips *is* since the biometric information would be more secure and virtually impossible to clone. Assuming that the answer has something to do with cost or available technology, some sort of encryption or authentication could be placed on the chip, though ADS has elected not to do this yet. The price would go up, as might the size of the chip. But, if cloned chips become a problem, these will be ready solutions. Similarly in the medical case, we might worry about someone's chip being cloned such that their medical information could be accessed; encryption or authentication would also solve these worries.

---

[63] See, for example, DeCew, *In Pursuit of Privacy*. See also the discussion of Prosser in §10.1.

RFID-chipped identification either is secure, in the passport case, or only offers access to an identification number; in this latter case, that number can only be used to access information if the end-user has access to a secure database. The databases already exist one way or the other, so their security is orthogonal to whether RFID poses privacy risks. But, as we discussed in §10.6, imagine that the identification numbers can be used for tracking. These would have to be isolated cases and, regardless, there are benefits to the new forms of identification. As we discussed, they help to enable the regular flow of commerce, and they are also harder to forge, thus increasing border security. These are surely both moral goods and therefore have to be factored into a discussion about the net benefits of using the technology. To our minds, these benefits can easily outweigh the costs. It is still not clear, though, why the information cannot be encrypted or require authentication, and none of the experts that we consulted was able to give an unequivocal answer on this issue. Protections would raise prices, but maybe the additional security would make people more comfortable with the technology; remember the outcry over the unprotected passports and the protections that thereafter ensued.

Item-level tagging, though, could facilitate ubiquitous tags that, by express purpose, are readable at various public outlets. We do worry that this information could be acquired and used for various purposes, though most of them seem fairly innocuous: companies using this information for marketing is more likely than private investigators gleaning important information from surveillance. Still, there are some fairly straightforward ways to deal with these issues. Lockton and Rosenberg mention three: aluminum foil, tag killing, and blocker tags, though they are not optimistic about any of these working.[64] We are more optimistic.[65]

If an RFID chip is covered in aluminum foil (or any other conductive metal), it cannot be read. This is not of much use with tags embedded in clothing, but it could be useful when we consider RFID-chipped identification: people could keep their license in foil except when they were required to show it. However, it would seem more convenient just to encrypt the identification codes or to require authentication.

Tag killing, though, could be quite effective with mitigating concerns over item-level tagging: tag killing renders a tag inoperative once an item is purchased, or whenever else it is meant to be deactivated. If stores wanted to use the tags to prevent shoplifting, they would not have any cause to

---

[64] Lockton and Rosenberg, "RFID," pp. 228–9. For the authors, these tactics are meant to address not merely concerns with item-level tagging but broader uses of RFID technology. We find them most appropriate for item-level tagging and, regardless, that is the focus of our primary moral concern as well.

[65] Another optimistic account, though from a different perspective, can be found in Alan R. Peslak, "An Ethical Exploration of Privacy and Radio Frequency Identification," *Journal of Business Ethics* 59 (2005): 327–45.

oppose tag killing since the deactivated tags would only be in purchased items. Those items would then be able to bypass security systems, but they would have been paid for, so the stores would have no reason to object.

Lockton and Rosenberg have three objections to the tag killing solution, though these all seem superable. First, tag killing does not stop in-store tracking. But in-store tracking of what? The items a customer is carrying around inside the store? We are not sure what is supposed to be objectionable about this, especially since the items are still the property of the store. There have been cases, like the Procter & Gamble and Gillette ones mentioned above, wherein hidden cameras were deployed for marketing purposes, but these practices are obviously inappropriate and have been discontinued regardless. Second, Lockton and Rosenberg object that stores could differentially interact with customers who did or did not have tags disabled: maybe customers with live tags could go to a special kiosk to return items faster. Given the size of most stores, this seems unlikely. And, even in bigger stores, is it really something to worry about? Why would a store care whether the tag had been disabled upon purchase? So it could keep tracking that person when she returned to the store? Maybe, though we would expect advocacy groups to put a quick end to this. And the store could just disable *all* the tags on purchase, not asking the customers whether to disable or not. Given some of the public relations issues they will face when item-level tagging becomes viable, it is hardly implausible to think that this might happen. Third, Lockton and Rosenberg argue that tag kills might be ineffective, or that the "kill" might just be a temporary disabling. Or else that tags might be disabled with zero overwrites (i.e., zeros would be written in as the new "EPC"), but that this could still be used for tracking, at least while the process was new or few stores were using it. But it just cannot be the case that, technologically, we are unable to come up with a way to effectively kill tags. They would not even have to be killed at all, but could simply be *removed* altogether. If it were too easy to remove tags, they would not be any good as security measures, but maybe the tags could be removed on purchase by some proprietary equipment at the register. These worries about effective tag killing all seem like they could be reasonably addressed.

The third idea Lockton and Rosenberg discuss is a "blocker tag."[66] A blocker tag is "able to effectively block the functioning of a reader by broadcasting every possible RFID identification code, thus preventing the reader from discerning valid information from garbage being sent by the blocker."[67] This blocker tag would therefore create a range of privacy around

---

[66] For the technical details, see Ari Juels, Ronald L. Rivest, and Michael Szydlo, "The Blocker Tag: Selective Blocking of RFID Tags for Consumer Privacy," in Vijay Atluri (ed.), *8th ACM Conference on Computer and Communications Security* (Washington, DC: ACM Press, 2003), pp. 103–11.

[67] Lockton and Rosenberg, "RFID," p. 229.

a consumer, and could be as cheap as 10 cents; stores wishing to look strong on privacy might even pass them out for free. Lockton and Rosenberg have various concerns with this proposal, the most plausible of which are that they would be banned and that they would shift the burden of protection to the consumer, who must be vigilant against surveillance. Regarding this latter, the ultimate burdens are always going to be on consumers, and it hardly seems like much of an imposition – or that it would require any technical expertise – for consumers to keep a small piece of paper, embedded with a tag blocker, in their wallets. The worry about tag blockers being banned, though, is a legitimate one, particularly since they would render security systems inoperable. We suspect that the way this would actually play out, though, is that consumers would threaten to use these unless stores implemented some sort of tag killing systems. Threatened with boycotts and negative media attention, we expect that stores would be forced to comply. This confrontation is still a few years away, though, so we will have to wait and see.

Our principal conclusions from this chapter are that RFID is not likely to be a huge imposition on privacy, or at least not an irrevocable one. As with the Benetton, Procter & Gamble, and Gillette cases, we have seen that consumers have been able to effect positive outcomes as pertains to privacy, and all of this without any dedicated regulations. Given our own views on regulation developed in Chapter 6, we are not inclined to support it in this case, but rather think that consumer advocacy groups are already sufficiently empowered. But we also wish to express some skepticism about whether these sorts of issues are really that important in the first place. As we saw in §10.3, for example, there are over a billion people in the world without access to clean water, and millions of them die per year from this lack of access. Many privacy alarmists grow incensed over the identification number of their sweater registering in a reader, but there really seem more substantive issues to worry about. This is not to trivialize the importance of privacy, or to deny that there are legitimate and substantive ways in which it should be protected; many of these were discussed in the court cases presented in §10.2. But it is to suggest that proper perspective is valuable, and we are not convinced that RFID is inherently perilous, though vigilance against its nefarious application is certainly warranted.

11

Medicine

Applications of nanotechnology to medicine are already under way and offer tremendous promise; these applications often go under the moniker of 'nanomedicine' or, more generally, 'bionanotechnology.'[1] In this chapter, we will present some general background on nanomedicine, particularly focusing on some of the investment that is being made in this emerging field (§11.1).[2] The bulk of the chapter, though, will consist in explorations of two areas in which the impacts of nanomedicine are likely to be most significant: diagnostics and medical records (§11.2) as well as treatment (§11.3), including surgery and drug delivery.[3] Under each discussion, we will survey some of the social and ethical issues that are likely to arise in these applications. Some other social and ethical issues pertaining to nanotechnology and medicine have been or will be covered elsewhere in the book. The subject of medical records and privacy, for example, receives discussion in §11.2, but it was already introduced in §10.5. Enhancement, closely related to medicine, is covered extensively in the following chapter, as well as in §9.2. For those reasons, and the one below, this chapter is somewhat shorter than some of the others.

Unlike some of the earlier chapters in the book, such as Chapter 7, "Equity and Access," or Chapter 10, "Privacy," there is not a lot of deep conceptual work to be done before moving straight into talk of nanomedicine.

[1] See, for example, Tuan Vo-Dinh, *Nanotechnology in Biology and Medicine Methods, Devices, and Applications* (Boca Raton, FL: CRC Press, 2007). See also Christof M. Niemeyer and Chad A. Mirkin, *Nanobiotechnology: Concepts, Applications, and Perspectives* (Weinheim, Germany: Wiley-VCH, 2004). Finally, see Christof M. Niemeyer and Chad A. Mirkin, *Nanobiotechnology II: More Concepts and Applications* (Weinheim, Germany: Wiley-VCH, 2007).

[2] This chapter is adapted from Fritz Allhoff, "The Coming Era of Nanomedicine," *American Journal of Bioethics* 9.10 (2009), 3–11.

[3] Some of that discussion will be adapted from Fritz Allhoff, "On the Autonomy and Justification of Nanoethics," *Nanoethics* 1.3 (2007): 185–210.

Whereas we had to clarify, for example, what justice and privacy even are, there are fewer conceptual challenges in thinking about medicine, or at least fewer challenges relevant for our discussions here. In this sense, the present chapter is more like Chapter 8, "Environment," or Chapter 9, "Military," insofar as we will assume some relatively uncontroversial conception of what medicine is. Some of the conceptual issues find themselves more at home in the following chapter on enhancement, so we will leave them for discussion there. Some other conceptual issues in medicine – e.g., the question of what is a disease[4] – do not have anything (directly) to do with discussions of nanomedicine in the way that, for example, conceptions of privacy had to do with RFID technology: in assessing whether that technology posed a threat to privacy, we had to think about what privacy was in the first place. The questions about nanomedicine, though, are different because most of the issues are not theoretically laden in the same way.

11.1   The Rise of Nanomedicine

Despite previous comments to the contrary, we do want to start with a simple conceptual worry, which is how to properly delimit the scope of nanomedicine. In each of the chapters for the third unit of this book, we consider specific applications of nanotechnology that raise social and ethical issues, and most of those applications straightforwardly incorporate nanotechnology. The status of nanomedicine, though, is somewhat different, in at least a couple of ways. First, as we will see in subsequent sections, nanomedicine does not always concern specific products, such as those considered in other chapters. Rather, nanomedicine can consist of certain techniques that we largely have not considered elsewhere. This challenges some of the classifications, in the sense that there is not always some specific *thing* that can be looked at and assessed as nanotechnology (or not). This point will become clearer in the remainder of the chapter.

Second, the non-scientific literature on nanomedicine is, at the time of writing, extremely undeveloped. In doing the research for this chapter, we were only able to find a small handful of papers that even countenanced some of the social and ethical implications that nanomedicine could have. This is in stark contrast to some of the topics considered in other chapters in the book: in a nanotechnology context, the developing world, environment, privacy, enhancement, and the military have all received more scholarly attention than nanomedicine. This is peculiar, especially in the

---

[4]   See, for example, Robert A. Freitas Jr, "Personal Choice in the Coming Era of Nanomedicine," in Fritz Allhoff et al., *Nanoethics: The Social and Ethical Implications of Nanotechnology* (Hoboken, NJ: John Wiley & Sons, 2007), pp. 161–72.

cases of enhancement and the military, which are far more speculative and, to some extent, futuristic than nanomedicine. One suggestion as to why this could be the case is that nanomedicine, unlike some of the other applications of nanotechnology, simply does not raise any social or ethical concerns that have not already appeared in other guises. But our view is closer to the claim that *none* of these applications raises any substantively new concerns; rather, they only change the context in which they are to be approached.[5] This sort of skepticism is not to cast aspersions on the present project, which can elucidate those contexts instead of trying to motivate some new theoretical approach altogether. Nevertheless, the asymmetry between the social and ethical discussion of nanomedicine and other applications of nanotechnology bears notice.

One of the few significant contributions to discussion of the social and ethical issues in medicine comes from Raj Bawa and Summer Johnson.[6] Bawa and Johnson, before moving to particular discussions about nanomedicine, offer some general comments about the pharmaceutical industry that warrant attention, particularly as is relevant for much of the investment that is being made in nanomedicine. Drug companies, for example, are always striving to increase the success rate of their products, as well as to decrease research and development (R&D) costs, including time to development. Getting a new drug to market is an extremely daunting task: the economic cost can be as high as $800 million;[7] the time to market is usually 10–15 years; and only one of every 8,000 compounds initially screened for drug development ultimately makes it to final clinical use.[8] Annual R&D investment by drug companies has climbed from $1 billion in 1975 to $40 billion in 2003, though the number of new drugs approved per year has not increased at all; it has stayed at a relatively constant 20–30 approvals per year.[9] (New drugs account for only approximately 25 percent of the approvals, with the other 75 percent coming from reformulations or combinations of already approved drugs.) And, due to these high costs, only about 30 percent of new drugs are recovering their R&D costs. International pressures are mounting on US pharmaceutical companies as

[5] See, for example, Allhoff, "On the Autonomy." See also Fritz Allhoff, "What Are Applied Ethics?" (unpublished).

[6] Raj Bawa and Summer Johnson, "Emerging Issues in Nanomedicine and Ethics," in Fritz Allhoff and Patrick Lin (eds), *Nanotechnology and Society: Current and Emerging Social and Ethical Issues* (Dordrecht: Springer, 2008), pp. 207–23.

[7] Joseph A. DiMasi, Ronald W. Hansen, and Henry G. Grabowski, "The Price of Innovation: New Estimates of Drug Development Costs," *Journal of Health Economics* 22 (2003): 151–85. See also Christopher P. Adams and Van V. Brantner, "Estimating the Cost of New Drug Development: Is it Really $802M?," *Health Affairs* 25.2 (2006): 420–8.

[8] Bawa and Johnson, "Emerging Issues," p. 211.

[9] Norman L. Sussman and James H. Kelly, "Saving Time and Money in Drug Discovery: A Pre-emptive Approach," in *Business Briefings: Future Drug Discovery 2003* (London: Business Briefings, Ltd).

well, especially with production being increased in low-cost countries like India and China, coupled with the expiration of some American patents.[10]

None of this is to lament the plights of pharmaceutical companies. The biggest ones – e.g., Johnson & Johnson, Pfizer, Bayer, and GlaxoSmithKline – have annual revenues at or close to $50 billion, and all have under $10 billion per year investments in R&D. Even factoring in total expenses, these biggest companies have annual profits of over $10 billion each.[11] But, of course, those companies always want to become more profitable, and nanotechnology offers promise in this regard. It is worth quoting Bawa and Johnson at length:

> Nanotechnology not only offers potential to address [the above] challenging issues but it can also provide significant value to pharma portfolios. Nanotechnology can enhance the drug discovery process via miniaturization, automation, speed and the reliability of assays. It will also result in reducing the cost of drug discovery, design and development and will result in the faster introduction of new cost-effective products to the market. For example, nanotechnology can be applied to current micro-array technologies, exponentially increasing the hit rate for promising compounds that can be screened for each target in the pipeline. Inexpensive and higher throughput DNA sequencers based on nanotechnology can rescue the time for both drug discovery and diagnostics. It is clear that nanotechnology-related advances represent a great opportunity for the drug industry as a whole.[12]

As expected, pharmaceutical companies are already investing in nanotechnology. Analysts have predicted that, by 2014, the market for pharmaceutical applications of nanotechnology will be close to $18 billion annually;[13] another report indicates that the US demand for medical products incorporating nanotechnology will increase over 17 percent per year to $53 billion in 2011 and $110 billion in 2016.[14,15]

The worry is that all of this investment and interest in nanotechnology, particularly as is motivated by a race to secure profitable patents, will lead to a neglect of important social and ethical issues that should be explored. Bawa and Johnson point out, correctly, that the time for such reflection is not once these technologies come to market, but before R&D even begins.[16] Insofar as the investments are large and growing and, as mentioned above, insofar as the social and ethical literature on nanomedicine is almost

---

[10]   Bawa and Johnson, "Emerging Issues," p. 212.
[11]   "The Top 50 Pharmaceutical Companies Charts & Lists," *MedAdNews* 13.9 (2007).
[12]   Bawa and Johnson, "Emerging Issues," p. 212.
[13]   Warren H. Hunt, "Nanomaterials: Nomenclature, Novelty, and Necessity," *Journal of Materials* 56 (2004): 13–19.
[14]   The Freedonia Group, Inc., *Nanotechnology in Healthcare* (Cleveland, OH: Freedonia, 2007). Available at www.freedoniagroup.com/brochure/21xx/2168smwe.pdf (accessed October 16, 2008).
[15]   Bawa and Johnson, "Emerging Issues," p. 212.
[16]   Ibid., pp. 212–13.

non-existent, this imperative has already been violated. The principal social and ethical issues will have to do with safety, especially toxicity: some of these issues have to do with the nanomaterials involved in nanomedicine, which have been poorly studied thus far, particularly as they interact with complex biological organisms. It is quite likely that some of these applications will have unwelcome results in their hosts, at least some of which might not be predicted ahead of time. These issues give rise to the notions of risk and precaution, which we considered in Chapter 5; therein, we articulated a general framework for thinking about risks, and this received some further articulation in Chapter 8, "Environment." For these reasons, we will not revisit those more general issues here, though we will develop some of the following issues within those contexts.

11.2   Diagnostics and Medical Records

Consider the trajectory of some particular health remediation. Effectively, there are two central steps: first, health care professionals have to assess the health of their patients, both through diagnostics and through pre-existing medical records; and, second, they have to choose some treatment plan to address whatever health issues those patients are having. This is surely an oversimplified view of medicine, particularly as it fails to include any of medicine's social aspects, including interaction with patients. But, for our purposes, it will work just fine: figure out what is wrong with the patient and then fix it. There are all sorts of diagnostic tools available to the medical community, ranging from simple patient interviews and examinations up to more sophisticated imaging tools like computerized tomography (CT), magnetic resonance imaging (MRI), and positron emission tomography (PET). Other diagnostic tools include blood work, DNA analysis, urine analysis, X-rays, and so on. And there are all sorts of ways to gain access to patients' medical histories. Again, simple patient interviews, if limited, can be effective, and medical records, prepared by other health care professionals, can offer a wealth of salient information. Once the diagnosis has been made, treatment can proceed through, for example, surgery or drugs, to which we will return in §11.3.

We start with diagnostics, deferring medical records to below. Many diagnostic techniques are limited, particularly given certain maladies that health care professionals would like to be able to diagnose effectively. Cancer, for example, is one arena in which nanotechnology is likely to have the biggest impact, and this impact will come in terms of both improved diagnostics and improved treatment.[17] Many cancer cells have a protein, epidermal growth factor receptor (EGFR), distributed on the outside of their

---

[17]   National Cancer Institute, "Targeted Gold Nanoparticles Detect Oral Cancer Cells," *Nanotech News* April 25, 2005. Available at http://nano.cancer.gov/news_center/nanotech_news_2005-04-25b.asp (accessed August 7, 2007).

membranes; non-cancer cells have much less of this protein. By attaching gold nanoparticles to an antibody for EGFR (anti-EGFR), researchers have been able to get the nanoparticles to bind to the cancer cells.[18] Once bound, the cancer cells manifest different light scattering and absorption spectra than benign cells.[19] Pathologists can thereafter use these results to identify malignant cells in biopsy samples.

This is just one example, but it has three noteworthy features: cost, speed, and effectiveness. Given some other traditional diagnostics that have been available, this one offers an improvement in all three regards. And, while cancer is a critical health problem that must be addressed, the general approach that this diagnostic uses can be generalized beyond just cancer. To wit, if nanoparticles can differentially bind to something – by which we mean that they bind more readily to that thing than to everything else – then it will be easier to identify. Improved diagnostics lead to improved treatments, which lead to better health outcomes.

Diagnostically, the concerns with these sorts of applications center around toxicity, particularly when the applications are manifest *in vivo* (as opposed to *in vitro*). Conventional diagnostic mechanisms, however, manifest the same structural features as nano-diagnostics (i.e., applications of nanotechnology to diagnostics); there do not seem to be any substantially new issues raised in this regard. Consider, for example, X-rays, which use electromagnetic radiation to generate images; these images can be used for medical diagnostics. But radiation, absorbed in large dosages, is carcinogenic, so health care professionals have to be judicious in their application thereof.[20] The radiation outputs for X-rays are reasonably well understood, as are their toxicities in regards to human biology.[21] As when considering treatment options, health care professionals must consider these toxicities,

---

[18]   Ivan El-Sayed, Xiaohua Huang, and Mostafa A. El-Sayed, "Selective Laser Photo-Thermal Therapy of Epithelial Carcinoma Using Anti-EGFR Antibody Conjugated Gold Nanoparticles," *Cancer Letters* 239.1 (2006): 129–35.

[19]   Ivan El-Sayed, Xiaohua Huang, and Mostafa A. El-Sayed, "Surface Plasmon Resonance Scattering and Absorption of anti-EGFR Antibody Conjugated Gold Nanoparticles in Cancer Diagnostics: Applications in Oral Cancer," *Nano Letters* 5.5 (2005): 829–34.

[20]   Note that one of the pioneers of radioactivity, Marie Curie, died from aplastic anemia, which was almost certainly caused by exposure to radiation. Rosalind Franklin, whose work on X-ray crystallography was critical to the discovery of the double helical structure of DNA, contracted ovarian cancer at a relatively young age; again, her work was almost certainly responsible.

[21]   There have been numerous studies of the effects of the use X-ray technology in diagnostic procedures. For a recent overview of data relating to risk of cancer see Amy Berrington de Gonzalez and Sarah Darby, "Risk of Cancer from Diagnostic X-rays: Estimates for the UK and 14 Other Countries," *The Lancet* 363.906 (January 31, 2004): 345–51. Also, see James G. Kereiakes and Marvin Rosenstein, *Handbook of Radiation Doses in Nuclear Medicine and Diagnostic X-Ray* (Boca Raton, FL: CRC Press, 1980); and National Research Council, *Health Effects of Exposure to Low Levels of Ionizing Radiation* (Washington, DC: National Academies Press, 1990).

as well as the benefits of this diagnostic mechanism (perhaps as contrasted with other options). Nano-diagnostics admits of a similar deliberative model, even if some of the risks are, at present, less well understood.

Another possibility is "lab-on-a-chip" technologies, which could detect "cells, fluids or even molecules that predict or indicate disease states."[22] These devices could also provide real-time monitoring of various biometric indicators, such as blood glucose levels, as might be useful to a diabetic.[23] Again, the issues could have to do with toxicity, depending on how these chips are deployed. They might be kept inside the body indefinitely or permanently, which could be different from some of the binding diagnostics that, if administered *in vivo*, would ideally leave the body soon thereafter (e.g., if the cancer cells to which they bound were destroyed; see below). Certainly clinical trials will be important, but it could be hard to get the sort of data that would be relevant for long-term effects. For example, perhaps the diagnostics only manifest toxicity some distant years in the future and only after they have been degraded; this information might not even be available for decades.

Depending on how this information is stored and shared, it could give rise to privacy issues, as we discussed in §10.5. In that section, we talked about how chips could serve as repositories for medical information, thus enabling quick access by health care professionals: a patient's medical history, for example, could be stored in a chip and then scanned when the patient checks into a medical facility. Strictly speaking, this seems to us not a diagnostic function, but one of medical record-keeping. In other words, these chips are "passive" in the sense that they are merely storing information. In the previous paragraph, though, we characterized chips that, in some sense, are "active": they could actually determine what is going on in the body and then make that informational available. This distinction is irrelevant socially and ethically insofar as there are privacy issues regardless, but the passive applications are likely to have more information. In terms of diagnostics, then, the focus should probably be on the active applications, but the passive ones, by virtue of their greater informational potential, give rise to greater privacy concerns. But, in §10.5, we expressed pessimism about some of the privacy worries since, in all likelihood, the chips would merely contain some identification number which would allow access to patients' records through some medical database; the chip itself would not actually contain any medical information, so there really is no privacy worry. (Even the identification number could require authentication or decryption as further protection and to prevent the possibility of tracking and surveillance.)

---

[22] Bawa and Johnson, "Emerging Issues," p. 218. See also Harold G. Craighead, "Future Lab-on-a-Chip Technologies for Interrogating Individual Molecules," *Nature* 442 (2006): 387–93.
[23] Bawa and Johnson, "Emerging Issues," p. 218.

But what are these databases and how are they supposed to work? First, we do not presently have the infrastructure to support this sort of databasing. The ideal is supposed to be something like the following: imagine that someone from California happens to be traveling to New York on business, when she falls seriously ill and is rushed, unconscious, to the emergency room. Maybe even, in the rush to get her to the emergency room, her identification is lost. The emergency room staff can simply scan her triceps, plug her identification number into the database and learn who she is, what previous conditions she has, what drugs she is allergic to, and so on. Treatment can then ensue given this access to her history and records. But, of course, no such database exists.[24] And several issues exist for its implementation.

First, it would be a huge endeavor and, presumably, very expensive. We would have to think about whether such a project would be worthwhile particularly given that, at least in many cases, there are fairly straight-forward ways of getting access to the relevant medical information (e.g., by asking the patient or by using identification to contact a relative). To be sure, this will not always work (e.g., with patients who are unconscious and lack relatives and/or identification), but we would have to consider how many cases could not be handled by more conventional means and whether it would be a worthwhile investment of resources to develop this new system. Second, even if the chips themselves did not pose any substantial privacy worries, the database itself might. If all the medical information really went on some sort of national server, then it could be hacked. If it were retained locally, but allowed for remote access (as it would have to be), the same worries would apply. Third, even if the privacy issues were mitigated and gave rise to improved outcomes, there could be social and ethical issues insofar as those outcomes would only be available to citizens of wealthier countries; as with other disparities, questions of distributive justice arise (see, e.g., §7.2).

There are some other aspects of nanomedicine that are promising and probably without any significant social and ethical worries. For example, "quantum dots have been used as an alternative to conventional dyes as contrast agents due to their high excitability and ability to emit light more brightly and over long periods of time."[25] If we can use quantum dots, especially *in vitro*, to allow for more sensitive detection, then this is a useful application. As mentioned above, *in vivo* uses raise issues about toxicity, but there are certainly benign applications of nanomedicine that should be pursued if they increase our diagnostic abilities. These other issues

---

[24]   Interestingly, we have not been able to find anything written on this issue, despite the fact that the media has often characterized these applications of nanotechnology to be revolutionary for medicine. The following discussion will therefore be inherently speculative.
[25]   Bawa and Johnson, "Emerging Issues," p. 218. See also A. Paul Alivisatos, "Less is More in Medicine," *Scientific American* 285.3 (2001): 66–73.

mentioned above, regarding toxicity and privacy, are quite likely superable, though social and ethical attention needs to be paid to how we move forward. (And, as discussed in §11.1, it is unlikely that this imperative has yet been recognized.)

Having offered discussion of how nanomedicine could affect diagnostics and medical record-keeping, let us move on to treatment.

11.3   Treatment

In this section, we discuss how nanomedicine can be used to improve treatment options, and we draw a distinction between two different kinds of treatment: surgery and drugs. 'Surgery' comes from the Greek χειρουργία and through the Latin *chirurgiae*, which is often translated as "hand work." Many cardiac techniques, for example, have to be done manually (literally, by hand) in the sense that a surgeon has to intervene physically on the heart of a patient. Some of these techniques can now be performed remotely, automatically, and so on, but the basic unifier is direct intervention on a physical system by a physician or proxy. Surgery can be contrasted with drug treatment, by which we mean the administration of some pharmacological substance that is prescribed for remediation of some physical ailment; this latter part of the definition serves to distinguish drug *treatment* from other uses of drugs. Drugs are often less direct than surgery in the sense that their introduction is not localized. For example, a patient may swallow some pills that have downstream physiological effects in his body, but those effects are mediated more by other processes than the effects of surgery would be. Though nothing substantial hangs on this distinction, it is useful for mode of presentation.[26]

Nanosurgery – by which we mean surgical applications of nanotechnology – enables techniques that are more precise and less damaging than traditional ones. Let us offer a few examples.[27] First, a Japanese group has performed surgery on living cells using atomic force microscopy with a nanoneedle (6–8 μm in length and 200–300 nm in diameter).[28] This needle was able to penetrate both cellular and nuclear membranes, and the thinness of the needle prevented fatal damage to those cells. In addition to

---

[26]   It is worth noting that there are probably non-surgical treatments that do not involve drugs: we do not mean to set up a false dichotomy between surgery and drugs. For example, rest might be a treatment for soreness, but it is neither surgical nor drug related. But this point is not relevant to nanomedicine either. For our purposes, the distinction between surgical and drug-related treatments will suffice, without implying that they are exhaustive.

[27]   See also Mette Ebbesen and Thomas G. Jensen, "Nanomedicine: Techniques, Potentials, and Ethical Implications," *Journal of Biomedicine and Biotechnology* (2006): 2–3.

[28]   Ikuo Obataya et al., "Nanoscale Operation of a Living Cell Using an Atomic Force Microscope with a Nanoneedle," *Nano Letters* 5.1 (2005): 27–30. Quoted in Ebbesen and Jensen, "Nanomedicine," p. 2.

ultra-precise and safe surgical needles, laser surgery at the nanoscale is also possible: femtosecond near-infrared (NIR) laser pulses can be used to perform surgery on nanoscale structures inside living cells and tissues without damaging them.[29] Because the energy for these pulses is so high, they do not destroy the tissue by heat – as conventional lasers would – but rather vaporize the tissue, preventing necrosis of adjacent tissue.[30] We already discussed in §11.2 that gold nanoparticles can be used for cancer diagnostics insofar as they can be attached to anti-EGFR which would then bind to EGFR;[31] once bound, cancer cells manifest different light scattering and absorption spectra than benign cells, thus leading to diagnostic possibilities. But this technology can be used for cancer treatment as well as diagnostics: since the gold nanoparticles differentially absorb light, laser ablation can then be used to destroy the attached cancer cells without harming adjacent cells.[32]

What these applications have in common is that they allow for direct intervention at the cellular level. While some traditional laser technologies, for example, have allowed for precision, the precision offered by surgery at the nanoscale is unprecedented. It is not just the promise of being able to act on individual cells, but even that of being able to act *within* cells without damaging them. Contrast some of this precision offered by nanosurgery with some more traditional surgical techniques. Just to take an extreme example, consider lobotomies, especially as were practiced in the first half of the twentieth century: these procedures were effectively carried out by inserting ice picks through the patient's eye socket, and the objective was to sever connections to and from the prefrontal cortex in the hopes of treating a wide range of mental disorders. Independently of whatever other social and ethical concerns attached to lobotomies – which have fallen out of practice since the introduction of anti-psychotics such as chlorpromazine – these procedures could hardly have any degree of precision. Even given the dark history of lobotomies, one of its highlights was the invention of more precise surgical devices; for example, António Egas Moniz's introduction of the leucotome, for which he won the Nobel Prize in 1949. However surgeries are practiced, precision is always important, by which we mean roughly that surgeons want to be able to access the damaged area of the patient's body without simultaneously compromising anything not damaged. The gains in precision from ice picks to nanosurgery

---

[29]   Uday K. Tirlapur and Karsten König, "Femptosecond Near-Infrared Laser Pulses as a Versatile Non-Invasive Tool for Intra-Tissue Nanoprocessing in Plants without Compromising Viability," *The Plant Journal* 31.2 (2003): 365–74.

[30]   Ebbesen and Jenson, "Nanomedicine," p. 2.

[31]   Recall that EGFR is a protein that is usually more often prevalent on cancer cells than on non-cancer cells; anti-EGFR is an antibody.

[32]   Some commentators talk about those applications for cancer treatment under the aegis of drug delivery but, for us, this does not sound quite right: it is not the *drug* that is being bound to the cancer cells, but rather a nanoparticle that is thereafter used to absorb light, heat the cancer cell, and destroy it. For this reason, we classify it as a surgical tool.

are multiple orders of magnitude, and even the gains in precision from recent surgical advances to nanosurgery could be a full order of magnitude.

So what are the worries? As was alluded to in §11.1 and will be discussed more in §11.4, the principal one is that these surgical techniques carry risks, and that in the excitement to rush them to market, those risks will not be adequately explored or assessed. We have already talked about risk in general in Chapter 5, "Risk and Precaution," as well as other places throughout the book, especially Chapter 8, "Environment," and will not review those discussions here. For now, though, it is worth noticing that there is nothing special about nanosurgery in this regard: whatever stance we otherwise adopt toward risk is equally transferrable to this context. That said, it is not obvious what the risks could be for nanosurgery, though dismissal of potential risks could certainly be one of them. Some of the applications with drug delivery do seem more risky than using high-precision surgical techniques, though we will argue below that those are not endemic to nanomedicine either. Of course there are the generic worries about distributive justice and health care such that only a few might have access to nanosurgical technologies, but those again have nothing in particular to do with nanosurgery, and we have already discussed this more general context in Chapter 7, "Equity and Access."

But let us say something about the use of nanoparticles; as discussed above, these might play a role in various cancer treatments. Of particular concern is the toxicity from nanoparticles that might be used, as well as other safety concerns. Whatever is to be said about these risks, though, it hardly follows that similar concerns do not attach to more traditional approaches. Consider, for example, chemotherapy, which uses cytotoxic drugs to treat cancer. The downside of chemotherapy is that these drugs are toxic to benign cells as well as to malignant ones, and there are side effects such as immunosuppression, nausea, vomiting, and so on. When physicians are prescribing chemotherapy, they therefore have to think about these risks and whether the risks are justified. But whether the treatment option involves nanoparticles or not, this basic calculus is unchanged: physicians must choose the treatment option that offers the best prognosis. Toxicity or side effects count against these outcomes, and improved health counts in favor of them. Obviously, there are epistemic obstacles to such forecasting, and physicians must be apprised of the relevant toxicity and side effect data, but there is nothing unique to nanomedicine in this regard.[33]

The point that we want to make is that there are *already* risks in treatments and there is no good reason to think that the risks of nanosurgery are any higher than the risks characteristic to conventional medicine. In fact, there are good reasons to think that the risks of nanosurgery are actually

---

[33] As mentioned above, lack of information about risks is more of an issue in nanomedicine than in some more traditional forms of medicine, though this does not affect general, formal deliberative models.

lower and that the benefits are higher. Compare laser ablation of malignant cells to chemotherapy, for example. Chemotherapy, as mentioned above, is a hard process and one with a lot of costs to the patient (including physical and psychological). If nanomedicine allows us, unlike chemotherapy, to destroy the malignant cells without harming the rest of the organism, it is a definite improvement. Are the nanoparticles that would be utilized toxic? Do they pass out of the body after the treatment? Even supposing that the answers are yes and no, respectively, it is quite probable that these new treatments would be improvements over the old. That is not to say that due diligence is not required, but we suspect that, in the end, there will not be that much to fear, at least at the appropriate comparative level.

Turning now to drug delivery, there are myriad advantages that nanomedicine will be able to confer. Per the discussion in §11.1, there is already a tremendous interest from the pharmaceutical companies in terms of incorporating nanotechnology into their product lines, and that interest is the primary driving force in the rise of nanomedicine; this is not to say that the above-discussed nanosurgical techniques are not impressive, just that they are probably not as profitable. Let us go through three traditional challenges in drug delivery, and explain how nanotechnology can mitigate these challenges.

First, consider absorption of drugs: when drugs are released into the body, they need to be absorbed as opposed to passed through. Nanotechnology can facilitate a reduction in the size of drugs (or, at least, their delivery mechanisms) and, therefore, an increase in their surface-to-volume ratios. The relevance of this was discussed in §1.3, but the basic idea is that it is the surfaces of materials that are most reactive and that, by increasing surface-to-volume ratios, greater reactivity is achieved. As the body absorbs drugs, it obviously has to act on the surfaces of those drugs and, by having more surface area per unit volume, the drugs can be dissolved faster, or even be rendered soluble at all. Speed of absorption is of critical importance to the success of drugs, and nanotechnology will make a difference in this regard. Also, it is worth noting that nanomedicine may obviate the need for oral (or other) administrations of drugs in some cases and allow for topical administration: because the drugs will be smaller, they will be more readily absorbed transdermally. Insofar as some patients would prefer this method of administration, it offers another advantage.

Second, it is not always the case that fast absorption is ideal since, in some cases, it would be better if the drug were released slowly over time. Consider, for example, time-release vitamins: because vitamins B and C are water soluble, they quickly flush from the body if not administered in some time-release manner. If the options were taking a vitamin every couple of hours or taking one that is slowly released over a longer period, there are obvious advantages to the latter. Nanotechnology could be used to create better time-release capacities insofar as it could allow for smaller apertures through which the pharmacological molecules would dissipate. In other words, nanotechnology could help to create lattices with openings through which, for example, single molecules would pass. If only single molecules could pass at any given time from the delivery system (e.g., the capsule), then it would take a longer time to disperse the drug supply and, depending on the application, this could lead to more effective treatment.

Third, because nanotechnology will be able to engineer smaller drugs, the associative drugs might be able to traverse various membranes or other biological barriers that had previously restricted their usefulness. For example, consider the blood–brain barrier, which is a membrane that restricts the passage of various chemical substances (and other microscopic entities, like bacteria) from the bloodstream into the neural tissue. Nano-particles are able to pass this barrier, thus opening up new possibilities for treatment of psychiatric disorders, brain injuries, or even the administration of neural anesthetic.[34] In this case, nanomedicine is not just improving existing treatment options, but perhaps even creating new ones.

Again tabling issues of distributive justice, the principal concern with bring-ing nanotechnology to bear on drug delivery has to do with toxicity and other risks: we simply do not know how these technologies will interact with the body, and there could be negative consequences. As with the dis-cussion of nanosurgery, we are inclined to think that the benefits conferred by the application of nanotechnology to drug delivery outweigh the risks, though the risks in drug delivery are probably greater than the risks with nanosurgery. It is worth reiterating, though, that risks pertaining to delivery are not unique to nanomedicine. Consider, for example, the celebrated case of Jesse Gelsinger, who died in a gene therapy trial.[35] Gelsinger had ornithine transcarbamylase deficiency: he lacked a gene that would allow him to break down ammonia (a natural byproduct of protein metabolism). An attempt to deliver this gene through adenoviruses was made, and Gelsinger suffered an immunoreaction that led to multiple organ failure and brain death. Whether talking about vectors for genetic interventions or nanoparticles, we surely have to think carefully about toxicity, immunoreac-tions, and other safety concerns; the point is merely that these issues are not unique to nanomedicine.

11.4 Moving Forward

In this last section, let us try to tie together various themes that have been developed in the preceding three and, in particular, make some comments about ways in which we hope that nanomedicine will move forward. Throughout, we have indicated skepticism about whether nanomedicine raises

---

[34] "Nanotechnology to Revolutionize Drug Delivery," *in-Pharmatechnologist.com* (March 7, 2005). Available at http://www.in-pharmatechnologist.com/Materials-Formulation/ Nanotechnology-to-revolutionise-drug-delivery (accessed August 27, 2008).
[35] Kristen Philipkowski, "Another Change for Gene Therapy," *Wired* October 1, 1999. Available at http://www.wired.com/science/discoveries/news/1999/10/31613 (accessed August 16, 2007).

any new social and ethical issues, or at least ones that have not already been manifest with existing technologies. This skepticism applies to various other areas of nanoethics as well, though perhaps more so to medicine. Issues in privacy might, for example, be transformed by the proliferation of RFID, though we doubt it. But the issues with nanomedicine seem to be, at most, risks (e.g., toxicity and safety) and distributive justice, and in fairly standard ways. Of course these considerations ought to be taken into account, but they should *always* be taken into account, and nothing inherent to nanomedicine makes us think differently about them.

What does make nanomedicine interesting, at least from a social and ethical perspective, is its extreme profitability. As mentioned in §11.1, pharmaceutical companies make tens of billions of dollars a year in profits, and this leads to a different motivational scheme than exists in other applications of nanotechnology. Let us quickly review the applications discussed in other chapters, starting with nanotechnology and the developing world (Chapter 7). The developing world, practically by definition, simply does not have tremendous amounts of money. Staying with pharmaceuticals, a primary ethical concern is that drugs critical for health in the developing world just are not developed because they would not be profitable; consider, for example, the billions of dollars spent in the US on erectile dysfunction drugs as against the lack of investment for lifesaving antimalarial medication. This is not to say that the developing world cannot be made profitable, but it will not be anywhere near as profitable as various domestic enterprises.

Applications of nanotechnology to RFID tags raise privacy worries, particularly given the profitability of the RFID industry and the expected proliferation of its products. For example, if Wal-Mart's principal suppliers use a billion tags a year, that is $50 million annually in revenues (see §10.3). But if Wal-Mart only used 1 percent of the produced tags, the annual *revenue* for the entire RFID industry (on tags) would be $5 billion, which is about half the annual *profit* of a single, large pharmaceutical company. There are other economic impacts of RFID technology, including scanners, manufacturing, training, and so on, but even aggregated, these have to fall well short of the economic impacts of nanomedicine.

The other three applications that we either considered or will consider at length – environment, enhancement, and military – are probably even less profitable still. It is always hard to make the environment profitable, particularly since it is more often governments than consumers or industry that have to produce the capital outlay. There will be profit in repairing or protecting the environment, but we think that it will be of a far lesser scale than that made by pharmaceutical companies through investment in nanomedicine. Enhancement, which we will consider in the next chapter, will only be available to some people, and we are skeptical about some of the widespread availability that the nanomedia has prognosticated. At least in the US, nearly everyone will take some pharmacological product at some stage of their lives, and many (or most) of those will be transformed by nanomedicine. Therefore the scope and, again, profitability are very high for nanomedicine. The third area is the military, and this one is a little trickier. The Department of Defense, for example, had a 2008 budget of $481 billion, but it is hard to get a sense for what implications this has for sellers of nanotechnology.[36] Obviously the Iraq War plays a large role in this expenditure, as do overall personnel. Will government military contractors be as motivated to pursue nanotechnology as pharmaceutical companies? Or, to put it a different way, will nanotechnology have a bigger economic impact on the military than on medicine? We suspect not, but will not pursue that suspicion here.

Regardless, the point is merely that nanomedicine is likely to be one of the most – if not *the* most – profitable applications of nanotechnology and, furthermore, is the one that is going to be primarily pursued by pharmaceutical companies committed to making profits. These features give rise to social and ethical concerns insofar as they are harbingers for market-first, ethics-last mantras. Imagine that Johnson and Johnson is competing with Pfizer to bring out the next greatest drug; a multi-year patent and billions of dollars hang in the balance. There is a lot of pressure to be first. And, potentially, a lot to be gained by cutting corners along the way, whether during pharmacological development, clinical trials, complete disclosure, or whatever. Organizations like the FDA will have to be extremely vigilant but, of course, they always have to be vigilant. The profits are there to be had whether the drugs incorporate nanotechnology or not; so, in that sense, this is nothing new. Our only point is that these pressures are likely to be more significant in nanomedicine than many or all other applications of nanotechnology.

Moving forward, we hope to see critical engagement on the promise of the technology with its social and ethical implications. In §11.1, we pointed out that there has been very little academic or public work done on these issues: at the time of writing, no more than a couple of papers.[37] Surely this needs to be remedied. To have the pharmaceutical companies developing products without an existing forum to discuss the potential effects of those products – including toxicity and other risks – is not good. Again, some of those effects might not be manifest for many years, as we simply do not have long-term research about how nanotechnology interacts with the body. The promise is certainly high, but it should be negotiated clearly and carefully with attention to social and ethical issues and an accompanying discourse.

[36] Office of Management and Budget, "Department of Defense." Available at http://www. whitehouse.gov/omb/budget/fy2008/defense.html (accessed August 29, 2008).
[37] A notable exception to this, particularly in terms of public engagement, is The European Working Group on Ethics in Science and New Technologies to the European Commission, "Opinion on the Ethical Aspects of Nanomedicine: Opinion No. 21" (2007). An abridged version is published as European Group on Ethics, "Ethical Aspects of Nanomedicine: A Condensed Version of the EGE Opinion 21," in Fritz Allhoff and Patrick Lin (eds), *Nanotechnology and Society: Current and Emerging Social and Ethical Issues* (Dordrecht: Springer, 2008), pp. 187–206.

12

Human Enhancement

This chapter is about the promise and perils of applying nanotechnology to human enhancement; it is a natural extension of the prior chapters on nanomedicine and military nanotechnology.[1] Clearly, nanotechnology's role in medicine is one of the most important and practical in both the near and the far term, perhaps even more so than nanotechnology's potential in humanitarian efforts which face discouraging political, profit, and public hurdles (see, e.g., §7.2). The other major drive for nanotechnology research is the need for national defense or military applications, both defensive and offensive. In this curious intersection of medicine and military, we see the issue of dual-use technologies emerge, where innovations can serve more than one purpose.

In medicine, drugs such as aspirin can both help alleviate headaches and prevent heart attacks. And in the military, a vehicle could be outfitted for both strike capabilities and transporting the wounded. But we do not mean 'dual use' in that limited sense: aspirin is still being used to treat some unwanted conditions, and vehicles still play a battlefield role. Rather, there is a more interesting challenge to policy and ethics. In medicine, we will focus on dual uses, as they follow the therapy–enhancement distinction (see §12.3): some treatments or procedures were originally developed as therapy for the sick, such as steroids to help muscular dystrophy patients regain lost strength, but steroid use by otherwise healthy persons – say, athletes – would increase their strength beyond what humans typically have, thereby enabling them to set new performance records in sports. While in both

---

[1] This chapter expands on Patrick Lin and Fritz Allhoff, "Untangling the Debate: The Ethics of Human Enhancement," *NanoEthics: Ethics for Technologies that Converge at the Nanoscale* 2.3 (2008): 251–64. See also Fritz Allhoff and Patrick Lin (eds), "Nanotechnology and Human Enhancement: A Symposium," *NanoEthics: Ethics for Technologies that Converge at the Nanoscale* 2.3 (2008): 251–327.

cases steroids are used to make an individual stronger, the purpose of the former is for therapy while the latter is for enhancement. Similarly for the military, some civilian technologies, such as speech-recognition software used by customer service call centers, can be used to monitor communications of interest.[2] Again, while the action may be the same – the use of software to monitor telecommunications – the aims are radically different, with the former for business purposes and the latter ostensibly for national security.

Human enhancement, then, is an issue that naturally arises from medicine; and given that defense organizations are always interested in creating stronger, more effective, more durable soldiers, it is an important focus for them as well (see §9.2 for more discussion of enhancement and the military). Nanotechnology is playing and will play an essential part in the development of human enhancement technologies, along with advances in other fields such as micro-electro-mechanical systems (MEMS), genetic engineering, robotics, cognitive science, information technology, and pharmacology.[3] For instance, designs have already been drawn for fantastical innovations such as a respirocyte: an artificial red blood cell that holds a reservoir of oxygen.[4] A respirocyte would come in handy for, say, a heart attack victim to continue breathing for an extra hour until medical treatment is available, despite a lack of blood circulation to the lungs or anywhere else. But in an otherwise healthy athlete or soldier, a respirocyte could boost performance by delivering extra oxygen to the muscles and brain, as if the person were breathing from a pure oxygen tank. We will provide more examples of how nanotechnology is connected to human enhancement throughout this chapter.

12.1 What is Human Enhancement?

"The story of humanity is the history of enhancement."[5] *Homo sapiens* has been such a prolific species, simply because we are very good at relentlessly adapting to our environment. At the most basic level, we have won control over fire and tools to forge a new world around us; we build shelter and

---

[2] In many cases for the military, technologies first developed for defense purposes were later found to have civilian applications, such as the Internet and nuclear power; though for medicine, it seems most dual uses have therapeutic origins.

[3] Mihail C. Roco and William Sims Bainbridge, *Converging Technologies for Improving Human Performance: Nanotechnology, Biotechnology, Information Technology and Cognitive Science* (Dordrecht: Kluwer Academic Publishers, 2003).

[4] Robert A. Freitas Jr, "Exploratory Design in Medical Nanotechnology: A Mechanical Artificial Red Cell," *Artificial Cells, Blood Substitutes, and Immobilization: Biotechnology* 26 (1998): 411–30.

[5] Henry T. Greely, "Regulating Human Biological Enhancements: Questionable Justifications and International Complications," *The Mind, the Body, and the Law: University of Technology, Sydney, Law Review* 7 (2005): 88 and *Santa Clara Journal of International Law* 4 (2006): 88 (joint issue).

weave clothes to repel the brutal elements; and we raise animals and crops for predictability in our meals. With our intellect and resourcefulness, we are thereby better able to survive this world.

However, it is not just the world around us that we desire to change. Since the beginning of history, we also have wanted to become more than human, to become *Homo superior*. From the godlike command of Gilgamesh, to the lofty ambitions of Icarus, to the preternatural strength of Beowulf, to the mythical skills of Shaolin monks, and to various shamans and shape-shifters throughout the world's cultural history, we have dreamt – and still dream – of transforming ourselves to overcome our all-too-human limitations.

In practice, this means that we improve our minds through education, disciplined thinking, and meditation; we improve our bodies with a sound diet and physical exercise; and we train with weapons and techniques to defend ourselves from those who would conspire to kill us or, more so in our past, to expand our empires. But today, something seems to be different. With ongoing work to unravel the mysteries of our minds and bodies, coupled with the art and science of emerging technologies, we are near the start of the Human Enhancement (or Engineering) Revolution.

Now we are not limited to "natural" methods to enhance ourselves or to merely wield tools such as a hammer or binoculars or a calculator. We are beginning to incorporate technology within our very bodies, which may hold moral significance that we need to consider. These technologies promise great benefits for humanity – such as increased productivity and creativity, longer lives, deeper serenity, stronger bodies and minds, and more – though, as we will discuss later, there is a question whether these things translate into *happier* lives, which many see as the point of it all.[6]

As examples of emerging technologies, in 2008, a couple of imaginative inventions in particular are closing the gap even more between science fiction and the real world. Scientists have conceptualized an electronic-packed contact lens that may provide the wearer with telescopic and night vision or act as an omnipresent digital monitor to receive and relay information.[7] Another innovation is a touch display designed to be implanted just under the skin that would activate special tattoo ink on one's arm to form images, such as telephone number keys to punch or a video to watch.[8] Together

---

[6] President's Council on Bioethics, *Beyond Therapy: Biotechnology and the Pursuit of Happiness* (Washington, DC: Government Printing Office, 2003); Raj Persaud, "Does Smarter Mean Happier?," in Paul Miller and James Wilsdon (eds), *Better Humans? The Politics of Human Enhancement and Life Extension* (London: Demos, 2006), pp. 129–36.

[7] Babak Parviz et al., "Contact Lens with Integrated Inorganic Semiconductor Devices," presentation at 21st IEEE International Conference on Micro Electro Mechanical Systems, Tucson, AZ, January 13–17, 2008.

[8] Jim Mielke, "Digital Tattoo Interface," entry at Greener Gadgets Design Competition 2008, New York, NY, February 2008. Available at http://www.core77.com/competitions/GreenerGadgets/projects/4673/ (accessed August 16, 2008).

with ever-shrinking computing devices, enabled by nanoelectronics and other advances, we appear to be moving closer to cybernetic organisms (or "cyborgs"), that is, where machines are integrated with our bodies or at least with our clothing in the near future. Forget about pocket PCs, mobile phones, GPS devices, and other portable gadgets; we might soon be able to communicate and access those capabilities without having to carry any external device, thus raising our productivity, efficiency, response time, and other desirable measures – in short, enabling us to survive our world even better.

Technology is clearly a game-changing field. The inventions of such things as the printing press, gunpowder, automobiles, computers, vaccines, and so on, have profoundly changed the world, for the better we hope. But at the same time, they have also led to unforeseen consequences, or perhaps consequences that might have been foreseen and addressed had we bothered to investigate them. Least of all, they have disrupted the *status quo*, which is not necessarily a terrible thing in and of itself; but unnecessary and dramatic disruptions, such as mass displacements of workers or industries, have real human costs to them.

Such may be the case as well with human enhancement technologies. For instance, on the issue of whether such technologies ought to be regulated or otherwise restricted, one position is that (more than minimal) regulation would hinder personal freedom or autonomy, infringing on some natural or political right to improve our own bodies, minds, and lives as we see fit.[9] Others, however, advocate strong regulation – and even a research moratorium – to protect against unintended effects on society, such as the presumably undesirable creation of a new class of enhanced persons who could outwit, outplay, and outlast "normal" or unenhanced persons for jobs, in schools, at sporting contests, and so on, among other reasons.[10] Still others seek a sensible middle path between stringent regulation and individual liberty.[11]

No matter where one is aligned on this issue, it is clear that the human enhancement debate is a deeply passionate and personal one, striking at the

---

[9] Ramez Naam, *More Than Human* (New York: Broadway Books, 2005); Ron Bailey, *Liberation Biology: The Scientific and Moral Case for the Biotech Revolution* (Amherst, NY: Prometheus Books, 2005); John Harris, *Enhancing Evolution: The Ethical Case for Making Ethical People* (Princeton: Princeton University Press, 2007).

[10] Francis Fukuyama, *Our Posthuman Future: Consequences of the Biotechnology Revolution* (New York: Picador, 2002); Francis Fukuyama, *Beyond Bioethics: A Proposal for Modernizing the Regulation of Human Biotechnologies* (Washington, DC: School of Advanced International Studies, Johns Hopkins University, 2006); Friends of the Earth, *The Disruptive Social Impacts of Nanotechnology: Issue Summary*, available at http://nano.foe.org.au/node/151 (accessed August 16, 2008).

[11] James Hughes, *Citizen Cyborg: Why Democratic Societies Must Respond to the Redesigned Human of the Future* (Cambridge, MA: Westview Press, 2004); Greely, "Regulating."

heart of what it means to be human. Some see it as a way to fulfill or even transcend our potential; others see it as a darker path toward becoming Frankenstein's monster. But before those issues are more fully presented, it will be helpful to lay out some background and context to better frame the discussion.

12.2 Defining Human Enhancement

First, we need to draw several important distinctions.[12] Strictly speaking, human enhancement includes any activity by which we improve our bodies, minds, or abilities – things we do to enhance our welfare. So reading a book, eating vegetables, doing homework, and exercising may count as enhancing ourselves, but these so-called natural human enhancements are morally uninteresting because it is difficult to see why we should not be permitted to improve ourselves through diet, education, physical training, and so on. Alternatively, it is an open question whether emerging, engineered enhancements are similarly unproblematic.

In any event, the natural-versus-artificial distinction may prove most difficult to defend as a way to identify human enhancements given the vagueness of the term 'natural.' For instance, if we consider X to be natural if X exists without any human intervention, then reading a book no longer qualifies as a natural activity since books do not exist without humans. Or else we might allow that some human-dependent things can be natural (e.g., a flint ax or book), depending on their level of complexity or the amount of engineering required. But this just seems an arbitrary requirement, difficult – if not impossible – to defend.[13] Even more contentious is resting the distinction on theological or teleological premises.

Rather, allow us to stipulate for the moment that human enhancement is about boosting our capabilities *beyond the species-typical level or statistically normal range* of functioning for an individual.[14] Similarly, human enhancement can be understood to be different from therapy, which is about

---

[12] We recognize that some advocates of human enhancement argue against making such a distinction, which seems to justify unrestricted human enhancement more easily; e.g., Nick Bostrom and Rebecca Roache, "Ethical Issues in Human Enhancement," in J. Ryberg, T.S. Petersen, and C. Wolf (eds), *New Waves in Applied Ethics* (New York: Palgrave Macmillan, 2008). Even if this position is tenable, we do not want to take that point for granted here, which we will discuss below.

[13] If we want to say that reading books is generally a form of natural enhancement (because books are not as complex to create as, say, a computer), would some books that are difficult to write or have taken years of research then count as *artificial* enhancements? What about reading a simple e-book on Amazon's Kindle™, which is a clearly a complex and engineered device?

[14] Norm Daniels, "Normal Functioning and the Treatment–Enhancement Distinction," *Cambridge Quarterly of Healthcare Ethics* 9 (2000): 309–22.

treatments aimed at pathologies that compromise health or reduce one's level of functioning below this species-typical or statistically normal level.[15] Another way to think about human enhancement technologies, as opposed to therapy, is that they change the structure and function of the body.[16] Admittedly, none of these definitions is immune to objections, but they are nevertheless useful as a starting point in thinking about the distinction, including whether there really is such a distinction.

Thus, corrective eyeglasses, for instance, would be considered therapeutic, rather than enhancement, since they serve to bring your vision back to normal; but strapping on a pair of night-vision binoculars would count as human enhancement, because they give you sight beyond the range of any unassisted human vision. We already mentioned the dual uses of steroids in medicine and sports. Applicable to both sports and soldiers, growing or implanting webbing between one's fingers and toes, or surgically altering one's knees to become double-jointed, to enable better swimming, changes the structure and function of those body parts; this counts then as a case of human enhancement and not therapy.

Likewise, as it concerns the mind, taking Ritalin to treat attention-deficit hyperactivity disorder (ADHD) is aimed at correcting the deficit; but taken by otherwise normal students to enable them to focus better in studying for exams, it is a form of human enhancement. And where reading a book may indeed make you more knowledgeable, it does not make you a great deal smarter than almost everyone else or push your intellect past natural limits; on the other hand, a computer chip implanted into your brain that gives you direct access to Google or spreadsheets would provide mental capabilities beyond the species-typical level.

The last example suggests a further distinction we should make. By 'human enhancement,' we do not mean the mere use of tools; that would render the concept impotent, turning nearly everything we do into cases of human enhancement. But if and when these tools are integrated into our bodies, rather than employed externally, then we consider them to be instances of human enhancement. Of course, this raises the question: what's so special about incorporating tools as part of our bodies, as opposed to merely using them externally to the body?[17] That is, why should the former count as human enhancement, but not the latter? A neural implant that gives access to Google and the rest of the online world does not seem to be different *in kind* to using a laptop computer or pocket PC to access the same; so why should it matter that we are embedding computing power into our heads rather than carrying the same capabilities with us by way of external devices?

---

[15] Eric Juengst, "Can Enhancement Be Distinguished from Prevention in Genetic Medicine?," *Journal of Medicine and Philosophy* 22 (1997): 125–42.

[16] Greely, "Regulating."

[17] Bostrom and Roache, "Ethical Issues."

Though it will be important to explore the issue further, we will not attempt to give a full discussion of that point here, except to suggest that integrating tools into our bodies (and perhaps with our everyday clothing to the extent that we are rarely without our clothes) appears to give us unprecedented advantages which may be morally significant. These advantages are that we would have easier, immediate, and "always-on" access to those new capabilities as if they were a natural part of our being; we would never be without those devices, as we might forget to bring a laptop computer with us to a meeting. And assimilating tools with our persons creates an intimate or enhanced connection with our tools that evolves our notion of personal identity, more so than simply owning things (as wearing name-brand clothes might boost our sense of self). This may translate into a *substantial* advantage for the enhanced person, more so than gained by purchasing an office computer or reading books or training with the best coaches.

Therefore, we might reasonably understand the distinction between human enhancement and mere tools by looking for an always-on (i.e., on-demand or permanent) feature, as opposed to the temporary or contingent access of our daily gadgets and tools (e.g., a mobile phone can be easily lost, stolen, or left behind). But even so, this attempt at a definition needs further elucidation as there are boundary cases. For instance, if permanence of a device, for all practical purposes, can be achieved by embedding or integrating it into our bodies, and permanence is relevant as it provides greater control over those extra capabilities, then it is not clear whether "smart" clothes[18] ought to count as forms of human enhancement. On one hand, clothes can be cast aside like an Apple iPhone™ since they are external to us, yet we are almost never without some clothes, if not particular ones. So while smart clothes may not be as permanent as, say, a computer chip implanted in one's brain, they may approximate the always-on characteristics that track the enhancement-versus-mere-tools distinction. As another example, eyeglasses that are not cumbersome to wear all day (cf. bulky night-vision goggles), and that give us super-vision or double as a computer display, might plausibly be called an enhancement. Therefore, we do not frame this particular distinction as one between internal or implanted tools and tools external to our bodies, which seems to carry less *prima facie* moral relevance and would need greater argument to establish.

This is not to say that our enhancement-versus-tools distinction is ultimately defensible. It is only to point out that it does not help an early investigation into the ethics of using such technological innovations – whatever we want to call them – to consider 'enhancement' so broadly that it obscures our intuitive understanding of the concept and makes everything that gives us an advantage in life into an enhancement.

---

[18] Consider, for example, clothing with devices embedded into it or clothing made from new, dynamic materials that may also serve as armor or for medical purposes.

12.3   The Therapy–Enhancement Distinction

We return to an issue previously raised, namely that some scholars have reasonably objected to, or at least raised difficulties with, the distinctions between therapy and enhancement. For instance, how should we think about vaccinations: are they a form of therapy, or are they an enhancement of our immune system?[19] On one hand, a vaccination seems to be an enhancement in that there is no existing pathology it is attempting to cure, merely a possible or likely pathology we wish to avoid; but are we drawn to declare it as some form of therapy – perhaps preventive therapy – given its close association with medicine? If enhancements in general are ultimately found to be socially or ethically problematic, then counting vaccinations as enhancement opens the possibility that they should be regulated or restricted, which would create a serious public health disaster as well as a counter-example to the claim that enhancements are problematic. Thus, even critics of human enhancement may be loath to put vaccinations in the enhancement bucket, though there does not seem to be an obviously superior reason to think otherwise.

Another dilemma: if a genius were to sustain a head injury, thereby reducing her IQ to merely the "average" or "species-normal" range, would raising her intelligence back to its initial "genius" level count as therapy or enhancement?[20] Either one would seem plausible, but is there a non-arbitrary reason for answering the question either way? If an enhancement, then how do we explain the difference between that and a clear (or clearer) case of therapy in which we return an "average" person who sustains a head injury back to the "normal" IQ range?

The therapy–enhancement distinction holds real stakes, beyond athletic and academic competition. Recent news reports show that the US military is increasingly prescribing anti-depressants to soldiers in combat to alleviate post-traumatic stress as well as stimulants to counteract sleep deprivation – actions which could be viewed as either creating a more effective, level-headed soldier or returning the soldier to the initial "normal" state of combat readiness, further blurring the distinction.[21,22] (Again, see §9.2 for more on military applications.)

---

[19]   Daniels, "Normal Functioning"; Harris, *Enhancing Evolution*; Bostrom and Roache, "Ethical Issues."

[20]   Bostrom and Roache, "Ethical Issues."

[21]   Mark Thompson, "America's Medicated Army," *Time* June 16, 2008. Available at http://www.time.com/time/nation/article/0,8599,1811858,00.html (accessed August 21, 2008); William Saletan, "Night of the Living Meds: The US Military's Sleep-Reduction Program," *Slate* July 16, 2008. Available at http://www.slate.com/id/2195466/ (accessed August 21, 2008).

[22]   If the military were to prescribe such medications prior or preemptively to combat or the targeted condition, then one could make the case for counting that as an enhancement; but this may take us full circle back to the vaccination question, particularly as soldiers are routinely vaccinated against bio-threats such as anthrax.

The above cases notwithstanding, we would agree that there are difficulties in precisely defining "human enhancement" (as there is with making clear definitions of nearly any other concept), but maintaining the enhancement–therapy distinction, at least until it can be more fully explored, is nonetheless important for several reasons.

First, to the extent that pro-enhancement advocates are primarily the ones arguing against the therapy–enhancement distinction, if a goal is to engage the anti-enhancement camp, then it would make for a far stronger case to meet those critics on their own ground (i.e., to grant the assumption that such a distinction exists). If it proves overly charitable to grant this assumption such that the pro-enhancement position is too difficult to defend without it, then perhaps more attention needs to be paid to arguing against the distinction in the first place, given that the debate may hinge on this fundamental issue.

Second, by not making these distinctions, specifically between therapy and enhancement, it may be too easy to argue that all forms of human enhancement are morally permissible given that the things we count as therapy are permissible. That is to say, we risk making a straw man argument that does not make a compelling case either for or against any aspect of human enhancement. Again, if the human enhancement debate turns on this distinction, then much more attention should be paid to defending or criticizing the distinction than has been to date.

Third, at least part of the reason that human enhancement is believed to be the most important issue in the twenty-first century by both sides of the debate[23] seems to be that it represents a collision between our intuitions and our actions. For instance, critics may believe that human enhancement technologies give an unfair advantage to some persons, fracturing local or global societies (even more) between the haves and have-nots.[24] Yet, at the same time, they seem to endorse – to the extent that they have not raised objections to – our use of existing technologies (e.g., mobile phones, computers, Internet) that also seem to countenance the same kind of division to which human enhancement technologies are said to lead us.

As another example, advocates of human enhancement may believe that individual autonomy should trump health and safety concerns, e.g., athletes should be permitted to take steroids or adults should be allowed to take mood-enhancing drugs at will.[25] Yet, at the same time, they do not offer

---

[23] William Hurlbut, Opening Remarks at Human Enhancement Technologies and Human Rights conference, Stanford University Law School, May 26–28, 2006.

[24] Fukuyama, *Our Posthuman Future*; Fukuyama, *Beyond Bioethics*; President's Council on Bioethics, *Beyond Therapy*; Michael J. Selgelid, "An Argument against Arguments for Enhancement," *Studies in Ethics, Law, and Technology* 1 (2007): Article 12, available at http://www.bepress.com/selt/vol1/iss1/art12/ (accessed August 21, 2008).

[25] Naam, *More Than Human*; Bennett Foddy and Julian Savulescu, "Ethics of Performance Enhancement in Sport: Drugs and Gene Doping," in Richard Ashcroft et al. (eds), *Principles of Health Care Ethics* (London: John Wiley & Sons, 2007).

objections to keeping some drugs illegal, such as crystal methamphetamine or crack cocaine, which becomes an even more complicated dilemma if they advocate legalizing other contraband such as marijuana.

This is not to say that these tensions with our intuitions are irresolvable, but only to suggest that common sense is at stake for both sides of the debate. There are strong intuitions that we should be unhindered in our quest to improve ourselves as we see fit, and that there *is* a therapy–enhancement distinction (since we understand 'therapy' and 'enhancement' as meaningfully discrete terms, even if some cases do not neatly fit into either category). So it would be more interesting for pro-enhancement advocates to reconcile their position with the latter intuition, if possible, rather than to reject the distinction, which is less satisfying. Or if the therapy–enhancement distinction really is untenable, then more vigorous argument seems to be needed before we are prepared to cast aside our intuition.

Fourth, the famous philosophical puzzle "The Paradox of the Heap" should be recalled here. Given a heap of N grains of sand, if we remove one grain of sand, we are still left with a heap of sand (that now only has $N - 1$ grains of sand). If we remove one more grain, we are again left with a heap of sand (that now has $N - 2$ grains). If we extend this line of reasoning and continue to remove grains of sand, we see that there is no clear point P where we can definitely say that a heap of sand exists on one side of P, but less than a heap exists on the other side. In other words, there is no clear distinction between a heap of sand and something less than a heap, or even no sand at all. However, the wrong conclusion to draw here is that there is no difference between them, or that the distinction between a heap and a no-heap should be discarded (or between being bald and having hair, as a variation of the paradox goes). Likewise, it would seem fallacious to conclude that there is no difference between therapy and enhancement or that we should dispense with the distinction. It may still be the case that there is no *moral* difference between the two, but we cannot arrive at it through the argument that there is no clear defining line or that there are some cases (such as vaccinations, etc.) that make the line fuzzy. As with 'heap,' the terms 'therapy' and 'enhancement' may simply be vaguely constructed and require more precision to clarify the distinction.

Therefore, at least for the time being and for the purposes of this chapter, we will assume that a therapy–enhancement distinction is defensible and illuminative, at least where it aligns with our intuitions. We allow that it may ultimately be the case that the distinction is shallow and/or that enhancements *per se* are not morally relevant, but we leave those as open questions to be explored in this stage of the debate. It is possible, and perhaps likely, that human enhancements need to be considered separately according to their type or application in an ethics investigation. So, for instance, even if we do not consider vaccines as preventive therapy – which may be the more natural way to see it – but rather view them as enhancements of our immune system, that does not imply that *all* enhancements are morally unproblematic.

What counts as an enhancement and whether it is morally relevant seems to be context dependent. For example, we can imagine a society in which strict equality is the all-important value, trumping individual rights to life, liberty, knowledge, and so on. Immunizing a person from a serious disease might be prohibited in such a world, so as to not upset egalitarian values or disrupt social institutions that strongly rely on a certain range of life expectancy (e.g., a social security system). As another example, consider that freely burning fossil fuels was less morally significant in 1910 than it is in 2010, given our current awareness of global warming, pollution, and their causes and effects. Therefore, context matters, and so it seems premature to say that all enhancements are morally worrisome, irrespective of context. On the other hand, it is also premature to declare all of them to be unproblematic, especially at the start of a debate exactly about those questions.

For a similar reason, it does not really help to dismantle or obviate the therapy–enhancement distinction by claiming that "everything good is an enhancement" as we first considered in §12.2 – i.e., that education, diet, exercise, and so on are all enhancements – and that because these instances are unobjectionable, then no enhancement is intrinsically problematic. Besides recognizing that this is a loose generalization, we could make an argument that even education, diet, etc. *do* have ethical implications (e.g., in that they may create inequities among individuals), but that these implications are outweighed by other considerations, such as liberty, the value of self-improvement, and so on. This, then, is why education, diet, exercise, and so on are unobjectionable, all things considered. Or we can make an argument that strict equality is not morally required in the first place, given the natural and manageable range of variations in our species; if some future vaccine takes us well beyond this "normal" range (e.g., into super-longevity or super-strength), our social systems are not equipped to account for those extra abilities, thereby bringing latent issues of equity, fairness, access, and so on back to the forefront. Similarly, we can imagine a world in which cognitive enhancements no longer hold as much controversy as they do today after social structures in the future have adapted to account for them. Again, context seems to matter.

12.4   Human Enhancement Scenarios

Given the above stipulations about what counts as human enhancement, let us lay out a few more scenarios – real, possible, and hypothetical – to further clarify what we mean by human enhancement. We can loosely group these scenarios into three categories: mental performance, physical performance, and other applications.

In the area of improving mental performance, individuals are already using pharmaceuticals available today to achieve such goals as increased productivity, creativity, serenity, and happiness. We previously mentioned Ritalin use, intended for ADHD patients, by otherwise normal students to boost concentration as a way to study more effectively. In sports, drugs such as beta-blockers, intended to treat high blood pressure and other disorders by slowing down the heart rate, have been used to reduce anxiety as a way to boost physical performance, such as in preparing for an important and nerve-racking putt in golf, or steadying an archer's hand to better release the arrow between heartbeats. In warfare, anti-depressants and stimulants have been used to treat post-traumatic stress and sleep deprivation, thereby creating better, more effective soldiers; see §9.2 for more discussion. And, of course, hallucinogenic and other recreational drugs, including alcohol, continue to be used (and used famously by some authors and artists) to achieve greater creativity, relaxation, and even enlightenment.

In the future, as technology becomes more integrated with our bodies, we can expect neural implants of the kind we mentioned above that effectively puts computer chips into our brains or allows devices to be plugged directly into our heads, giving us always-on access to information as well as unprecedented information-processing powers. Again, nanotechnology is expected to play a significant role here with its ability to shrink electronic devices and batteries. New and future virtual reality programs are much better able to simulate activities, for instance, to train law enforcement officers and soldiers in dangerous situations so that they can respond better to similar events in the real world; and augmented reality applications can help capture more information about a dynamic environment in real time, such as the infrared signature of a sniper hiding in bushes or even a deer on a dark highway.

In the area of physical performance, the use of steroids by athletes is one of the most obvious examples. Cosmetic surgery has also grown in popularity, not for corrective purposes but to increase (perceived) attractiveness. Prosthetic limbs have improved to such a degree that they are already enabling their wearers greater than normal strength and capabilities, sparking a debate on whether athletes with those artificial limbs might participate in the Olympics.[26] Nanomedicine clearly can help advance cosmetic surgery techniques, e.g., to minimize scarring or to perform surgery non-invasively (even from the inside). And nanomaterials, as well as nanoelectronics in robotics, can create stronger, lighter, self-powered prosthetic limbs and other body parts.

In the future, we can expect continuing advances in robotics and bio-nanotechnology to give us cybernetic body parts, from bionic arms to artificial

[26] Steven D. Edwards, "Should Oscar Pistorius Be Excluded from the 2008 Olympic Games?," *Sports, Ethics, and Philosophy* 2 (2008): 112–25.

noses and ears, that surpass the capabilities of our natural body. Today, research organizations such as MIT's Institute for Soldier Nanotechnologies are working on an exoskeleton to give the wearer superhuman strength as well as flexible battlesuits that can, for instance, harden when needed to create a splint or tourniquet to attend to injuries more quickly and effectively; see also §9.2 for more on MIT's work.[27] And we previously mentioned innovative designs such as a contact lens that enables us to see in the dark or receive information from a miniature digital monitor.

And perhaps as an example of both mental and physical enhancement, we should also consider life extension, whether it comes by curing fatal pathologies (such as cancer) or rejuvenating the body/mind or developing anti-aging medicine, and whether it enables us to live another 20 or 100 or 1,000 years (radical life extension). This is a particularly contentious issue in the human engineering debate, not just for obvious concerns related to the burden of overpopulation on quality of life or loss of meaning in life, but also because it seems that we are already – and presumably unproblematically – extending our lives through better nutrition, medicine, exercise, sanitation, and so forth; yet there is something troubling to many about the prospect of radical life extension, even if we can all agree that, in principle, more life is better than less life. We will return to this below.

Other applications include enhancements that may seem gratuitous, such as attempting to physically transform into a lizard by tattooing scales all over one's body and forking one's tongue, or into a cat by implanting whiskers, sharpening teeth and clipping one's ears, or into something other than human by implanting horns in one's forehead; all of these procedures have been done already. In the future, we can envision the possibility that prosthetic flippers, designed today for dolphins, along with artificial gills, etc., might be requested by humans who want to transform into an aquatic animal. These types of enhancement, of course, bring to the forefront the question whether 'enhancement' is the right word to use in the debate in the first place, as opposed to simply 'human engineering' or a more neutral term that does not imply improvement. Indeed, even in cases where technology boosts mental and physical capabilities, it seems that we cannot predict with any accuracy whether there will be any negative psychological or physiological side effects that will offset the intended benefits of a particular enhancement. For instance, in drinking alcohol as a mood enhancer of sorts, we already know that it can hold the unintended effect of a painful hangover; and steroids taken by athletes can have disastrous health consequences.

Moreover, if human enhancement can be ultimately defended, then *un*-enhancements may seem to be morally permissible as well, if individual

---

[27] MIT, Institute for Soldier Nanotechnologies (2008). Available at http://web.mit.edu/ISN/research/index.html (accessed August 21, 2008).

autonomy is the most important value to consider in the debate. There are already medical cases in which: individuals want to amputate some healthy limb from their bodies;[28] parents want to stunt the growth of their bedridden child to keep her portable and easier to care for;[29] and deaf parents specifically want a deaf baby in selecting embryos for *in vitro* fertilization.[30] Un-enhancements aside, we will continue to use 'enhancement' in our discussion here for the most part, since there is a presumption that whatever technology is integrated with our bodies will be expected to deliver some net benefit, real or perceived (otherwise, why do it?). Further, we will limit our discussion here primarily to those technologies that enhance human cognitive and physical abilities, rather than seemingly gratuitous procedures or un-enhancements.

12.5   Untangling the Issues in Human Enhancement

Now, given the above understanding of 'human enhancement,' let us tease apart the myriad issues that arise in the debate. These too are loose non-exclusive categories that may overlap with one another, but perhaps are still useful in providing an overview of the debate: (1) freedom and autonomy; (2) health and safety; (3) fairness and equity; (4) societal disruption; and (5) human dignity.

Let us make a few preliminary notes. First, just as no one could predict with much accuracy how the Internet Revolution would unfold, raising policy issues from privacy to piracy and beyond, the same is likely true with the Human Enhancement Revolution; that is, the framework presented below will undoubtedly evolve over time. However, this does not mean that we should not attempt to address the issues we are able to anticipate. Second, the objective of this chapter is neither to anticipate nor to fully address any given issue, but simply to broadly sketch the major issues. Therefore, the following discussion will raise more questions than it answers in constructing that framework. Finally, in the following, we abstract away from nanotechnology, since we have previously described its role in human enhancement. Also, many if not all of these issues have been raised prior to nanotechnology and so need not be linked to nanotechnology; however, nanotechnology accelerates the urgency of these issues and adds greater depth, since it is enabling breakthroughs in such fields as electronics, robotics, and science research generally, which all drive the development of human enhancement applications.

---

[28]   Clare Dyer, "Surgeon Amputated Healthy Legs," *British Medical Journal* 320 (2000): 332.
[29]   Steven D. Edwards, "The Ashley Treatment: A Step Too Far, or Not Far Enough?," *Journal of Medical Ethics* 34 (2008b): 341–3.
[30]   Carina Dennis, "Genetics: Deaf by Design," *Nature* 431 (2004): 894–6.

Freedom and autonomy

There is perhaps nothing more valued, at least in democracies, than the cherished concepts of freedom and autonomy. (The distinction between the two is not critical to this discussion, so we will not take the space to give precise definitions here; but allow us to stipulate that, at minimum, both concepts are about negative liberty, or the absence of constraints.) But because freedom and autonomy are central to the issue of human enhancement, they add much fuel to the impassioned debate.

Pro-enhancement advocates have argued against regulating enhancements on the grounds that it would infringe on our fundamental ability to choose how we want to live our own lives.[31] Or, in other words, if enhancing our bodies does not hurt anyone (other than possibly ourselves; more on this in the next section), then why should we be prevented from doing so? This is a common objection – arguing especially against governmental intervention – to any number of proposals that involve regulation, from hiring practices to home improvements to school clothing and so on.

Though freedom and autonomy may be viewed in democracies as "sacred cows" that ought not to be corralled, the reality is that we do not have complete freedom or autonomy in the areas of life that we think we do anyway. As examples, freedom of the press and freedom of speech do not protect the individual from charges of libel, slander, or inciting panic by yelling "Fire!" in a crowded theater; our privacy expectations quietly give way to security measures, such as searches of our property and persons at airports or eavesdropping on our communications; and even ancestral homes built by the hands of one's forefathers could be unilaterally seized (and demolished) by the state under eminent domain laws. This is to say that whatever rights we have also imply responsibilities and exist within some particular political system. Therefore it is not unreasonable to expect or define certain limits for those rights, especially where they conflict with other rights and obligations.

Maximal freedom is a hallmark of a *laissez-faire* or minimal state, but a democratic society is not compelled to endorse such a stance, as some political philosophers have suggested.[32] Nor would reasonable people necessarily want unrestricted freedom anyway, e.g., no restrictions or background checks for gun ownership. Even the most liberal democracy today understands the value of regulations as a way to enhance our freedom. For instance, our economic system is not truly a free market: though we may advocate freedom in general, regulations exist not only to protect our rights, but also to create an orderly process that greases the economic wheel, accelerating both innovations and transactions. As a simpler example, by imposing laws on traffic, we can actually *increase* our freedom: by driving

---

[31]   Naam, *More Than Human*; Bailey, *Liberation Biology*; Harris, *Enhancing Evolution*.
[32]   Robert Nozick, *Anarchy, State, and Utopia* (New York: Basic Books, 1974).

forward on only one side of the road, for instance, we can be (more) assured that we will not be a victim of a head-on collision, which makes driving faster a more sensible proposition.

There is another sense, related to free will, in which cognitive enhancements may be infringing: if an enhancement, such as a mood-altering drug or neural implant, interferes with or alters our deliberative process, then it is an open question whether or not we are truly acting freely while under the influence of the enhancement. For instance, a "citizen chip" embedded in the brain might cause us to be unswervingly patriotic and hold different values from what we would otherwise have. Further, external pressure by or from peers, employers, competitors, national security, and others also may unduly influence one's decision making.[33]

Finally, as it relates to autonomy, we should acknowledge the issue of parental rights and responsibilities associated with decisions to enhance children, whether directly (after the child is born) or indirectly through germline enhancements (i.e., interventions to the parent's transmittable genes, prior to the child's birth).[34] Insomuch as parents generally strive to provide the best for their children – e.g., quality of school/education, a proper diet, moral guidance, and so on – do enhancements overstep any bounds in that effort, which might be either obligatory or supererogatory? Prevailing wisdom suggests that parents do not have an unlimited right to raise their children however the parents want; that right, if one exists, seems to be limited by health and safety concerns related to the child, which are reasons similar to those that limit our right to free speech, as mentioned above. Yet, in bioethics, real-world cases exist: for example, deaf parents have been permitted to deliberately select embryos for *in vitro* fertilization that would lead to deaf babies, as previously mentioned in our discussion about un-enhancements. (See also the subsection "Societal disruption" below on paternalistic responsibilities.)

Health and safety

To justify restrictions on our freedom and autonomy, of course, we would need strong, compelling reasons to offset that *prima facie* harm; specifically, we need to identify conflicting values that ought to be factored into our policy making. One possible reason is that human enhancement technologies may pose a health risk to the person operated upon, similar to illegal or unprescribed steroids use by athletes: given how precious little we still

---

[33] David Guston, John Parsi, and Justin Tosi, "Anticipating the Ethical and Political Challenges of Human Nanotechnologies," in Fritz Allhoff et al. (eds), *Nanoethics: The Ethical and Social Implications of Nanotechnology* (Hoboken, NJ: John Wiley & Sons, 2007), pp. 185–97.

[34] For discussion of some ethical issues in germline enhancements, see Fritz Allhoff, "Germ-Line Genetic Enhancement and Rawlsian Primary Goods," *Kennedy Institute of Ethics Journal* 15.1 (2005): 39–56.

know about how our brains and other biological systems work, any tinkering with those systems would likely give rise to unintended effects, from mild to most serious.[35] Even drinking pure water – perhaps the safest thing we can do to our own bodies – may have some harms. For instance, maybe we become dependent on fluoridated water to prevent tooth decay, or we drink too much water, which dilutes sodium in the body to dangerously low or fatal levels. Or consider that many of the foods we eat every day are suspected to have some causal connection to disease or unwanted conditions. It is therefore quite likely that making radical changes to our bodies undoubtedly will have surprising side effects.

Is this reason enough to restrict human enhancement technologies, for the sake of protecting the would-be patient? The answer is not clear. Even if such technologies prove to be so dangerous or risky that we strongly believe we need to protect individuals from their own decisions to use those technologies (through paternalistic regulations), the well-informed individual might circumvent this issue by freely and knowingly consenting to those risks, thereby removing this reason for restricted use.

But even this case does not solve the conflict between freedom/ autonomy and health/safety. First, it is not always clear whether a person's consent is sufficiently informed or not. For instance, consider a partygoer who may have heard that smoking cigarettes can be addictive and harmful but nonetheless begins to smoke anyway; this seems to be a less-informed decision than one made by a person with a parent whose smoking caused a specific and horrible illness (and associated expenses). Furthermore, the partygoer may be unduly influenced by peers or movies that glamorize smoking. So paternalistic regulations could be justified under some circumstances, e.g., where risks are not adequately communicated or understood, for children, and so on.

Second, the assumption that a procedure to implant some human enhancement technology may affect the health and safety of *only* that patient appears to be much too generous. Indeed, it is rare to find any human activity that has absolutely no impact on other persons, either directly or indirectly, such that our own freedom or autonomy is the only value at stake and clearly should be protected. For instance, opponents to regulating such activities as gambling, recreational drugs (including smoking tobacco), prostitution, segregation, and so forth commonly cite the need to protect their freedom or rights as objections to those regulations. Yet, this objection ignores the opposing argument, which is that such activities may harm *other persons*, either actually or statistically.

To look at just one of many examples, at first glance unfettered gambling seems to affect only the gambler (it is his money to win or lose, so the argument goes); but a broader analysis would point out that many gamblers have families whose bank accounts are being risked and that desperate

---

[35]   President's Council on Bioethics, *Beyond Therapy*.

gamblers may commit crimes to finance their addiction, never mind harms to the out-of-control gambler himself. Even marijuana use, which in many cases may be justified and allegedly harms no one, might be traced back to dangerous cartels that terrorize or bully the local population. Furthermore, irresponsible use of the drug could cause accidents or cause the user to neglect his or her obligations, family, etc. Notice here that we are not arguing that activities such as gambling and recreational drug use should be completely banned, but only that some measure of oversight seems to be appropriate for the sake of others, if not also for the welfare of the individual.

Relating back to the human enhancement debate, it seems premature to say that only the would-be enhanced person assumes any risk, even if the procedure does not affect his or her germline (i.e., cannot be passed on to the next generation). The harm or risk to others could also be indirect: where steroid use by athletes sets a presumably wrong example for children whose bodies and minds are still developing, we can anticipate a similar temptation to be created with human enhancement technologies among children. As we suggested above, even parents may feel pressure – or even an obligation – to enhance their children, which arises from the natural desire to want the best for our children or, in this case, make them the best they can be.

Third, even if the harm that arises from any given instance of human enhancement is so small as to be practically negligible, the individual choice to enhance oneself can lead to aggregate harms that are much larger and substantial. For instance, in today's environmental debate, calls are increasing to limit activities from lawn care or drinking bottled water: on one hand, the amount of extra water needed to keep one's lawn green seems small, as is also the amount of fertilizer or pesticide that might leach into the groundwater, but the cumulative effect of millions of homeowners caring for a pristine patch of grass can be disastrous for a nation's water supply and health.

Likewise, as human enhancement technologies improve and are adopted by more people, the once-negligible harms that arise from individual cases may metastasize into very real harms to large segments of society.[36] Life extension, as one case, may appear to be a great benefit for the individual, but on an aggregate scale it could put pressure or burdens on families, retirement programs, overpopulation, and so on; we will return to this below.

Fairness and equity

Even if we can understand why there would be pressure to enhance one's self or children, it is important to note the following: advantages gained by enhanced persons also imply a relative *disadvantage* for the unenhanced, whether in sports, employment opportunities, academic performance, or any other area. That is to say, fairness is another value to consider in the debate.

---

[36] Derek Parfit, *Reasons and Persons* (New York: Oxford University Press, 1986).

A related worry is that the wealthy would be the first adopters of human enhancement technologies, given that they can best afford such innovations (like LASIK eye surgery), thus creating an even wider gap between the haves and the have-nots.[37]

In considering the issue of fairness, we need to be careful not to conflate it with equity. Under most economic theories, fairness does not require that we need to close the gap entirely between economic classes, even when justice is defined as fairness.[38] Indeed, there are good reasons to think that we want some gap to exist, for example, to provide incentives for innovations, in order to move up the economic ladder, and to allow flexibility in a workforce to fill vacancies and perform a wide range of tasks. At least some competition seems to be desirable, especially when resources to be allocated are limited or scarce and when compared to the historically unsuccessful alternative of the state attempting to equalize the welfare of its citizens.

Thus, inequality itself is not so much the point, though any poverty or decline in welfare related to increased inequality may be a serious concern. We do not want people to stop striving to improve their own lives, even if the situation for others is not improved at the same time or ever. Natural advantages and inequities already exist unproblematically anyway; Hobbes recognized that these organic differences did not give any individual or group of individuals so much *net* advantage that they would be invulnerable to the "nasty, brutish, and short" conditions that mark human life.[39]

Yet if human enhancement technologies develop as predicted, they can afford us a tremendous advantage in life, e.g., over others in a competition for resources, so much so that it overstretches the natural range of equality to the point where inequality becomes a more salient issue. This is where the gap between enhanced and unenhanced persons may be too wide to bridge, making the latter into dinosaurs in a hypercompetitive world. If we assume that the benefits of being an enhanced person must be largely paid from the welfare of others, e.g., a job gain by one person is a job loss by another, since the others are now at a relative disadvantage, this may impoverish the unenhanced, which would limit their access to such things as health care, legal representation, political influence, and so on.

Related to the notion of equity is that of fairness. Even if pronounced inequality is morally permissible, there is still a question of *how* an individual accesses or affords a human enhancement technology, which may be unfair or unacceptably magnify the inequality. If the distribution of or access to enhancement technologies is not obviously unfair, e.g., illegally discriminatory, then perhaps we can justify the resulting inequities. But

---

[37] Bill McKibben, *Enough: Staying Human in an Engineered Age* (New York: Henry Holt & Co., 2004).

[38] John Rawls, *A Theory of Justice*, rev. edn. (Cambridge, MA: Belknap Press, 1999).

[39] Thomas Hobbes, *Leviathan* [1651] (New York: Penguin Group, 1982), p. 186.

what would count as a fair distribution of those technologies? A scheme based on need or productivity or any other single dimension would be easily defeated by the standard arguments that they overlook other relevant dimensions.[40] Even if a market system is considered to be fair or an acceptable approximation of it – which is highly contestable, especially after a fresh round of job layoffs and mortgage defaults – many still object to the unfairness of our starting points, which may date back to monarchies, aristocracies, "robber barons" (recall the saying that behind every great fortune there is a great crime), bad luck, and other arbitrary circumstances.[41] Even if the starting points were fair, the subsequent market processes would need to be fair in order for the results to be declared fair (e.g., so that it does not happen that only the wealthy, who can afford human enhancement technologies, gain significant advantages over the unenhanced).[42]

Societal disruption

Fairness and equality are not just theoretical values, but they have practical effects. Gross inequality itself, whether fair or not, can motivate the worse-off masses to revolt against a state or system. But societal disruption need not be so extreme to be taken seriously. Entire institutions today – as well as the lack thereof – are based on a specific range of abilities and rough equality of natural assets. Sports, for instance, would change dramatically if enhanced persons were permitted to compete to the clear disadvantage of unenhanced athletes, smashing their previous records. (This is not to say that sports should ban enhanced competitors, only that allowing them would have a significant effect on careers, and require the expense of valuable resources to adjust sporting programs and contests; and in the end, it is not clear that sports are better off because of it.)

Other institutions and systems include the economic (e.g., jobs), privacy, communications, pensions, security, and many other areas of society. For instance, if life-extension technologies can increase our average lifespan by 20 years (let alone the 100+ years predicted by some scientists and futurists,[43] and assuming that the extra 20 years will be a good life, not one burdened with the illness and unproductivity that afflicts many elderly today), then we would need to adjust retirement programs radically. Would we move the retirement age to 85, which would have negative consequences for job seekers such as new tenure-track academic faculty, or would we

---

[40] Nicholas Rescher, "The Canons of Distributive Justice," in J. Sterba (ed.), *Justice: Alternative Political Perspectives* (Belmont: Wadsworth Publishing Co., 1980).

[41] Honoré de Balzac, *Père Goriot* (1835), trans. Henry Reed (New York: Signet Classics, 2004).

[42] Nozick, *Anarchy*.

[43] Aubrey de Grey, *Ending Aging: The Rejuvenation Breakthroughs That Could Reverse Human Aging in Our Lifetime* (New York: St Martin's Press, 2007); Ray Kurzweil, *The Singularity is Near: When Humans Transcend Biology* (New York: Viking, 2005).

increase contributions to pension plans, which would put pressure on household budgets and employers? Or both? Also, assuming birth rates do not
decline (which causes problems of its own), longer lives will mean more
pressure on resources such as energy and food, in addition to jobs, so this
could disrupt society in negative ways.

Looking more into the distance, if enhancement technologies enable
us to adapt our bodies to, say, underwater living (with implantable gills,
flippers, echolocation, new skin, etc.), then we would need to construct new
institutions to govern that lifestyle, from underwater real estate to pollution rules to law enforcement to handling electronic devices to currency
(replacing the paper money of non-waterworlds). Or if this sounds too far-
fetched, consider humanity's rush into outer space that will require similar
attention to be paid to such issues in the near future.[44]

There are other nearer-term scenarios that may cause social disruption:
a job candidate with a neural implant that enables better information
retention and faster data processing would consistently beat out unenhanced
candidates; a person with superhuman hearing or sight could circumvent
existing privacy protections and expectations by easily and undetectably
eavesdropping or spying on others; more students (and professors) using
Ritalin may grab admission or tenure at all the best universities, reducing
those opportunities for others; and so on.

So societal disruption is a significant concern and seems to be something
we want to mitigate where we can, though this does not imply that we
should resist change in general. Minimizing disruption might be achieved
by transitioning laid-off workers immediately to a new job or job-training
program, rather than allowing the layoffs to come unexpectedly which leaves
the newly unemployed with few options but to fend for themselves. Today,
without this kind of preparation, we trust that these social and economic
disruptions eventually will be handled, but there is still a real cost to those
affected by layoff that could have been better mitigated. The typewriter
industry, as an example, was blindsided by the fast-growing word-processing
industry in the 1980s, leading to the displacement of thousands of workers,
on both the manufacturing and the end-user sides. (Similar situations exist
for the spreadsheet industry that displaced countless accountants and book-
keepers, the computer-aided design industry that displaced graphic artists,
and so on.)

But, unless it will be clearly and seriously harmful, social disruption by
itself does not seem enough to count as a strong reason against regulating
enhancement technologies. After all, we do not wish that typewriters were
never replaced with word-processing programs, though we hope the affected
employees readily found gainful jobs elsewhere. Human enhancement
technologies, likewise, do not necessarily need to be halted or regulated,

---

[44] Patrick Lin, "Space Ethics: Look Before Taking Another Leap for Mankind,"
Astropolitics 4 (2006): 281–94.

but it seems more prudent and responsible to anticipate and prepare for any disruptive effects.

To be clear, there presumably will be benefits to society from enhanced persons. We can expect greater productivity or more creative and intellectual breakthroughs, which is why individuals would want to be enhanced in the first place. But what remains difficult to calculate is whether these gains outweigh the costs or risks, or even the likelihood of either gains or costs – which is needed if we do find it sensible to use a precautionary principle to guide our policy making.

A final, more distant thought related to disruption: if future enhancements give us much greater abilities than we have now, would that allow us to have greater rights or liberties than unenhanced persons? For instance, would the enhanced then have some duty to care for the unenhanced, just as the better-informed and capable parent has a duty to care for her child? If so, then human enhancement may create a moral chasm between the haves and have-nots, just as a theist might believe exists between gods and mortals. In such a case, societal disruption would not be so much a concern by itself, but a more important issue of rights and responsibilities might arise to take precedence over disruptive effects. An analogous case would be if we stepped into a world occupied only by children who, in our adult opinion, are not living the way they ought to live in order to maximize health and happiness: we might take control of that world for paternalistic reasons, no matter how disruptive that change might be.

Human dignity

The fiercest resistance to human enhancement technologies is perhaps a concern about their effect on "human dignity" and what it means to be human.[45] For instance, does the desire for enhancement show ingratitude for what we have and (further) enable an attitude of unquenchable dissatisfaction with one's life? Some researchers suggest that discontent is hardwired into the genetic makeup of humans, which is why we constantly innovate, strive to achieve and gain more, etc.[46] However, even if this is true, it seems to be not so much an argument to promote human enhancement technologies, but more a worry that those technologies are not the panacea or Holy Grail of happiness we might believe them to be; that is, we will still be dissatisfied with ourselves no matter how much we enhance ourselves (unless, of course, we somehow eradicate that part of our DNA that causes discontent).

---

[45] President's Council on Bioethics, *Beyond Therapy*; Michael Sandel, *The Case Against Perfection: Ethics in the Age of Genetic Engineering* (Cambridge, MA: Belknap Press, 2007).
[46] Sarah E. Hill, *Dissatisfied by Design: The Evolution of Discontent* (dissertation) (Austin: University of Texas, 2006); Jack Woodall, "Programmed Dissatisfaction: Does One Gene Drive All Progress in Science and the Arts?," *The Scientist* 21.6 (2007): 63.

Would human enhancement technologies hinder moral development? Many believe that "soul making" is impossible without struggle,[47] and achievements ring hollow without sacrifice or effort;[48] so if technology makes life and competition easier, then we may lose opportunities to feed and grow our moral character. On the other hand, compare our lives today with pre-Internet days: increased connectivity to friends, work, information, etc. is often a double-edged proposition that also increases stress and decreases free time. This, then, raises the related concern of whether enhancement technologies will actually make our lives happier. (If the research mentioned above about discontent in our genes is accurate, then we might have a psychobiological reason to think not.)

Is the frailty of the human condition necessary to best appreciate life? There is something romantic about the notion of being mortal and fallible. But with existing pharmacology, we could eliminate the emotion of sadness today, and work is continuing for drugs that repress memories; but it is not clear that sadness (at least in the normal range, as opposed to clinical depression) is a "pathology" we should want to eliminate, rather than a poignant human experience that we should preserve.[49] Other critics have suggested that life could be too long, leading to boredom after one's life goals are achieved.[50]

Finally, we will mention here the related, persistent concern that we are playing God with world-changing technologies, which is presumably bad.[51] But what exactly counts as "playing God," and why is that morally wrong, i.e., where exactly is the proscription in religious scripture? If we define the concept as manipulating nature, then we all have been guilty of that since the first person picked up a stick. Making life-and-death decisions is a plausible definition, but then physicians as well as soldiers (even in holy wars) could be accused of this charge.

12.6   Restricting Human Enhancement Technologies?

Given the preceding discussion, it should be clear that human enhancement concerns more than just the individual's freedom or autonomy, and that there are plausibly negative consequences for others and society that need to be considered. At least an argument needs to be made that freedom/

---

[47]   John Hick, *Evil and the God of Love* (New York: Harper and Row Publishers, 1966).
[48]   President's Council on Bioethics, *Beyond Therapy*.
[49]   Ibid.
[50]   See, for example, Bernard Williams, *Problems of the Self* (Cambridge: Cambridge University Press, 1973).
[51]   Ted Peters, "Are We Playing God with Nanoenhancement?," in Fritz Allhoff et al. (eds), *Nanoethics: The Ethical and Social Implications of Nanotechnology* (Hoboken, NJ: John Wiley & Sons, 2007), pp. 173–83.

autonomy trumps all other values, but such a position seems unnecessarily dogmatic. These issues point to a policy dilemma of whether we should have regulations or restrictions on human enhancement technologies, so to prevent or mitigate some of the negative impacts considered. (See Chapter 6 for a broader discussion on regulating nanotechnology and its applications, especially §6.3.) Three answers suggest themselves: no restrictions, some restrictions, or a moratorium (i.e., full ban).

A moratorium seems unrealistic to the extent that a worldwide ban would be needed to truly stem the use of human enhancement technologies, and that no worldwide moratorium on anything has yet to work, including on (alleged) attempts to clone a human being. A local moratorium would send patients to "back-alley" enhancement clinics or to more liberal regions of the world, as is the case with "cosmetic surgery vacations" in which those medical procedures are less expensive in other nations. Further, a ban on enhancement research seems to be much too premature – an over-reaction to perceived, future risks – as well as a real threat to therapy-related research today.

At the other end of the spectrum, the idea of having no restrictions on human enhancement technologies seems to be reckless or at least unjustifiably optimistic, given that there are plausible risks. As pointed out earlier, complete freedom or autonomy may be a recipe for disaster and chaos in any case; we do not want to grant the right to yell "Fire!" in a crowded venue or the right for dangerous felons to own firearms.[52]

So what about finding some middle ground with some non-Draconian regulations? Critics have argued that any regulation would be imperfect and likely ineffectual, much like laws against contraband or prostitution.[53] But it is not clear that eliminating these laws would improve the situation, all things considered. Also, as a society, we believe that we ought at least to try to solve social ills, even if we cannot ultimately fix the entire problem; e.g., we cannot stop any given crime from ever occurring again, yet we still have laws against such acts. And even if there are practical reasons to not pursue regulations, would that send the wrong message, e.g., to children, that we countenance or support enhancement without reservations?

The issue of regulating human enhancement technologies (and other issues discussed in this book) will surely not be settled here, nor do we intend it to be. Yet it is important to keep in mind that the human enhancement debate is not just a theoretical discussion about ethics, but has bearing on the real world with policy decisions that may affect not just the would-be enhanced, but also researchers, manufacturers, and social institutions, as well as our ideals of freedom and human dignity. The drive to improve ourselves is strong if not irresistible, and nanotechnology presents itself as a key enabler toward that goal.

---

[52] Perhaps even the right to be happy may be inappropriately exercised, say, at a funeral?
[53] Naam, *More Than Human*.

13

Conclusion

In Greek mythology, Prometheus, the titan of forethought, gave fire to mankind as a gift (though he was summarily punished by the gods with unending torture). In nanotechnology, we also have a rare gift that can enable us to profoundly change our world, from human enhancement to environmental remediation to military technology and more. But just as we should not play with fire before we learn how to control it or its risks – or at least if we do play with fire we should be cautious and keep a pail of water on hand – common sense requires the same for nanotechnology. By understanding what nanotechnology is and why it matters, we can begin to exercise some forethought in developing this important emerging technology and its related policies to ensure that we do not carelessly burn ourselves or burn down our world.

It is not surprising that nanotechnology is still shrouded in mystery. As an emerging technology, it is immensely complex, especially given its wide array of applications and far-reaching consequences. And, where knowledge and policy are lacking, hope and hype rush in to fill the gap. Our discussions in this book were intended to promote a balanced understanding of what nanotechnology really is and why it matters – particularly since nanotechnology seems poised to impact virtually all areas of life, if not usher in the next technological revolution.

Meanwhile, we have resisted the urge to make overarching normative conclusions, beyond the advice to guard against both the extreme of rushing toward nanotechnology without caution, and the opposite of fearing its possible changes and risks so much that we are afraid to make a move at all. It is simply too early for any sweeping conclusions, especially given how little is known and how much is unknown. In any event, we intend for this book to catalyze further dialogue and spark the investigations needed for further progress. Having identified the associative issues throughout the book, let us now retrace our steps to see where we have been and where we are headed.

13.1 Chapter Summaries

This book was organized in three units. In the first unit (Chapters 1–4), we became familiar with the nanoscale and some of nanotechnology's basic features. In the second unit (Chapters 5–7), we attended to some basic issues and principles shared across the various areas of nanotechnology's applications. And, in the third unit (Chapters 8–12), we examined some of the most important areas of applications – environment, privacy, military, medicine, and human enhancement – as well as their ethical and social implications. Let us revisit these chapters individually.

In Chapter 1, we introduced the basic concepts of nanotechnology, including the basic scale at which it operates. Furthermore, we provided some historical context, from Richard Feynman's prizes through Eric Drexler's *Engines of Creation* to the modern explosion of nanoscale materials development. To date, the development of nanotechnology has focused on these novel materials, properties of which we can predict with computer simulation and modeling. With that background, we could better survey the nanotechnologies that already exist in nature, as well as in current applications; the former is informative since nature is an inspiration for many other technologies.

With Chapter 2, we recognized that any full understanding of a scientific discipline or a technology requires familiarity with the basic tools used to develop it. Thus we looked at both the physical tools that are used for developing nanotechnology, and the corresponding theoretical tools. An essential tool today is the electron microscope, since even our best optical microscopes are unable to image structures at the nanoscale. It is also important to understand both the scientific limits to development and the engineering issues, including physics at the nanoscale, such as quantum mechanics, chemical bonding, and material crystal structure.

In Chapter 3, we discussed the large amount of research under way in the field of nanomaterials. Nanomaterials are a critical dimension of nanotechnology because the composition of these materials is nearly inseparable from the associative functions and devices. We introduced the basic materials and the differences between top-down technology and bottom-up technology formation. For instance, carbon nanotubes are one of the most studied nanoscale materials, holding significant promise for a wide variety of applications. But inorganic nanomaterials are also important, such as zero-dimensional materials (e.g., quantum dots), one-dimensional materials (e.g., silicon nanowires and zinc oxide nanobelts), and two-dimensional materials (e.g., photovoltaic thin films).

In Chapter 4, we presented various current applications of nanotechnology and examined how they work, why nanotechnology is vital to their function, and what insight these applications offer into the future of nano-technology. Nanomaterials are pushing the boundaries of physical law to achieve novel technologies, using new methods to place individual nano-structures into higher-order architectures, such as self-assembly. We next explored future technologies that nanotechnology is predicted to enable – such as more powerful computing and robotics – and how nanotechnology provides a pathway to continue the exponential growth predicted by Moore's law. Nanotechnology promises to have a great impact in a wide range of industries, from military to medicine to manufacturing, and we went on to discuss some of these more contentious fields in the third unit of the book.

Chapter 5 marked the start of the second unit of the book; this chapter focused on risk and precaution as fundamental concepts, given the promise and (potentially unknown) hazards of nanotechnology. A thoughtful road-map for nanotechnology requires us to think carefully about these risks and rewards; though such consideration is inescapably complicated by the fact that we will need to make some decisions under uncertainty. Much of our discussion here abstracted away from nanotechnology in particular since the point was to arrive at the conceptual apparatus that would allow us to evaluate the specific applications considered in the third unit of the book.

Chapter 6, in a similar vein, considered issues of regulation, particularly as the policy debate rages on about whether stricter laws and regulations are needed; this debate applies to many if not all areas of nanotechnology and throughout the entire product lifecycle. We discussed conceptual argu-ments both for and against stricter laws, and we proposed an interim solution as this debate is sorted out at local, state, and federal levels.

In Chapter 7, we were concerned with equity and access in nano-technology, and especially what fair distribution requires with regard to developing countries. Looking at cases related to water purification, solar energy, and medicine, we observed that countries lacking capital might not have *access* to nanotechnology, making the distributions of those goods *unequal*. However, we did not take a stance on whether such a distribution would be *inequitable*, which is to say unfair. Rather, this is a substantive question for social and political philosophy; our purposes in this chapter were not to defend some particular position, but rather to show what some of the possible positions were and how those positions bore on the dis-tribution of nanotechnology.

In Chapter 8, we began the third unit of the book by examining several areas of nanotechnology's impact and applications. In this endeavor, our focus was on the current and near term as opposed to, say, more distant or speculative applications such as molecular manufacturing, life extension, and space exploration. The focus of this first chapter was nanotechnology's effect on the environment, which may also affect the health and safety of both humans and animals. We looked at the oft-conflicting relationships among society, technology, and the environment throughout history, and we discussed some of the reasons why nanotechnology might pose unique problems (e.g., increased toxicity and reactivity). But we also presented ways in which nanotechnology can help the environment (e.g., remediation, prevention, "green" manufacturing). Finally, we offered some possible frameworks for assessing and proceeding with developing new technologies, including considerations of efficiency and our obligations to future generations.

In Chapter 9, we recognized the military as one of the largest and primary drivers of new technologies throughout history, resulting in significant changes to society and international relations. Nanotechnology is no exception to this trend since the military aggressively pursues research in the field. We examined a range of projects under way in introducing nanotechnology to the modern military and defense systems – such as lightweight body armor and the enhancement of physical and cognitive abilities – as well as how these might impact society.

In Chapter 10, we discussed privacy; this is a persistent concern that arises in applications of nanotechnology from medicine to civil security to everyday consumer electronics. We began by discussing the conceptual foundations of privacy and whether we have a legal or moral right to it. While privacy has been discussed at least since Plato, it has come under more intense focus in a post-9/11 world that places a greater emphasis on security. For these historical and contemporary reasons, it is important to think about how nanotechnology will bear on privacy. In doing so, we can draw lessons from radio frequency identification (RFID) chips; nanotechnology will be used to make this conventional technology smaller, less expensive, and widely disseminated. After explaining the basics of RFID, we went on to consider emerging social and ethical issues.

Chapter 11 looked at nanomedicine, perhaps the most promising area of nanotechnology for individual and societal benefits. We started with a discussion of the investments being made in this emerging field and then focused on two areas in which the impacts of nanomedicine are likely to be most significant: the first was diagnostics and medical records, and the second was treatment (including surgery and drug delivery). While we argued that no substantively new social or ethical issues arose in nanomedicine, we nevertheless held that this is an importantly new context in which to refocus more traditional debates.

Finally, in Chapter 12, we addressed a highly contentious issue with roots in our military and medical chapters: the application of nanotechnology to human enhancement. This topic is complicated by the fact that some uses of nanotechnology could be therapeutic rather than enhancing; this issue, though, begs the question of whether the therapy–enhancement distinction is intelligible in the first place. We presented some of the basic conceptual issues involved in thinking about enhancement, and we considered some of the applications of enhancement that nanotechnology portends.

13.2   Final Thoughts and Future Investigations

Given the preceding discussions, what conclusions can we draw? First, we do not presume to have exhausted or given a full discussion of the areas covered; to do so would require a complete book dedicated to the material in any given chapter. Rather, our intention has been to provide a broad survey of the basics of nanoscience, as well as the ways in which this technology will have social and ethical implications. Our hope has been to provide the basis for further dialogue and to invite others to help expand and develop these debates further.

In these closing remarks, let us try to tie various themes together through another issue that we have not yet addressed much: intellectual property (IP) and patents. Consider, for example, the litigious environment in which we live today. Plaintiffs are quick to file injunctions and lawsuits, to assert strong rights to their creations and discoveries, and so on, especially when it comes to technological innovation. While previous scientific revolutions did not take place in such an environment, nanotechnology does, for better or worse. IP rights may help to incentivize and safeguard the hard work of scientists and engineers, but there is also a troubling effect in overzeal-ously exercising and enforcing IP rights, which serve to stifle a nascent field like nanotechnology. The basic tools and processes that nanoscientists and engineers have created – *any* tool to help us navigate these uncharted waters – seem to be critical in moving forward, so if researchers unduly hinder others from using or building upon their inventions and discoveries (e.g., by charging exorbitant licensing fees) then progress may be slowed.

To explain, imagine if the basic optical microscope (invented hundreds of years ago) – or its predecessor in eyeglasses (more than a thousand years ago) – had been patented, and its inventors denied the invention's use to other scientists who could not pay hefty licensing fees: our knowledge of biology and medicine would not be nearly what it is now, and countless human and animal lives would have been lost as a result. This appears to be the situation we are now facing in nanotechnology, and one that needs investigation by those with more legal and regulatory expertise than we have.

There have been many substantial discussions related to IP and nano-technology,[1] and just in time, as applications of nanotechnology are tran-scending the more mundane stain-resistant pants and longer-lasting tennis balls. But these emerging and future applications also raise serious social and ethical issues that need to be considered, and certainly we have not

---

[1]   See the following papers, as a very small sampling, from the journal *Nanotechnology, Law & Business* (Berkeley Electronic Press): Raj Bawa, "Nanotechnology Patenting in the US," 1.1 (2004): 31–50; Ted Sabety, "Nanotechnology Innovation and the Patent Thicket: Which IP Policies Promote Growth?," 1.3 (2004): 262–83; Sean O'Neill et al., "Broad Claiming in Nanotechnology Patents: Is Litigation Inevitable?," 4.1 (2007): 29–40. See also Drew L. Harris, "Carbon Nanotube Patent Thickets," in Fritz Allhoff and Patrick Lin (eds), *Nanotechnology and Society: Current and Emerging Ethical Issues* (Dordrecht: Springer, 2007), pp. 163–86.

mentioned or anticipated the entire range of current and possible applications. For instance, imagine that nanotechnology can enable the creation of "smart dust" – millions of dust- or sand-sized sensors that can be carried by the wind to, say, cover a battlefield and locate enemy movement and buildings, or create a broad wi-fi network for a large city, or map the terrain of inhospitable planets. If this is possible, then we need to measure those benefits against possible risks related to the environment, privacy, and other relevant ethical areas.

The challenge with emerging technologies is that we *know* that there are plenty of things that we do not know, let alone the innumerable things we do not know that we do not know. This makes risk assessment and policy setting an inexact science at best: it is fraught with possible mistakes that include an over- or under-abundance of caution as we develop nanotechnology, either unnecessarily retarding progress and economic growth or else foolishly rushing toward harms. Therefore, a conceptual understanding of what risk is and what risks are tolerable can help inform the policy or regulatory issue.

Our discussion of the regulatory debate needs to be tied to an evaluation of actual laws and regulations applicable to nanotechnology; these are continuing to evolve and they require more legal expertise than we can provide here. Further, the full range of possible remedies and their efficacy needs to be considered, from mandatory data reporting to product labeling to global regulations.

Regarding law in a global context, there is not a mandate that some percentage of nanotechnology investments, for example, should go toward humanitarian efforts, such as creating solutions to provide clean water, affordable energy, and so on. This is notable (and ironic) since so much talk in nanotechnology is about how it can help society and alleviate suffering, especially in impoverished parts of the world. But given that business interests are driving nanotechnology research and development – e.g., in pharmacology, which also holds the potential to save millions of lives in impoverished communities – as well as military interests, it is an open question to what extent these nanotechnology solutions will ultimately trickle down to those who most desperately need them. Equity and access, then, are socially and ethically salient issues.

And other areas of nanotechnology's applications reveal their own social and ethical dimensions. We are already concerned about the apparently declining health of the planet; nanotechnology adds a new element to this distress in offering both possible and significant benefits and harms. As our military tactics and tools enter the next generation – e.g. the rise of the war robot[2] – we are again confronted with the issue of arms proliferation,

---

[2] For instance, see Patrick Lin, Keith Abney, and George Bekey, "Autonomous Military Robotics: Risk, Ethics, and Design," report for the US Department of Navy, Office of Naval Research, 2008. Available online at http://ethics.calpoly.edu/ONR_report.pdf (accessed April 27, 2009).

which we have yet to solve, even more than 60 years after the birth of the atomic bomb. Privacy continues to be a top public or consumer concern, as evidenced by furious backlashes directed at technology innovators, such as Facebook or Google, whose policies are perceived to overreach into our private affairs. Perhaps the most valuable application for nanotechnology in the near term is in revolutionizing medicine and saving lives. But even this application can elicit ethical dilemmas, such as those related to the morality of human enhancement.

From our investigations here, we are reminded of Alan Turing, one of the founding fathers of the modern computing age on which so much of nanotechnology is dependent. In describing the coming era of computing, Turing said: "We can only see a short distance ahead, but we can see plenty there that needs to be done."[3] To attend effectively to nanotechnology's social and ethical implications, we surely need more than a nanoscientist and a pair of philosophers to station the floodgates: we need a dialogue with sociologists, economists, historians, biologists, toxicologists, lawyers, policymakers, business minds, and other experts. We need the participation of the global public who stand to win – or lose – much from this powerful, potentially world-changing technology. Whereas we have let our irrational exuberance for science blind us to ethical pitfalls in the past, we now have a chance to get it right. Instead of fumbling around in the darkness, we can, and must, walk responsibly and confidently into a bright future.

[3] Alan Turing, "Computing Machinery and Intelligence," *Mind* 59.236 (1950): 460.

References

Ackerman, Frank and Lisa Heinzerling. 2003. *Priceless: On Knowing the Price of Everything and the Value of Nothing* (New York: New Press).

Adams, Christopher P. and Van V. Brantner. 2006. "Estimating the Cost of New Drug Development: Is it Really $802M?" *Health Affairs* 25.2: 420–8.

Adams, Douglas. 2002. *The Salmon of Doubt: Hitchhiking the Galaxy One Last Time* (New York: Macmillan Books).

Adams, John Couch. 1846. "Explanation of the Observed Irregularities in the Motion of Uranus, on the Hypothesis of Disturbance by a More Distant Planet," *Monthly Notices of the Royal Astronomical Society* November 13, 7: 149.

Agarwal, Ashok, Fnu Deepinder, and Kartikeya Makker. 2007. "Cell Phones and Male Infertility: Dissecting the Relationship," *Reproductive Bio-Medicine Online* 15.3: 266–70. Available at www.clevelandclinic.org/reproductiveresearchcenter/docs/agradoc250.pdf (accessed August 21, 2008).

Aitken, R.J., K.S. Creely, and C.L. Tran. 2004. *Nanoparticles: An Occupational Hygiene Review*. Report prepared by the Institute of Occupational Medicine for the Health and Safety Executive. Available at www.hse.gov.uk/research/rrhtm/rr274.htm (accessed September 23, 2008).

Alivisatos, A. Paul. 2001. "Less is More in Medicine," *Scientific American* 285.3: 66–73.

Alivisatos, A. Paul et al. 1988. "Electronic States of Semiconductor Clusters: Homogeneous and Inhomogeneous Broadening of the Optical Spectrum," *The Journal of Chemical Physics* 89.7: 4001–11.

Allhoff, Fritz. 2005. "Germ-Line Genetic Enhancement and Rawlsian Primary Goods," *Kennedy Institute of Ethics Journal* 15.1: 39–56.

Allhoff, Fritz. 2007. "On the Autonomy and Justification of Nanoethics," *Nanoethics* 1.3: 185–210.

Allhoff, Fritz. 2009. "The Coming Era of Nanomedicine," *American Journal of Bioethics* 9.10, 3–11.

Allhoff, Fritz. 2009. "Risk, Precaution, and Emerging Technologies," *Studies in Ethics, Law, and Society* 3.2: 1–27.

Allhoff, Fritz. "What Are Applied Ethics?" (unpublished).

Allhoff, Fritz et al. (eds). 2007. *Nanoethics: The Social and Ethical Implications of Nanotechnology* (Hoboken, NJ: John Wiley and Sons).

Allhoff, Fritz and Patrick Lin (eds). 2008. "Nanotechnology and Human Enhancement: A Symposium," *Nanoethics: The Ethics of Technologies that Converge at the Nanoscale* 2.3: 251–327.

Allhoff, Fritz and Patrick Lin (eds). 2008. *Nanotechnology and Society: Current and Emerging Ethical Issues* (Dordrecht: Springer).

Amendment 14, US Constitution. Available at www.usconstitution.net/const.html#Am14 (accessed August 5, 2008).

Aristotle. 1992. *Politics*, trans. T.A. Sinclair, rev. Trevor J. Saunders (London: Penguin Books).

Aristotle. 1999. *Nichomachean Ethics*, 2nd edn, trans. Terence Irwin (Indianapolis, IN: Hackett Publishing Company).

Arnold, Michael S. et al. 2003. "Field-Effect Transistors Based on Single Semiconducting Oxide Nanobelts," *Journal of Physical Chemistry B* 107: 659–63.

Associated Press. 2004. "FDA Approves Computer Chips for Humans: Devices Could Help Doctors Store Medical Information." October 13. Available at www.msnbc.msn.com/id/6237364/ (accessed August 14, 2008).

Australian Associated Press. 2007. Australian Green Party: "Call for Moratorium on Nanotechnology," press release, March 17.

Autumn, Kellar et al. 2002. "Evidence for Van der Waals Adhesion in Gecko Setae," *Proceedings of the National Academy of Sciences U.S.A.* 99.19: 12252–6.

Axelrod, Robert. 1984. *The Evolution of Cooperation* (New York: Basic Books).

Bailey, Ron. 2005. *Liberation Biology: The Scientific and Moral Case for the Biotech Revolution* (Amherst, NY: Prometheus Books).

Bainbridge, William Sims and Mihail C. Roco (eds). 2001. *Societal Implications of Nanoscience and Nanotechnology* (Washington, DC: US National Science Foundation).

Barker, Todd F. et al. 2008. "Nanotechnology and the Poor Opportunities and Risks for Developing Countries," in Fritz Allhoff and Patrick Lin (eds), *Nanotechnology and Society: Current and Emerging Ethical Issues* (Dordrecht: Springer), pp. 243–63.

Battelle Memorial Institute and Foresight Nanotech Institute. 2007. "Productive Nanosystems: A Technology Roadmap." Available at www.foresight.org/Roadmaps/Nanotech_Roadmap_2007_main.pdf (accessed November 8, 2008).

Batzner, Derk Leander et al. 2004. "Stability Aspects in CdTe/CdS Solar Cells," *Thin Solid Films* 451–2: 536–43.

Baum, Rudy. 2003. "Nanotechnology: Drexler and Smalley Make the Case For and Against 'Molecular Assemblers'," *Chemical & Engineering News* 81.48: 37–42.

Bawa, Raj. 2004. "Nanotechnology Patenting in the US," *Nanotechnology, Law & Business* 1.1: 31–50.

Bawa, Raj and Summer Johnson. 2008. "Emerging Issues in Nanomedicine and Ethics," in Fritz Allhoff and Patrick Lin (eds), *Nanotechnology and Society: Current and Emerging Social and Ethical Issues* (Dordrecht: Springer), pp. 207–23.

Bawendi, Moungi G. et al. 1992. "Luminescence Properties of CdSe Quantum Crystallites: Resonance between Interior and Surface Localized States," *The Journal of Chemical Physics* 96.2: 946–54.

BBC News. 2004. "Barcelona Clubs Get Chipped." September 29. Available at http://news.bbc.co.uk/2/hi/technology/3697940.stm (accessed August 14, 2008).

BBC News. 2007. "The Tech Lab: Gordon Frazer." September 7. Available at http://news.bbc.co.uk/1/hi/technology/6981704.stm (accessed November 13, 2008).

BBC News. 2007. "World's Tiniest RFID Tag Unveiled." February 23. Available at http://news.bbc.co.uk/2/hi/technology/6389581.stm (accessed October 16, 2008).

Bear, Greg. 1998. *Slant* (New York: Tor Books).

Bernal, John Desmond. 1964. "The Bakerian Lecture, 1962: The Structure of Liquids," *Proceedings of the Royal Society* 208: 299–322.

Berrington de Gonzalez, Amy and Sarah Darby. 2004. "Risk of Cancer from Diagnostic X-rays: Estimates for the UK and 14 Other Countries," *The Lancet* January 31, 363.906: 345–51.

Berube, David. 2006. "Regulating Nanoscience: A Proposal and a Response to J. Clarence Davies," *Nanotechnology Law & Business* 3.4: 485–506.

The Bible. 1918. King James Version, ed. David Norton (New York: Penguin Group).

Bill of Rights, US Constitution. Available at www.usconstitution.net/const.html (accessed August 5, 2008).

Bloustein, Edward J. 1964. "Privacy as an Aspect of Human Dignity: A Response to Dean Prosser," *New York University Law Review* 39: 962–1007.

Bolt, Kristen Millares. 2007. "New Driver's License Ok'd for Border: Gregoire Signs Test Program to Allow Non-Passport Travel," *Seattle Post-Intelligencer* March 23. Available at http://seattlepi.nwsource.com/local/308864_border24.html (accessed August 15, 2008).

Boot, Max. 2006. *War Made New: Technology, Warfare, and the Course of History: 1500 to Today* (New York: Gotham Books).

Borenstein, Seth and Jennifer C. Yates. 2008. "Pittsburgh Cancer Center Warns of Cell Phone Risks," Associated Press, July 23. Available at http://ap.google.com/article/ALeqM5hwzQ6Jsq3cSWa721yR84l99_pnlAD923R4DO1 (accessed August 21, 2008).

Bork, Robert. 1997. *The Tempting of America: The Political Seduction of the Law* (New York: Simon and Schuster).

Bostrom, Nick and Rebecca Roache. 2008. "Ethical Issues in Human Enhancement," in Thomas S. Petersen, Jesper Ryberg, and Clark Wolf (eds), *New Waves in Applied Ethics* (New York: Palgrave Macmillan).

Bowers v. Hardwick 478 US 186. 1986. Available at http://caselaw.lp.findlaw.com/scripts/getcase.pl?court=US&vol=410&invol=113 (accessed August 8, 2008).

Broderick, Damien. 2002. *The Spike: How Our Lives Are Being Transformed by Rapidly Advancing Technologies* (New York: Forge Books).

Bronson, Rachel. 2006. *Thicker than Oil: America's Uneasy Partnership with Saudi Arabia* (New York: Oxford University Press).

Brown, Marilyn A. 2005. "Nano-Bio-Info Pathways to Extreme Efficiency." Presented at the AAAS Annual Meeting, Washington, DC.

Bruchez Jr, Marcel et al. 1998. "Semiconductor Nanocrystals as Fluorescent Biological Labels," *Science* 281.5385: 2013–16.

Brus, Louis E. 1984. "Electron–Electron and Electron–Hole Interactions in Small Semiconductor Crystallites: The Size Dependence of the Lowest Excited Electronic State," *The Journal of Chemical Physics* 80.9: 4403–9.

Bull, Hedley. 1977. *The Anarchical Society* (New York: Columbia University Press).

Burgess, Adam. 2004. *Cellular Phones, Public Fears, and a Culture of Precaution* (Cambridge: Cambridge University Press).

Castellani, John W. et al. 2006. "Cognition during Sustained Operations: Comparison of a Laboratory Simulation to Field Studies," *Aviation, Space and Environmental Medicine* 77.9: 929–35.

Caswell, K.K. et al. 2003. "Preferential End-to-End Assembly of Gold Nanorods by Biotin-Streptavidin Connectors," *Journal of the American Chemical Society* 125.46: 13914–15.

Chan, Warren C.W. and Shuming Nie. 1998. "Quantum Dot Bioconjugates for Ultrasensitive Nonisotopic Detection," *Science* 281.5385: 2016–18.

Chappel, Shlomit and Arie Zaban. 2002. "Nanoporous SnO2 Electrodes for Dye-Sensitized Solar Cells: Improved Cell Performance by the Synthesis of 18 nm SnO_2 Colloids," *Solar Energy Materials and Solar Cells* 71.2: 141–52.

Chappel, Shlomit, Si-Guang Chen, and Arie Zaban. 2002. "TiO_2-coated Nanoporous SnO_2 Electrodes for Dye-Sensitized Solar Cells," *Langmuir* 18.8: 3336–42.

Chen, Si-Guang et al. 2001. "Preparation of Nb_2O_5 Coated TiO_2 Nanoporous Electrodes and Their Application in Dye-Sensitized Solar Cells," *Chemistry of Materials* 13.12: 4629–34.

Chico, Leonor et al. 1996. "Quantum Conductance of Carbon Nanotubes with Defects," *Physical Review B* 54.4: 2600–6.

Churchill, Winston S. 1986. *The Gathering Storm* (New York: Mariner Books).

Cientifica. 2006. "Where Has My Money Gone? Government Nanotechnology Funding and the $18 Billion Pair of Pants," January. Available at www.cientifica.eu/index.php?option=com_content&task=view&id=33&Itemid=63 (accessed August 21, 2008).

Clarke, Arthur C. 2000. *2001: A Space Odyssey* [1968] (New York: Penguin Group).

Clean Air Act. US Environmental Protection Agency. Available at www.epa.gov/air/caa/ (accessed September 1, 2008).

Cohen, Jean L. 2002. *Regulating Intimacy: A New Legal Paradigm* (Princeton, NJ: Princeton University Press).

Colvin, Vicki and Mark Wiesner. 2002. *Environmental Implications of Nanotechnology: Progress in Developing Fundamental Science as a Basis for Assessment.* Keynote presentation delivered at the US EPA's Nanotechnology and the Environment: Applications and Implications STAR Review Progress Workshop, Arlington, VA, August 28.

Cooke, W.E. 1924. "Fibrosis of the Lungs Due to the Inhalation of Asbestos Dust," *British Medical Journal* 2: 147–50.

Craighead, Harold G. 2006. "Future Lab-on-a-Chip Technologies for Interrogating Individual Molecules," *Nature* 442: 387–93.

Cranor, Carl F. 2004. "Toward Understanding Aspects of the Precautionary Principle," *Journal of Medicine and Philosophy* 29.3: 259–79.

Crichton, Michael. 2002. *Prey* (New York: HarperCollins).

Cui, Yi and Charles M. Lieber. 2001. "Functional Nanoscale Electronic Devices Assembled Using Silicon Nanowire Building Blocks," *Science* 291.5505: 851–3.

Cui, Yi et al. 2001. "Diameter-Controlled Synthesis of Single-Crystal Silicon Nanowires," *Applied Physics Letters* 78.15: 2214–16.

Cui, Yi et al. 2001. "Nanowire Nanosensors for Highly Sensitive and Selective Detection of Biological and Chemical Species," *Science* 293: 1289–92.

Daar, Abdallah S., Anisa Mnyusiwalla, and Peter A. Singer. 2003. " 'Mind the Gap': Science and Ethics in Nanotechnology," *Nanotechnology* 14.3: R9–R13.

Daniels, Norm. 2000. "Normal Functioning and the Treatment–Enhancement Distinction," *Cambridge Quarterly of Healthcare Ethics* 9: 309–22.

Davies, J. Clarence. 2006. *Managing the Effects of Nanotechnology* (Washington, DC: Woodrow Wilson International Center for Scholars).

De Balzac, Honoré. 2004. *Père Goriot*, trans. Henry Reed (New York: Signet Classics).

DeCew, Judith. 1997. *In Pursuit of Privacy: Law, Ethics, and the Rise of Technology* (Ithaca, NY: Cornell University Press).

DeCew, Judith. 2008. "Privacy," *Stanford Encyclopedia of Philosophy*. Available at http://plato.stanford.edu/archives/sum2008/entries/privacy/ (accessed August 8, 2008).

De Grey, Aubrey. 2007. *Ending Aging: The Rejuvenation Breakthroughs that Could Reverse Human Aging in Our Lifetime* (New York: St Martin's Press).

Dennis, Carina. 2004. "Genetics: Deaf by Design," *Nature* 431: 894–6.

Derfus, Austin M., Warren C.W. Chan, and Sangeeta N. Bhatia. 2004. "Probing the Cytotoxicity of Semiconductor Quantum Dots," *Nano Letters* 4.1: 11–18.

Diamond, Jared. 1997. *Guns, Germs, and Steel: The Fate of Human Societies* (New York: W.W. Norton).

DiMasi, Joseph A., Ronald W. Hansen, and Henry G. Grabowski. 2003. "The Price of Innovation: New Estimates of Drug Development Costs," *Journal of Health Economics* 22: 151–85.

Dong, Lifeng et al. 2003. "Catalytic Growth of CdS Nanobelts and Nanowires on Tungsten Substrates," *Chemical Physics Letters* 376: 653–8.

Doyle, Charles. 2004. "Patriot Act: Sunset Provisions that Expire on December 31, 2005" (Washington, DC: Congressional Research Service). Available at www.fas.org/irp/crs/RL32186.pdf (accessed August 8, 2008).

Doyle, Charles. 2005. "USA PATRIOT Act Reauthorization in Brief" (Washington, DC: Congressional Research Service). Available at http://fpc.state.gov/documents/organization/51133.pdf (accessed August 8, 2008).

Drexler, K. Eric. 1987. *Engines of Creation: The Coming Era of Nanotechnology* (New York: Broadway Books). Available at www.e-drexler.com/d/06/00/EOC/EOC_Table_of_Contents.html (accessed September 26, 2008).

Drexler, K. Eric. 1993. "Appendix A: Machines of Inner Space," in B.C. Crandall and James Lewis (eds), *Nanotechnology: Research and Perspectives* (New York: MIT Press), pp. 325–35.

Drexler, K. Eric and Richard Smalley. 2003. "Point–Counterpoint: Nanotechnology," *Chemical and Engineering News* December 1, 81.48: 37–42.

Dyer, Clare. 2000. "Surgeon Amputated Healthy Legs," *British Medical Journal* 320: 332.

Earman, John. 1992. *Bayes or Bust: A Critical Examination of Bayesian Confirmation Theory* (Cambridge, MA: MIT Press).

Ebbesen, Mette and Thomas G. Jensen. 2006. "Nanomedicine: Techniques, Potentials, and Ethical Implications," *Journal of Biomedicine and Biotechnology* 2006: 1–11.

Edwards, Steven D. 2008. "The Ashley Treatment: A Step Too Far, or Not Far Enough?" *Journal of Medical Ethics* 34: 341–3.

Edwards, Steven D. 2008. "Should Oscar Pistorius Be Excluded from the 2008 Olympic Games?" *Sports, Ethics, and Philosophy* 2: 112–25.

Eigler, Donald M. and Erhard K. Schweizer. 1990. "Positioning Single Atoms with a Scanning Tunneling Microscope," *Nature* 344: 524–6.

Einstein, Albert. 1905. "Über einen die Erzeugung und Verwandlung des Lichtes betreffenden heuristischen Gesichtspunkt (On a Heuristic Viewpoint Concerning the Production and Transformation of Light)," *Annalen der Physik* 17: 132–48. Available at www.physik.uni-augsburg.de/annalen/history/papers/1905_17_132-148.pdf (accessed October 24, 2008).

Einstein, Albert. 1971. "Letter to Max Born, written December 4, 1926," in Irene Born (trans.), *The Born–Einstein Letters* (New York: Walker and Company).

Einstein, Albert. 1981. "Letter to Ilse Rosenthal-Schneider, written May 11, 1945," in Ilse Rosenthal-Schneider (trans.), *Reality and Scientific Truth: Discussions with Einstein, Van Laue, and Planck* (Detroit, MI: Wayne State University Press).

Elliott, Daniel W. and Wei-Xian Zhang. 2002. "Field Assessment of Nanoscale Bimetallic Particles for Groundwater Treatment," *Environmental Science Technology* 35.24: 4922–6.

El-Sayed, Ivan, Xiaohua Huang, and Mostafa A. El-Sayed. 2005. "Surface Plasmon Resonance Scattering and Absorption of Anti-EGFR Antibody Conjugated Gold Nanoparticles in Cancer Diagnostics: Applications in Oral Cancer," *Nano Letters* 5.5: 829–34.

El-Sayed, Ivan, Xiaohua Huang, and Mostafa A. El-Sayed. 2006. "Selective Laser Photo-Thermal Therapy of Epithelial Carcinoma Using Anti-EGFR Antibody Conjugated Gold Nanoparticles," *Cancer Letters* 239.1: 129–35.

Encyclopædia Britannica. 2008. "English Longbow." Available at www.britannica.com/EBchecked/topic/188247/English-longbow (accessed August 19, 2008).

Encyclopædia Britannica. 2008. "Hydroelectric Power." Standard Edition (Digital) (Chicago, IL: Encyclopædia Britannica).

Encyclopædia Britannica. 2008. "Waterwheel." Standard Edition (Digital) (Chicago, IL: Encyclopædia Britannica).

EPCglobal Inc. 2008. "EPCglobal Tag Data Standards Version 1.4." Available at www.epcglobalinc.org/standards/tds/tds_1_4-standard-20080611.pdf (accessed August 15, 2008).

Erwin, Steven C. et al. 2005. "Doping Semiconductor Nanocrystals," *Nature* 436: 91–4.

ETC Group. 2003. "No Small Matter II: The Case for a Global Moratorium," *Occasional Paper Series* 7.1, April. Available at www.etcgroup.org/upload/publication/165/01/occ.paper_nanosafety.pdf (accessed October 13, 2008).

Etzioni, Amitai. 1999. *The Limits of Privacy* (New York: Basic Books).

Etzioni, Amitai. 2005. "The Limits of Privacy," in Andrew I. Cohen and Christopher Heath Wellman (eds), *Contemporary Debates in Applied Ethics* (Malden, MA: Blackwell Publishing, 2005), pp. 253–62.

Etzioni, Amitai. 2007. "Are New Technologies the Enemy of Privacy?" *Knowledge, Technology, and Privacy* 20: 115–19.

European Group on Ethics. 2008. "Ethical Aspects of Nanomedicine: A Condensed Version of the EGE Opinion 21," in Fritz Allhoff and Patrick Lin (eds), *Nanotechnology and Society: Current and Emerging Social and Ethical Issues* (Dordrecht: Springer), pp. 187–206.

European Working Group on Ethics in Science and New Technologies to the European Commission. 2007. "Opinion on the Ethical Aspects of Nanomedicine," Opinion No. 21, January 17 (Brussels: European Commission).

Feynman, Richard P. 1992. "There's Plenty of Room at the Bottom," *Journal of Microelectromechanical Systems* 1: 60–6.

Feynman, Richard P. 2002. *The Pleasure of Finding Things Out: The Best Short Works of Richard P. Feynman* (New York: Perseus Books Group).

Filley, T.R. et al. 2005. "Investigations of Fungal Mediated (C60-C70) Fullerene Decomposition," *Preprints of Extended Abstracts Presented at ACS* 45: 446–50.

Finkbeiner, Ann. 2006. *The Jasons: The Secret History of Science's Postwar Elite* (New York: Penguin Group).

Foddy, Bennett and Julian Savulescu. 2007. "Ethics of Performance Enhancement in Sport: Drugs and Gene Doping," in Richard Ashcroft et al. (eds), *Principles of Health Care Ethics* (Oxford: John Wiley & Sons). Available at www.wiley.com/WileyCDA/WileyTitle/productCd-0470027134,descCd-tableOfContents.html.

Food, Drug and Cosmetic Act. US Food and Drug Administration. Available at www.fda.gov/opacom/laws/fdcact/fdctoc.htm (accessed September 1, 2008).

Freedonia Group, Inc. 2007. *Nanotechnology in Healthcare* (Cleveland, OH). Available at www.freedoniagroup.com/brochure/21xx/2168smwe.pdf (accessed October 16, 2008).

Freitas Jr, Robert A. 1998. "Exploratory Design in Medical Nanotechnology: A Mechanical Artificial Red Cell," *Artificial Cells, Blood Substitutes, and Immobilization Biotechnology* 26: 411–30.

Freitas Jr, Robert A. 1999. *Nanomedicine, Volume I: Basic Capabilities* (Austin, TX: Landes Bioscience). Available at www.nanomedicine.com/NMI.htm (accessed November 23, 2008).

Freitas Jr, Robert A. 2007. "Personal Choice in the Coming Era of Nanomedicine," in Fritz Allhoff et al. (eds), *Nanoethics: The Social and Ethical Implications of Nanotechnology* (Hoboken, NJ: John Wiley & Sons), pp. 161–72.

Fried, Charles. 1970. *An Anatomy of Values* (Cambridge, MA: Harvard University Press).

Friedlander, Sheldon K. and David Y.H. Pui. 2003. *Emerging Issues in Nanoparticle Aerosol Science and Technology* (Washington, DC: US National Science Foundation, November). Available at www.nano.gov/html/res/NSFAerosolParteport.pdf# (accessed October 14, 2008).

Friedman, David D. 2005. "The Case for Privacy," in Andrew I. Cohen and Christopher Heath Wellman (eds), *Contemporary Debates in Applied Ethics* (Malden, MA: Blackwell Publishing), pp. 264–75.

Friends of the Earth. 2006. "Nano-Ingredients Pose Big Risks in Beauty Products: Friends of the Earth Report Highlights Unregulated Risks of Nanoparticles in Cosmetics and Sunscreens," May 16.

Friends of the Earth. 2008. *The Disruptive Social Impacts of Nanotechnology: Issue Summary*. Available at http://nano.foe.org.au/node/151 (accessed August 16, 2008).

Fukada, Toshio, Fumihito Arai, and Lixin Dong. 2003. "Assembly of Nanodevices with Carbon Nanotubes through Nanorobotic Manipulations," *Proceedings of the IEEE* 91: 1803–18.

Fukuyama, Francis. 2002. *Our Posthuman Future: Consequences of the Biotechnology Revolution* (New York: Picador).

Fukuyama, Francis. 2006. *Beyond Bioethics: A Proposal for Modernizing the Regulation of Human Biotechnologies* (Washington, DC: School of Advanced International Studies, Johns Hopkins University).

Gao, Xiaohu et al. 2004. "In Vivo Cancer Targeting and Imaging with Semi-conductor Quantum Dots," *Nature Biotechnology* 22.8: 969–76.

Gardiner, Stephen M. 2006. "A Core Precautionary Principle," *Journal of Political Philosophy* 14.1: 33–60.

Gerstein, Robert S. 1978. "Intimacy and Privacy." *Ethics* 89: 76–81.

Gibbs, Lawrence and Mary Tang. 2004. "Nanotechnology: Safety and Risk Management Overview." Presented at the NNIN Nanotechnology Safety Workshop at Georgia Institute of Technology, Atlanta, GA, December 2.

Gimpel, Jean. 1976. *The Medieval Machine* (New York: Henry Holt and Company LLC).

Global Development Research Center. 2008. "EU's Communication on Precautionary Principle." Available at www.gdrc.org/u-gov/precaution-4.html (accessed June 23, 2008).

Godfrey-Smith, Peter. 2003. *Theory and Reality: An Introduction to the Philosophy of Science* (Chicago, IL: University of Chicago Press).

Goldblatt, Michael. 2002. "Office Overview." Presented at DARPATech 2002 Symposium. Available at www.darpa.mil/darpatech2002/presentations/dso_pdf/speeches/GOLDBLAT.pdf (accessed September 7, 2008).

Gore, Al. 2006. *An Inconvenient Truth* (New York: Rodale Books).

Greely, Henry T. 2005/2006. "Regulating Human Biological Enhancements: Questionable Justifications and International Complications," *The Mind, the Body, and the Law: University of Technology, Sydney, Law Review* 7: 87–110 and *Santa Clara Journal of International Law* 4: 87–110 (joint issue).

Green, Martin A. 2000. "Photovoltaics: Technology Overview," *Energy Policy* 28.14: 989–98.

Greene, Thomas C. November 30, 2004. "Anti-RFID Outfit Deflates Mexican VeriChip Hype: Only 18 Volunteers to Date," *The Register*. Available at www.theregister.co.uk/2004/11/30/mexican_verichip_hype/ (accessed August 14, 2008).

Griswold v. Connecticut 381 US 469. 1965. Available at http://supreme.justia.com/us/381/479/case.html (accessed August 5, 2008).

Gubrud, Mark. 1997. "Nanotechnology and International Security." Presented at Fifth Foresight Conference on Molecular Nanotechnology at Palo Alto, CA. Available at www.foresight.org/Conferences/MNT05/Papers/Gubrud/ (accessed September 4, 2008).

Gulliford, James B. 2008. "Toxic Substances Control Act Inventory Status of Carbon Nanotubes" (Washington, DC: Environmental Protection Agency, October 27). Available at http://nanotech.lawbc.com/uploads/file/00037805.PDF (accessed April 21, 2009).

Guston, David, John Parsi, and Justin Tosi. 2007. "Anticipating the Ethical and Political Challenges of Human Nanotechnologies," in Fritz Allhoff et al. (eds), *Nanoethics: The Ethical and Social Implications of Nanotechnology* (Hoboken, NJ: John Wiley & Sons, 2007), pp. 185–97.

Hale, Sir Matthew. 1677. *The Primitive Origination of Mankind* (London: William Godbid).

Halford, Bethany. 2005. "Dendrimers Branch Out," *Chemical and Engineering News* 83.24: 30–6. Available at http://pubs.acs.org/cen/coverstory/83/8324dendrimers.html (accessed July 31, 2008).

Hall, J. Storrs. 2005. *Nanofuture: What's Next for Nanotechnology* (New York: Prometheus Books).

Hansson, Sven Ove. 1996. "Decision Making under Great Uncertainty," *Philosophy of the Social Sciences* 26.3: 369–86.

Hansson, Sven Ove. 1996. "What Is Philosophy of Risk?" *Theoria* 62: 169–86.

Hansson, Sven Ove. 2004. "Philosophical Perspectives on Risk," *Techné* 8.1: 10–35.

Hardin, Garrett. 1974. "Lifeboat Ethics: The Case against Helping the Poor," *Psychology Today* 38–40, 123–4, 126.

Harremoes, Poul et al. (eds). 2002. *The Precautionary Principle in the 20th Century: Late Lessons from Early Warnings* (London: Earthscan).

Harris, Drew L. 2007. "Carbon Nanotube Patent Thickets," in Fritz Allhoff and Patrick Lin (eds), *Nanotechnology & Society: Current and Emerging Ethical Issues* (Dordrecht: Springer), pp. 163–86.

Harris, John. 2007. *Enhancing Evolution: The Ethical Case for Making Ethical People* (Princeton, NJ: Princeton University Press).

Hayward, Tim. 2008. "On the Nature of Our Debt to the Global Poor," *Journal of Social Philosophy* 39.1: 1–19.

Hick, John. 1966. *Evil and the God of Love* (New York: Harper and Row Publishers).

Hill, Sarah E. 2006. "Dissatisfied by Design: The Evolution of Discontent" (unpublished doctoral dissertation) (Austin: University of Texas).

Hillie, Mohan Munasinghe, Mbhuti Hlope, and Yvani Deraniyagala. 2006. "Nanotechnology, Water, and Development" (Meridian Institute). Available at www.merid.org/nano/waterpaper/ (accessed July 28, 2008).

Hobbes, Thomas. 1982. *Leviathan* [1651] (New York: Penguin Group).

Hooke, Robert. 2005. *Micrographia* [1665]. Available at www.gutenberg.org/etext/15491 (accessed October 22, 2008).

Hughes, James. 2004. *Citizen Cyborg: Why Democratic Societies Must Respond to the Redesigned Human of the Future* (Cambridge, MA: Westview Press).

Hughes, Jonathan. 2006. "How Not to Criticize the Precautionary Principle," *Journal of Medicine and Philosophy* 31: 447–64.

Hunt, Warren H. 2004. "Nanomaterials: Nomenclature, Novelty, and Necessity," *Journal of Materials* 56: 13–19.

Hurlbut, William. 2006. Opening Remarks at the Human Enhancement Technologies and Human Rights Conference, Stanford University Law School, Stanford, CA, May 26–28.

Huygens, Christian. 2005. *Treatise on Light* [1690]. Available at www.gutenberg.org/files/14725/14725-h/14725-h.htm (accessed October 14, 2008).

Huynh, Wendy U., Xiaogang Peng, and A. Paul Alivisatos. 1999. "CdSe Nanocrystal Rods/Poly(3-hexylthiophene) Composite Photovoltaic Devices," *Advanced Materials* 11.11: 923–7.

Inness, Julie C. 1992. *Privacy, Intimacy, and Isolation* (Oxford: Oxford University Press).

In-PharmaTechnologist. 2005. "Nanotechnology to Revolutionize Drug Delivery," March 7. Available at www.in-pharmatechnologist.com/Materials-Formulation/Nanotechnology-to-revolutionise-drug-delivery (accessed August 27, 2008).

Intel. 2007. "Moore's Law." Available at www.intel.com/technology/mooreslaw/ (accessed August 17, 2007).

International Technology Roadmap for Semiconductors (ITRS). Available at www.itrs.net/ (accessed November 7, 2008).

Jain, Prashant K. et al. 2006. "Calculated Absorption and Scattering Properties of Gold Nanoparticles of Different Size, Shape, and Composition: Applications in Biological Imaging and Biomedicine," *Journal of Physical Chemistry B* 110.14: 7238–48.

Jasanoff, Sheila. 2005. *Designs on Nature: Science and Democracy in Europe and the United States* (Princeton, NJ: Princeton University Press).

Jeong, Heejun, Albert M. Chang, and Michael R. Melloch. 2001. "The Kondo Effect in an Artificial Quantum Dot Molecule," *Science* 293.5538: 2221–3.

Jiang, Yang et al. 2003. "Hydrogen-Assisted Thermal Evaporation Synthesis of ZnS Nanoribbons on a Large Scale," *Advanced Materials* 15: 323–7.

Jiang, Yang et al. 2004. "Zinc Selenide Nanoribbons and Nanowires," *Journal of Physical Chemistry B* 108.9: 2784–7.

Jiang, Yong-Hou et al. 2005. "SPL7013 Gel as Topical Microbicide for Prevention of Vaginal Transmission of $SHIV_{89.6P}$ in Macaques," *AIDS Research and Human Retroviruses* 21.3: 207–13.

Johnsen, Mads G. et al. 2005. "Chitosan-Based Nanoparticles for Biomedicine," *Journal of Biotechnology* 118: S34.

Joy, Bill. 2000. "Why the Future Doesn't Need Us," *Wired Magazine* April, 8.04.

Juels, Ari, Ronald L. Rivest, and Michael Szydlo. 2003. "The Blocker Tag: Selective Blocking of RFID Tags for Consumer Privacy," in Vijay Atluri (ed.), *8th ACM Conference on Computer and Communications Security* (Washington, DC: ACM Press), pp. 103–11.

Juengst, Eric. 1997. "Can Enhancement Be Distinguished from Prevention in Genetic Medicine?" *Journal of Medicine and Philosophy* 22: 125–42.

Karuppuchamy, Subbian et al. 2002. "Cathodic Electrodeposition of Oxide Semiconductor Thin Films and Their Application to Dye-Sensitized Solar Cells," *Solid State Ionics* 151.1–4: 19–27.

Kay, Alan. 1989. "Predicting the Future," *Stanford Engineering* 1.1: 1–6.

Keis, Karin et al. 2002. "Nanostructured ZnO Electrodes for Dye-Sensitized Solar Cell Applications," *Journal of Photochemistry and Photobiology A: Chemistry* 148.1–3: 57–64.

Kereiakes, James G. and Marvin Rosenstein. 1980. *Handbook of Radiation Doses in Nuclear Medicine and Diagnostic X-Ray* (Boca Raton, FL: CRC Press).

Keren, Kinneret et al. 2003. "DNA-Templated Carbon Nanotube Field-Effect Transistor," *Science* 302: 1380–2.

Krupke, Ralph et al. 2003. "Separation of Metallic from Semiconducting Single-Walled Carbon Nanotubes," *Science* 301.5631: 344–7.

Kukowska-Latallo, Jolanta F. et al. 2005. "Nanoparticle Targeting of Anticancer Drug Improves Therapeutic Response in Animal Model of Human Epithelial Cancer," *Cancer Research* June 15, 65.12: 5317–24.

Kurzweil, Ray. 2001. "The Law of Accelerating Returns." March 7. Available at www.kurzweilai.net/articles/art0134.html (accessed April 25, 2007).

Kurzweil, Ray. 2005. *The Singularity Is Near: When Humans Transcend Biology* (New York: Viking).

Lawrence et al. v. Texas 539 US 558. 2003. Available at http://supreme.justia.com/us/539/558/case.html (accessed October 16, 2008).

Lederman, Michael M., Robin E. Offord, and Oliver Hartley. 2006. "Microbicides and Other Topical Strategies to Prevent Vaginal Transmission of HIV," *Nature Reviews Immunology* 6: 371–82.

Li, Jing et al. 2003. "Carbon Nanotube Sensors for Gas and Organic Vapor Detection," *Nano Letters* 3: 929–33.

Li, Ning et al. 2006. "Toxic Potential of Materials at the Nanolevel," *Science* 311.5761: 622–7.

Li, Quan and Chunrui Wang. 2003. "Fabrication of Wurtzite ZnS Nanobelts via Simple Thermal Evaporation," *Applied Physics Letters* 83: 359–61.

Li, Xiaoqin et al. 2003. "An All-Optical Quantum Gate in a Semiconductor Quantum Dot," *Science* 301.5634: 809–11.

Li, Y.Q. et al. 2006. "Manganese Doping and Optical Properties of ZnS Nanoribbons by Postannealing," *Applied Physics Letters* 88: 013115.

Liang, Changhao et al. 2001. "Catalytic Growth of Semiconducting In$_2$O$_3$ Nanofibers," *Advanced Materials* 13.17: 1330–3.

Lieber, Charles M. 1998. "One-Dimensional Nanostructures: Chemistry, Physics and Applications," *Solid State Communications* 107.11: 607–16.

Lin, M.F. and Kenneth W.K. Shung. 1995. "Magnetoconductance of Carbon Nanotubes," *Physical Review B* 51.12: 7592–7.

Lin, Patrick. 2006. "Space Ethics: Look Before Taking Another Leap for Mankind," *Astropolitics* 4: 281–94.

Lin, Patrick. 2007. "Nanotechnology Bound: Evaluating the Case for More Regulation," *NanoEthics: Ethics for Technologies that Converge at the Nanoscale* 2: 105–22.

Lin, Patrick and Fritz Allhoff. 2008. "Untangling the Debate: The Ethics of Human Enhancement," *NanoEthics: Ethics for Technologies that Converge at the Nanoscale* 2: 251–64.

Lin, Patrick, Keith Abney, and George Bekey. 2008. "Autonomous Military Robotics: Risk, Ethics, and Design." Report for the US Department of Navy, Office of Naval Research. Available at http://ethics.calpoly.edu/ONR_report.pdf (accessed April 27, 2009).

Locke, John. 1986. *Second Treatise on Civil Government* (New York: Prometheus Books).

Lockton, Vance and Richard S. Rosenberg. 2005. "RFID: The Next Serious Threat to Privacy," *Ethics and Information Technology* 7: 221–31.

Lockwood, Julie, Martha Hoopes, and Michael Marchetti. 2006. *Invasion Ecology* (Hoboken, NJ: Wiley-Blackwell).

Low, Tim. 2002. *Feral Future: The Untold Story of Australia's Exotic Invaders* (Chicago, IL: University of Chicago Press).

Lux Research Inc. 2008. "Nanomaterials State of the Market Q3 2008." Available at www.luxresearchinc.com/ and www.industryweek.com/ReadArticle.aspx?ArticleID=16884&SectionID=4 (accessed August 21, 2008).

Ma, Christopher and Zhong Lin Wang. 2005. "Road Map for the Controlled Synthesis of CdSe Nanowires, Nanobelts, and Nanosaws: A Step Towards Nanomanufacturing," *Advanced Materials* 2005.17: 1–6.

Ma, Christopher et al. 2003. "Nanobelts, Nanocombs, and Nano-Windmills of Wurtzite ZnS," *Advanced Materials* 15: 228–31.

Ma, Christopher et al. 2004. "Single-Crystal CdSe Nanosaws," *Journal of the American Chemical Society* 126: 708–9.

Ma, D.D.D. et al. 2003. "Small-Diameter Silicon Nanowire Surfaces," *Science* 299: 1874–7.

Macoubrie, Jane. 2005. *Informed Public Perceptions of Nanotechnology and Trust in Government* (Washington, DC: Woodrow Wilson International Center for Scholars).

Madan, T. et al. 2005. "Biodegradable Nanoparticles as a Sustained Release System for the Antigens/Allergens of Aspergillus Fumigatus: Preparation and Characterization," *International Journal of Pharmaceutics* 159: 135–47.

Manson, Neil A. 2002. "Formulating the Precautionary Principle," *Environmental Ethics* 24: 263–72.

Manson, Neil A. 2007. "The Concept of Irreversibility: Its Use in the Sustainable Development and Precautionary Principle Literatures," *Electronic Journal of Sustainable Development* 1.1: 1–15.

Markoff, John. 1999. "An Internet Pioneer Ponders the Next Revolution," *The New York Times on the Web* December 20. Available at http://partners.nytimes.com/library/tech/99/12/biztech/articles/122099outlook-bobb.html (accessed September 1, 2008).

Marx, Karl. 1949. "Critique of the Gotha Program," in *Marx–Engels Selected Works*, vol. II (Moscow: Foreign Languages Publishing House). Available at www.marxists.org/archive/marx/works/1875/gotha/ch01.htm (accessed July 25, 2008).

McKeon, Richard and C.D. Reeve. 2001. *The Basic Works of Aristotle* (New York: Modern Library).

McKibben, Bill. 2004. *Enough: Staying Human in an Engineered Age* (New York: Henry Holt & Co.).

Meadows, Donella H. 1974. *The Limits to Growth* (New York: University Books).

MedAdNews. 2007. "The Top 50 Pharmaceutical Companies Charts and Lists," *MedAdNews* 13.9.

Merewether, E.R.A. and C.W. Price. 1930. *Report on Effects of Asbestos Dust on the Lung* (London: HM Stationery Office).

Mielke, Jim. 2008. "Digital Tattoo Interface," entry at Greener Gadgets Design Competition 2008, February (New York). Available at www.core77.com/competitions/GreenerGadgets/projects/4673/ (accessed August 16, 2008).

Mill, John Stuart. 1989. "On Liberty," in *On Liberty and Other Writings* (Cambridge: Cambridge University Press), pp. 1–116.

MIT, Institute for Soldier Nanotechnologies. 2008. Available at http://web.mit.edu/ISN/research/index.html (accessed August 21, 2008).

Moore, Daniel et al. 2004. "Wurtzite ZnS Nanosaws Produced by Polar Surfaces," *Chemical Physics Letters* 385: 8–11.

Moore, Gordon. 1965. "Cramming More Components onto Integrated Circuits," *Electronics* 38.8: 56–9.

Murray, C.B., D.J. Norris, and Moungi G. Bawendi. 1993. "Synthesis and Characterization of Nearly Monodisperse CdE (E = Sulfur, Selenium, Tellurium) Semiconductor Nanocrystallites," *Journal of the American Chemical Society* 115.19: 8706–15.

Naam, Ramez. 2005. *More Than Human* (New York: Broadway Books).

Nagel, Thomas. 1998. "Concealment and Exposure," *Philosophy & Public Affairs* 27.1: 3–30.

Nanoscale, Science, Engineering and Technology (NSET) Subcommittee of the National Science and Technology Council. 2003. *Nanotechnology and the*

Environment: Report of a National Nanotechnology Initiative Workshop. May 8–9, 2003. Available at www.nano.gov/NNI_Nanotechnology_and_the_ Environment.pdf (accessed August 15, 2008).

Narendran, Nadarajah. 2005. "Improved Performance White LED." Presented at the Fifth International Conference on Solid State Lighting, *Proceedings of SPIE* 5941: 45–50.

National Cancer Institute. 2005. "Targeted Gold Nanoparticles Detect Oral Cancer Cells," *Nanotech News* April 25. Available at http://nano.cancer.gov/ news_center/nanotech_news_2005-04-25b.asp (accessed August 7, 2007).

National Nanotechnology Institute. 2008. "Nanotech Facts." Available at www.nano.gov/html/facts/The_scale_of_things.html (accessed September 23, 2008).

National Nanotechnology Initiative. 2008. "What Is Nanotechnology?" Available at www.nano.gov/html/facts/whatIsNano.html (accessed October 11, 2008).

National Research Council. 1990. *Health Effects of Exposure to Low Levels of Ionizing Radiation* (Washington, DC: National Academies Press).

Nel, Andre et al. 2006. "Toxic Potential of Materials at the Nanolevel," *Science* 311.5761: 622–7.

Nguyen, Pho et al. 2003. "Epitaxial Directional Growth of Indium-Doped Tin Oxide Nanowire Arrays," *Nano Letters* 3.7: 925–8.

Nichols, Mary P. 1991. *Citizens and Statesmen: A Study of Aristotle's Politics* (Savage, MD: Rowman & Littlefield).

Niemeyer, Christof M. and Chad A. Mirkin. 2004. *Nanobiotechnology: Concepts, Applications, and Perspectives* (Weinheim, Germany: Wiley-VCH).

Niemeyer, Christof M. and Chad A. Mirkin. 2007. *Nanobiotechnology II: More Concepts and Applications* (Weinheim, Germany: Wiley-VCH).

Nobel Prize in Chemistry. 1996. Available at http://nobelprize.org/nobel_prizes/ chemistry/laureates/1996/illpres/discovery.html (accessed October 12, 2008).

Nordqvist, Christian. 2006. "Extensive Cell Phone Use Linked to Brain Tumors, Swedish Study," *Medical News Today* April 1. Available at www.medicalnewsday. com/articles/40764.php (accessed August 21, 2008).

Nozick, Robert. 1974. *Anarchy, State, and Utopia* (New York: Basic Books).

Obataya, Ikuo et al. 2005. "Nanoscale Operation of a Living Cell Using an Atomic Force Microscope with a Nanoneedle," *Nano Letters* 5.1: 27–30.

Oberdörster, Eva. 2004. "Manufactured Nanomaterials (Fullerenes, C60) Induce Oxidative Stress in the Brain of Juvenile Largemouth Bass," *Environmental Health Perspectives* 112.10: 1058–62.

Occupational Safety and Health Act. US Department of Labor Occupational Safety and Health Administration. Available at www.osha.gov/pls/oshaweb/ owasrch.search_form?p_doc_type=oshact (accessed September 1, 2008).

Odom, Teri Wang et al. 1998. "Atomic Structure and Electronic Properties of Single-Walled Carbon Nanotubes," *Nature* 391.1: 62–4.

Office of Management and Budget. 2008. "Department of Defense." Available at www.whitehouse.gov/omb/budget/fy2008/defense.html (accessed August 29, 2008).

Ohnishi, Hideomi. 1989. "Electroluminescent Display Materials," *Annual Review of Materials Science* 19: 83–101.

O'Neill, Sean et al. 2007. "Broad Claiming in Nanotechnology Patents: Is Litigation Inevitable?" *Nanotechnology, Law & Business* 4.1: 29–40.

Ono, Tomoya, Shigeru Tsukamoto, and Kikuji Hirose. 2003. "Magnetic Orderings in A1 Nanowires Suspended between Electrodes," *Applied Physics Letters* 82.25: 4570–2.

Oppenheimer, Andrew. 2005. "Nanotechnology Paves Way for New Weapons," *Jane's* August 1. Available at www.janes.com/defence/news/jcbw/jcbw050801_1_n. shtml (accessed September 9, 2008).

Osakabe, Nobuyuki et al. 1997. "Time-Resolved Observation of Thermal Oscillations by Transmission Electron Microscopy," *Applied Physics Letters* 70.8: 940–2.

Overbye, Dennis. 2008. "Gauging a Collider's Odds of Creating a Black Hole," *New York Times*. Available at www.nytimes.com/2008/04/15/science/15risk. html?em (accessed October 7, 2008).

Pan, Zheng Wei, Zu Rong Dai, and Zhong Lin Wang. 2001. "Nanobelts of Semiconducting Oxides," *Science* 291: 1947–9.

Pankhurst, Quentin A. et al. 2003. "Applications of Magnetic Nanoparticles in Biomedicine," *Journal of Physics D: Applied Physics* 36.13: R167–R181.

Parak, Wolfgang J., Teresa Pellegrino, and Christian Plank. 2005. "Labelling of Cells with Quantum Dots," *Nanotechnology* 16.2: R9–R25.

Parent, William A. 1983. "Privacy, Morality and the Law," *Philosophy & Public Affairs* 12.4: 269–88.

Parfit, Derek. 1986. *Reasons and Persons* (New York: Oxford University Press).

Parviz, Babak et al. 2008. "Contact Lens with Integrated Inorganic Semiconductor Devices," presented at the 21st IEEE International Conference on Micro Electro Mechanical Systems, Tucson, AZ, January 13–17.

Peirce, Charles Sanders. 1934. "The Fixation of Belief," in Charles Hartshorne and Paul Weiss (eds), *Collected Papers of Charles Peirce* (Cambridge, MA: Harvard University Press), pp. 223–47.

Pengfei, Qui et al. 2003. "Toward Large Arrays of Multiplex Functionalized Carbon Nanotube Sensors for Highly Sensitive and Selective Molecular Detection," *Nano Letters* 3: 347–51.

Persaud, Raj. 2006. "Does Smarter Mean Happier?" in Paul Miller and James Wilsdon (eds), *Better Humans? The Politics of Human Enhancement and Life Extension* (London: Demos), pp. 129–36.

Peslak, Alan R. 2005. "An Ethical Exploration of Privacy and Radio Frequency Identification," *Journal of Business Ethics* 59: 327–45.

Peters, Ted. 2007. "Are We Playing God with Nanoenhancement?" in Fritz Allhoff et al. (eds), *Nanoethics: The Ethical and Social Implications of Nanotechnology* (Hoboken, NJ: John Wiley & Sons, 2007), pp. 173–83.

Philipkowski, Kristen. 1999. "Another Change for Gene Therapy," *Wired* October 1. Available at www.wired.com/science/discoveries/news/1999/10/31613 (accessed August 16, 2007).

Plato. 1968. *Laws, Plato in Twelve Volumes*, trans. R.G. Bury (Cambridge, MA: Harvard University Press), 739c–d.

Plato. 1992. *The Republic*, 2nd edn, trans. G.M.A. Grube, rev. C.D.C. Reeve (Indianapolis, IN: Hackett Publishing Company).

Plato. 2008. *Critias*, trans. Benjamin Jowett. The Internet Classics Archive. Available at http://classics.mit.edu/Plato/critias.html (accessed August 1, 2008).

Pogge, Thomas. 2002. *World Poverty and Human Rights* (Malden, MA: Blackwell Publishing).

Pogge, Thomas. 2005. "Severe Poverty as a Violation of Negative Duties," *Ethics and International Affairs* 19.1: 55–83.

Poland, Craig A. et al. 2008. "Carbon Nanotubes Introduced into the Abdominal Cavity of Mice Show Asbestos-Like Pathogenicity in a Pilot Study," *Nature Nanotechnology* 3: 423–8.

Poncharal, Philippe et al. 1999. "Electrostatic Deflections and Electromechanical Resonances of Carbon Nanotubes," *Science* 283.5407: 1513–16.

Posner, Richard. 2004. *Catastrophe: Risk and Response* (New York: Oxford University Press).

Postma, Henk W.C., Allard Sellmeijer, and C. Dekker. 2000. "Manipulation and Imaging of Individual Single-Walled Carbon Nanotubes with an Atomic Force Microscope," *Advanced Materials* 12.17: 1299.

President's Council on Bioethics. 2003. *Beyond Therapy: Biotechnology and the Pursuit of Happiness* (Washington, DC: Government Printing Office).

Prosser, William L. 1960. "Privacy," *California Law Review* 48.3: 383–423.

Prosser, William L. 1964. *Handbook of the Law of Torts*, 3rd edn (St Paul, MN: West Publishing Co.).

Qi, Hang, Cuiying Wang, and Jie Liu. 2003. "A Simple Method for the Synthesis of Highly Oriented Potassium-Doped Tungsten Oxide Nanowires," *Advanced Materials* 15.5: 411–14.

Quinten, M. et al. 1998. "Electromagnetic Energy Transport via Linear Chains of Silver Nanoparticles," *Optics Letters* 23.17: 1331–3.

Rachels, James. 1975. "Why Privacy Is Important," *Philosophy & Public Affairs* 4: 323–33.

Ravindran, Sathyajith et al. 2005. "Quantum Dots as Bio-Labels for the Localization of a Small Plant Adhesion Protein," *Nanotechnology* 16.1: 1–4.

Rawls, John. 1993. *Political Liberalism* (New York: Columbia University Press).

Rawls, John. 1999. *A Theory of Justice*, rev. edn (Cambridge, MA: Belknap Press).

Rawls, John. 1999. *The Law of Peoples* (Cambridge, MA: Harvard University Press).

Rawls, John. 2001. *Justice as Fairness: A Restatement* (Cambridge, MA: Belknap Press).

Rees, William E. 1992. "Ecological Footprints and Appropriated Carrying Capacity: What Urban Economics Leaves Out," *Environment and Urbanization* 2: 121–30.

Report of the United Nations Conference on Environment and Development. 2008. Available at www.un.org/documents/ga/conf151/aconf15126-1annex1.htm (accessed June 23, 2008).

Rescher, Nicholas. 1980. "The Canons of Distributive Justice," in James P. Sterba (ed.), *Justice: Alternative Political Perspectives* (Belmont, CA: Wadsworth Publishing Co.).

Resnik, David B. 2003. "Is the Precautionary Principle Unscientific?" *Studies in the History and Philosophy of Biological and Biomedical Sciences* 34: 329–44.

Reuters. 2006. "Testing Cell Phone Radiation on Human Skin," March 3. *CNET News*. Available at www.sarshield.com/news/Testing%20cell%20phone%20radiation%20on%20human%20skin%20%20CNET%20News_com.htm (accessed August 21, 2008).

Ringel, Steven A. et al. 2002. "Single-Junction InGaP/GaAs Solar Cells Grown on Si Substrates with SiGe Buffer Layers," *Progress in Photovoltaics* 10.6: 417–26.

Robinson, David. 1983. *Chaplin: The Mirror of Opinion* (Bloomington: Indiana University Press).

Roco, Mihail C. and William Sims Bainbridge. 2003. *Converging Technologies for Improving Human Performance: Nanotechnology, Biotechnology, Information Technology and Cognitive Science* (Dordrecht: Kluwer Academic Publishers).

Roco, Mihail C., R. Stanley Williams, and Paul Alivisatos (eds). 1999. *Nanotechnology Research Directions: IWGN Workshop Report. Vision for Nanotechnology Research and Development in the Next Decade* (Baltimore, MD: Loyola College). Available at www.wtec.org/loyola/nano/IWGN.Research.Directions/chapter10.pdf (accessed July 28, 2008).

Roe v. Wade 410 US 113. 1973. Available at http://caselaw.lp.findlaw.com/scripts/getcase.pl?court=US&vol=410&invol=113 (accessed August 8, 2008).

Rojas-Chapana, Jose A. and Michael Giersig. 2006. "Multi-Walled Carbon Nanotubes and Metallic Nanoparticles and Their Application in Biomedicine," *Journal of Nanoscience and Nanotechnology* 6.2: 316–21.

Romeo, Alessandro et al. 2000. "Recrystallization in CdTe/CdS," *Thin Solid Films* 361: 420–5.

Ross, Brian. 2006. "Government Moves to Curb Use of Chemical in Teflon," *ABC News*, January 25. Available at http://abcnews.go.com/WNT/story?id=1540964 (accessed August 21, 2008).

Rothemund, Paul. 2006. "Folding DNA to Create Nanoscale Shapes and Patterns," *Nature* 440: 297–302.

Sabety, Ted. 2004. "Nanotechnology Innovation and the Patent Thicket: Which IP Policies Promote Growth?" *Nanotechnology, Law & Business* 1.3: 262–83.

Sagoff, Mark. 1981. "At the Shrine of Our Lady of Fatima; Or, Why All Political Questions Are Not All Economic," *Arizona Law Review* 23.4: 1283–98.

Salamanca-Buentello, Fabio et al. 2005. "Nanotechnology and the Developing World," *PLoS Medicine* 2.5: e97, doi:10.1371/journal.pmed.0020097. Available at http://medicine.plosjournals.org/archive/1549-1676/2/5/pdf/10.1371_journal.pmed.0020097-L.pdf (accessed October 13, 2008).

Saletan, William. 2008. "Night of the Living Meds: The US Military's Sleep-Reduction Program," *Slate* July 16. Available at www.slate.com/id/2195466/ (accessed August 21, 2008).

Salvetat, Jean Paul et al. 1999. "Mechanical Properties of Carbon Nanotubes," *Applied Physics A: Materials Science & Processing* 69.3: 255–60.

Sandel, Michael. 2007. *The Case Against Perfection: Ethics in the Age of Genetic Engineering* (Cambridge, MA: Belknap Press).

Sandin, per. 1999. "Dimensions of the Precautionary Principle," *Human and Ecological Risk Assessment* 5.5: 889–907.

Scanlon, Thomas M. 1975. "Thomson on Privacy," *Philosophy & Public Affairs* 4: 315–22.

Schaller, Richard D. et al. 2006. "Seven Excitons at a Cost of One: Redefining the Limits for Conversion Efficiency of Photons into Charge Carriers," *Nano Letters* 6.3: 424–9.

Schrödinger, Erwin. 1935. "Die gegenwärtige Situation in der Quantenmechanik," *Naturwissenschaften* 23: 807–12, 823–8, 844–9. Trans. by John D. Trimmer as "The Present Situation in Quantum Mechanics," *Proceedings of the American Philosophical Society* 124 (1980): 323–38.

Schummer, Joachim. 2007. "Impact of Nanotechnology on Developing Countries," in Fritz Allhoff et al. (eds), *Nanoethics: The Ethical and Social Implications of Nanotechnology* (Hoboken, NJ: John Wiley & Sons), pp. 291–307.

Science and Environmental Health Network. 2008. "Precautionary Principle." Available at www.sehn.org/wing.html (accessed June 23, 2008).

Selgelid, Michael J. 2007. "An Argument against Arguments for Enhancement," *Studies in Ethics, Law, and Technology* 1.1: Article 12. Available at www.bepress.com/selt/vol1/iss1/art12/ (accessed August 21, 2008).

Shah, Arvind et al. 1999. "Photovoltaic Technology: The Case for Thin-Film Solar Cells," *Science* 285.5428: 692–8.

Shim, Moonsub et al. 2002. "Functionalization of Carbon Nanotubes for Biocompatibility and Biomolecular Recognition," *Nano Letters* 2: 285–8.

Short, R.V. 2006. "New Ways of Preventing HIV Infection: Thinking Simply, Simply Thinking," *Philosophical Transactions of the Royal Society B* 361.1469: 811–20.

Shrader-Frechette, Kristen. 2007. *Taking Action, Saving Lives: Our Duties to Protect Environmental and Public Health* (New York: Oxford University Press).

Singer, Peter. 1972. "Famine, Affluence, and Morality," *Philosophy and Public Affairs* 1.1: 229–43.

Smalley, Richard. 1997. "Discovering the Fullerenes," *Review of Modern Physics* 69: 723–30.

Small Times. 2005. "Vermont's Seldon Labs Want to Keep Soldier's Water Pure," *Small Times* April 26. Available at www.smalltimes.com/articles/article_display.cfm?Section=ARCHI&C=Profi&ARTICLE_ID=269416&p=109 (accessed July 28, 2008).

Smith, Andrew M., Xiaohu Gao, and Shuming Nie. 2004. "Quantum Dot Nanocrystals for *In Vivo* Molecular and Cellular Imaging," *Photochemistry and Photobiology* 80.3: 377–85.

Solar Panel Estimator. Available at http://files.blog-city.com/files/M05/102402/b/solarcalc3.html (accessed July 30, 2008).

Sparrow, Rob. 2007. "Killer Robots," *Journal of Applied Philosophy* 24.1: 62–77.

Stenton, Doris Mary. 1951. *English Society in the Early Middle Ages* (Baltimore, MD: Penguin Books).

Sunkara, Mahendra K. et al. 2001. "Bulk Synthesis of Silicon Nanowires Using a Low-Temperature Vapor–Liquid–Solid Method," *Applied Physics Letters* 79.10: 1546–8.

Sunstein, Cass R. 2002. *The Cost–Benefit State* (Washington, DC: American Bar Association).

Sunstein, Cass R. 2003. "Beyond the Precautionary Principle," *Pennsylvania Law Review* 151: 1003–58.

Sunstein, Cass R. 2004. *Risk and Reason: Safety, Law, and the Environment* (Cambridge: Cambridge University Press).

Sunstein, Cass R. 2005. "Cost–Benefit Analysis and the Environment," *Ethics* 115.2: 351–85.

Sunstein, Cass R. 2005. *Laws of Fear: Beyond the Precautionary Principle* (Cambridge: Cambridge University Press).

Sunstein, Cass R. 2008. "Two Conceptions of Irreversible Environmental Harm," *University of Chicago Law & Economics, Olin Working Paper No. 407*. Available at http://ssrn.com/abstract=1133164 (accessed October 2008).

Sussman, Norman L. and James H. Kelly. 2003. "Saving Time and Money in Drug Discovery: A Pre-emptive Approach," in *Business Briefings: Future Drug Discovery 2003* (London: Business Briefings, Ltd), pp. 46–9.

Sutton, Robert I. (ed.). 2001. *Weird Ideas That Work: 11¹/₂ Practices for Promoting, Managing and Sustaining Innovation* (New York: Free Press).

Swanson, Judith A. 1992. *The Public and Private in Aristotle's Political Philosophy* (Ithaca, NY: Cornell University Press).

Taniguchi, Norio. 1974. "On the Basic Concept of Nanotechnology." *Proceedings of the International Conference of Production Engineering, London, Part II.* British Society of Precision Engineering.

Tennakone, Kirti et al. 2002. "Dye-Sensitized Composite Semiconductor Nano-structures," *Physica E* 14.1–2: 190–6.

Thelakkat, Mukundan, Christoph Schmitz, and Hans-Werner Schmidt. 2002. "Fully Vapor-Deposited Thin-Layer Titanium Dioxide Solar Cells," *Advanced Materials* 14.8: 577–81.

Thompson, Mark. 2008. "America's Medicated Army," *Time* June 16. Available at www.time.com/time/nation/article/0,8599,1811858,00.html (accessed August 21, 2008).

Thucydides. 1903. *History of the Peloponnesian War*, trans. Richard Crawley (London: J.M. Dent & Sons). Available at http://ebooks.adelaide.edu.au/t/thucydides/crawley/ (accessed October 15, 2008).

Tirlapur, Uday K. and Karsten König. 2003. "Femtosecond Near-Infrared Laser Pulses as a Versatile Non-Invasive Tool for Intra-Tissue Nanoprocessing in Plants without Compromising Viability," *The Plant Journal* 31.2: 365–74.

Tiwari, Ayodhya N. et al. 2004. "CdTe Solar Cell in a Novel Configuration," *Progress in Photovoltaics* 12.1: 33–8.

Topol, Anna W. et al. 2004. "Chemical Vapor Deposition of ZnS : Mn for Thin-Film Electroluminescent Display Applications," *Journal of Materials Research* 19.3: 697–706.

Toxic Substances Control Act. Available at http://epw.senate.gov/tsca.pdf (accessed September 1, 2008).

Trager, Rebecca. 2008. "EPA Nanosafety Scheme Fails to Draw Industry," *Chemistry World*, August 5. Available at www.rsc.org/chemistryworld/News/2008/August/05080801.asp (accessed August 21, 2008).

Tran, P.T. et al. 2002. "Use of Luminescent CdSe-ZnS Nanocrystal Bioconjugates in Quantum-Dot-Based Nanosensors," *Physica Status Solidi B* 229.1: 427–32.

Trouwborst, Arie. 2002. *Evolution and Status of the Precautionary Principle in International Law* (London: Kluwer Law International).

Tsuji, Joyce S. et al. 2006. "Research Strategies for Safety Evaluation of Nano-materials, Part IV: Risk Assessment of Nanoparticles," *Toxicological Sciences* 88.1: 12–17.

Turing, Alan. 1950. "Computing Machinery and Intelligence," *Mind* 59.236: 434–60.

Uldrich, Jack. 2006. *Investing in Nanotechnology* (New York: Adams Media Corporation).

Uldrich, Jack. 2006. "Nanotech of the North," *The Motley Fool.* Available at www.fool.com/investing/high-growth/2006/03/06/nanotech-of-the-north.aspx (accessed August 21, 2008).

UNAIDS. 2006. *Report on the Global AIDS Epidemic.* Available at http://data.unaids.org/pub/EpiReport/2006/2006_EpiUpdate_en.pdf (accessed October 13, 2008).

UNAIDS. 2008. *Report on the Global AIDS Epidemic.* Available at www.unaids.org/en/KnowledgeCentre/HIVData/GlobalReport/2008 (accessed July 30, 2008).

United Nations Development Programme. 2006. *Human Development Report 2006. Beyond Scarcity: Power, Water, and the Global Water Crisis* (New York: Palgrave Macmillan). Available at http://hdr.undp.org/en/media/hdr06-complete.pdf (accessed July 27, 2008).

Uniting and Strengthening America by Providing Appropriate Tools Required to Intercept and Obstruct Terrorism Act (USA PATRIOT Act) of 2001. Pub. L. No. 107-56, 115 Stat. 272. 2001. Available at http://frwebgate.access.gpo.gov/cgi-bin/getdoc.cgi?dbname=107_cong_bills&docid=f:h3162enr.txt.pdf (accessed October 16, 2008).

University of Pittsburgh Cancer Institute. 2008. Cancer warning. Available at www.upci.upmc.edu/news/ (accessed August 21, 2008).

UN Millennium Project. 2004. *Interim Full Report of Task Force 7 on Water and Sanitation.* Available at www.unmillenniumproject.org/documents/tf7interim.pdf (accessed October 13, 2008).

UN Millennium Project. 2005. *Innovation: Applying Knowledge in Development.* Available at www.unmillenniumproject.org/documents/Science-complete.pdf (accessed October 13, 2008).

USA PATRIOT Act Additional Reauthorizing Amendments Act of 2006, 109–170, S.2271. Available at http://thomas.loc.gov/cgi-bin/query/z?c109:s2271: (accessed August 8, 2008).

USA PATRIOT Improvement and Reauthorization Act of 2005, 109–333, H.R.3199. Available at http://thomas.loc.gov/cgi-bin/cpquery/R?cp109:FLD010:@1(hr333) (accessed August 8, 2008).

US Department of Energy: Pacific Northwest National Laboratory. 2008. "Description of the Technology: SAMMS Assembly." Available at http://samms.pnl.gov/tech_descrip.stm (accessed August 20, 2008).

US Department of State: Bureau of Consular Affairs. 2008. "Western Hemisphere Travel Initiative." Available at http://travel.state.gov/travel/cbpmc/cbpmc_2223.html (accessed August 15, 2008).

US Department of State: Bureau of Consular Affairs. 2008. "The US Electronic Passport: Frequently Asked Questions #12." Available at http://travel.state.gov/passport/eppt/eppt_2788.html#Twelve (accessed August 15, 2008).

US Environmental Protection Agency. 2006. "EPA Seeking PFOA Reductions," press release, January 25. Available at http://yosemite.epa.gov/opa/admpress.nsf/a543211f64e4d1998525735900404442/fd1cb3a075697aa485257101006afbb9!OpenDocument (accessed August 21, 2008).

US Environmental Protection Agency. 2008. Nanosafety Scheme. Available at www.epa.gov/oppt/nano/stewardship.htm (accessed August 21, 2008).

US Environmental Protection Agency: National Center for Environmental Research. 2008. "Nanotechnology Toxicity." Available at http://es.epa.gov/ncer/nano/research/nano_tox.html (accessed August 14, 2008).

US National Nanotechnology Infrastructure Network. 2008. Explanatory text. Available at www.nnin.org/nnin_what.html (accessed August 21, 2008).

US National Science and Technology Council, Committee on Technology. 2000. *National Nanotechnology Initiative: Leading to the Next Industrial Revolution* (Washington, DC).

Van den Hoven, Jeroen. 2007. "Nanotechnology and Privacy: Instructive Case of RFID," in Fritz Allhoff et al. (eds), *Nanoethics: The Ethical and Social Implications of Nanotechnology* (Hoboken, NJ: John Wiley & Sons), pp. 253–66.

Van den Hoven, Jeroen. 2008. "The Tangled Web of Tiny Things: Privacy Implications of Nano-electronics," in Fritz Allhoff and Patrick Lin (eds), *Nanotechnology and Society: Current and Emerging Ethical Issues* (Dordrecht: Springer), pp. 147–62.

Van den Hoven, Jeroen and Pieter E. Vermaas. 2007. "Nano-Technology and Privacy: On Continuous Surveillance outside the Panopticon," *Journal of Medicine and Philosophy* 32: 283–97.

Viscusi, W. Kip. 1993. *Fatal Tradeoffs* (New York: Oxford University Press).

Vo-Dinh, Tuan. 2007. *Nanotechnology in Biology and Medicine: Methods, Devices, and Applications* (Boca Raton, FL: CRC Press).

Waltzer, Michael. 2006. *Just and Unjust Wars: A Moral Argument with Historical Illustrations*, 4th edn (New York: Basic Books).

Wang, Zhong Lin. 2005. "Nanotechnology and Nanomanufacturing," in Hwaiyu Geng (ed.), *Semiconductor Manufacturing Handbook* (New York: McGraw-Hill Professional).

Wang, Zhong Lin and Z.C. Kang. 1998. *Functional and Smart Materials: Structural Evolution and Structure Analysis*, 1st edn (New York: Plenum Press).

Warren, Samuel and Louis Brandeis. 1890. "The Right to Privacy," *Harvard Law Review* 4: 193–220.

Weckert, John and James Moor. 2006. "The Precautionary Principle in Nano-technology," *International Journal of Applied Philosophy* 2.2: 191–204.

Weissert, Will. 2004. "Microchips Implanted in Mexican Officials: Attorney General, Prosecutors Carry Security Pass under Their Skin," *Associated Press* July 14. Available at www.msnbc.msn.com/id/5439055/ (accessed August 14, 2008).

Wenar, Leif. 2007. "Rights," *The Stanford Encyclopedia of Philosophy*. Available at http://plato.stanford.edu/entries/rights/ (accessed August 21, 2008).

Westwater, John et al. 1997. "Growth of Silicon Nanowires via Gold/Silane Vapor–Liquid–Solid Reaction," *Journal of Vacuum Science Technology B* 15.3: 554–7.

Whitney, M.T. 2007. "Mobile Phones Boost Brain Tumor Risk by Up to 270 Percent on Side of Brain Where Phone is Held," *Natural News* February 22. Available at www.naturalnews.com/021634.html (accessed August 21, 2008).

Wilder, Jeroen W.G. et al. 1998. "Electronic Structure of Atomically Resolved Carbon Nanotubes," *Nature* 391.1: 59–62.

Williams, Bernard. 1973. *Problems of the Self* (Cambridge: Cambridge University Press).

Williams, David H. 2004. "The Strategic Implications of Wal-Mart's RFID Mandate," *Directions Magazine* July 29. Available at www.directionsmag.com/article.php?article_id=629&trv=1 (accessed August 13, 2008).

Williams, E. et al. 2008. "Human Performance," *JASON: The MITRE Corporation* March. Available at www.fas.org/irp/agency/dod/jason/human.pdf (accessed September 4, 2008).

Witteveen, William J. 2005. "A Self-Regulation Paradox: Notes Towards the Social Logic of Regulation," *Electronic Journal of Comparative Law* 9.1. Available at www.ejcl.org/91/art91-2.html (accessed August 21, 2008).

Wohlstadler, Jacob N. et al. 2003. "Carbon Nanotube-Based Biosensor," *Advanced Materials* 15: 1184–7.

Wong, Eric W., Paul E. Sheehan, and Charles M. Lieber. 1997. "Nanobeam Mechanics: Elasticity, Strength, and Toughness of Nanorods and Nanotubes," *Science* 277.5334: 1971–5.

Woodall, Jack. 2007. "Programmed Dissatisfaction: Does One Gene Drive All Progress in Science and the Arts?" *The Scientist* 21.6: 63.

World Energy Council. 1999. "The Challenge of Rural Energy Poverty in Developing Countries." Available at http://217.206.197.194:8190/wec-geis/publications/reports/rural/exec_summary/exec_summary.asp (accessed July 30, 2008).

World Health Organization (WHO). 2008. "Water, Sanitation, and Hygiene Links to Health: Facts and Figures." Available at www.who.int/entity/water_sanitation_health/factsfigures2005.pdf (accessed July 28, 2008).

Yantasee, Wassana. 2005. "Nanostructured Electrochemical Sensors Based on Functionalized Nanoporous Silica for Voltammetric Analysis of Lead, Mercury and Copper," *Journal of Nanoscience and Nanotechnology* 5.9: 1537–40.

Yi, Jeong W. et al. 2003. "*In Situ* Cell Detection Using Piezoelectric Lead Zirconate Titanate Stainless-Steel Cantilevers," *Journal of Applied Physics* 93.1: 619–25.

Yu, D.P. et al. 1998. "Amorphous Silica Nanowires: Intensive Blue Light Emitters," *Applied Physics Letters* 73.21: 3076–8.

Yu, D.P. et al. 1998. "Nanoscale Silicon Wires Synthesized Using Simple Physical Evaporation," *Applied Physics Letters* 72: 3458–60. Erratum in *Applied Physics Letters* 85 (2004): 5104.

Yu, Min-Feng et al. 2000. "Strength and Breaking Mechanism of Multi-Walled Carbon Nanotubes Under Tensile Load," *Science* 287.5453: 637–40.

Zhang, Xin-Yi et al. 2001. "Synthesis of Ordered Single Crystal Silicon Nanowire Arrays," *Advanced Materials* 13.16: 1238–41.

Zhang, Y. et al. 2000. "Bulk-Quantity Si Nanowires Synthesized by SiO Sublimation," *Journal of Crystal Growth* 212: 115–18.

Index